Mathematical Reasoning
Writing and Proof

Mathematical Reasoning
Writing and Proof

Ted Sundstrom

Grand Valley State University

Prentice
Hall

Pearson Education, Inc., Upper Saddle River, New Jersey 07458

Library of Congress Cataloging-in-Publication Data

Sundstrom, Theodore A.
    Mathematical reasoning: writing and proof / Ted Sundstrom
  p. cm.
    Includes index.
    ISBN 0-13-061815-2
      1. Proof theory.   I. Title
QA9.54.S86 2003
511.3--dc21   2002066764

Acquisition Editor: *George Lobell*
Editor-in-Chief: *Sally Yagan*
Editorial/Production Supervision: *Bayani Mendoza de Leon*
Vice President/Director of Production and Manufacturing: *David W. Riccardi*
Senior Managing Editor: *Linda Mihatov Behrens*
Executive Managing Editor: *Kathleen Schiaparelli*
Manufacturing Buyer: *Alan Fischer*
Manufacturing Manager: *Trudy Pisciotti*
Marketing Assistant: *Rachel Beckman*
Art Director: *Jayne Conte*
Cover Designer: *Bruce Kenselaar*

Printed in the United States of America

10   9   8   7   6   5   4   3   2   1

ISBN 0-13-061815-2

Pearson Education LTD, *London*
Pearson Education Australia PTY, Limited, *Sydney*
Pearson Education Singapore, Pte. Ltd
Pearson Eudcation North Asia Ltd, *Hong Kong*
Pearson Education Canada, Ltd., *Toronto*
Pearson Educatión de Mexico, S.A. de C.V.
Pearson Education—Japan, *Tokyo*
Pearson Education Malaysia, Pte. Ltd

To my family and the memory of my parents,
Ted and Bertha Sundstrom

Contents

Preface

Mathematical Reasoning: Writing and Proof is designed to be a text for the first course in the college mathematics curriculum that focuses on the formal development of mathematics. The primary goals of the text are as follows:

- To help students learn how to read and understand mathematical definitions and proofs;

- To help students learn how to construct mathematical proofs;

- To help students learn how to write mathematical proofs according to accepted guidelines so that their work and reasoning can be understood by others; and

- To provide students with some mathematical material that will be needed for their further study of mathematics.

This type of course is becoming a standard part of the mathematics major at most colleges and universities. It is often referred to as a "transition course" from the calculus sequence to the upper level courses in the major. The transition is from the problem-solving orientation of calculus to the more abstract and theoretical upper-level courses. This is needed today because the principal goals of most calculus courses are developing students' understanding of the concepts of calculus and improving their problem-solving skills. Consequently, most students complete their study of calculus without seeing a formal proof or having constructed a proof of their own. This is in contrast to many upper-level mathematics courses, where the emphasis is on the formal development of abstract mathematical ideas, and the expectations are that students will be able to read and understand proofs and to construct and write coherent, understandable mathematical proofs.

Important Features of the Book

Mathematical Reasoning was written to assist students with the transition from calculus to upper level mathematics courses. Students should be able to use this text with a background of one semester of calculus. Following are some of the important ways this text will help with this transition.

1. **Emphasis on Writing in Mathematics**

   The issue of writing mathematical exposition is addressed throughout the book. Guidelines for writing mathematical proofs are incorporated into the text. These guidelines are introduced as needed and begin in Chapter 1. Appendix A contains a summary of all the guidelines for writing mathematical proofs that are introduced in the text. In addition, every attempt has been made to ensure that each proof presented in this text is written according to these guidelines in order to provide students with examples of well-written proofs.

2. **Instruction in the Process of Constructing Proofs**

   One of the primary goals of this book is to develop students' abilities to construct mathematical proofs. Another goal is to develop their abilities to write the proof in a coherent manner that conveys an understanding of the proof to the reader. These are two distinct skills.

   Instruction on how to write proofs begins in Section 1.2 and is developed further in Chapter 3. In addition, Chapter 5 is devoted to developing students' abilities to construct proofs using mathematical induction. Students are taught to organize their thought processes when attempting to construct a proof with a so-called know-show table. (See Sections 1.2 and 3.1.) Students use this table to work backward from what it is they are trying to prove while at the same time working forward from the assumptions of the problem.

3. **Emphasis on Active Learning**

   One of the underlying premises of this text is that the best way to learn and understand mathematics is to be actively involved in the learning process. However, it is unreasonable to expect students to go out and learn mathematics on their own. Students actively involved in learning mathematics need appropriate materials that will provide guidance and support in their learning of mathematics. This text provides these by incorporating two or three Preview Activities for each section and some activities within each section based on the material in that section. These activities can be done individually or in a collaborative learning setting, where students work in groups to brainstorm, make conjectures, test each others' ideas, reach consensus, and, it is hoped, develop sound mathematical arguments to support their work.

The Preview Activities at the beginning of each section should be completed by the students prior to the classroom discussion of the section. The purpose of the Preview Activities is to prepare students to participate in the classroom discussion of the section. Some Preview Activities will review prior mathematical work that is necessary for the new section. This prior work may contain material from previous mathematical courses or it may contain material covered earlier in this text. Other preview activities will introduce new concepts and definitions that will be used when that section is discussed in class.

In addition to the Preview Activities, each section of the text contains two or three activities related to the material contained in that section. These activities can be used for in-class group work or can be assigned as homework in addition to the exercises at the end of each section.

Content and Organization

Mathematical content is needed as a vehicle for learning how to construct and write proofs. The mathematical content for this text is drawn primarily from elementary number theory, including congruence arithmetic; elementary set theory; functions, including injections, surjections, and the inverse of a function; relations and equivalence relations; further topics in number theory such as greatest common divisors and prime factorizations; and cardinality of sets, including countable and uncountable sets. This material was chosen because it can be used to illustrate a broad range of proof techniques and it is needed as a prerequisite for many upper level mathematics courses.

The chapters in the text can roughly be divided into the following (possibly overlapping) classes:

- Constructing and Writing Proofs: Chapters 1, 3, and 5

- Content: Chapters 4, 6, 7, 8, and 9

- Logic: Chapter 2

The first chapter sets the stage for the rest of the book. It introduces the writing guidelines, discusses conditional statements, and begins instruction in the process of constructing a direct proof of a conditional statement. This is not meant to be a thorough introduction to methods of proof. Before this is done, it is necessary to introduce the students to the parts of logic that are needed to aid in the construction of proofs. This is done in Chapter 2.

Students need to learn some logic and gain experience in the traditional language and proof methods used in mathematics. Since this is a text that deals with constructing and writing mathematical proofs, the logic that is presented in Chapter 2 is intended to aid in the construction of proofs. The goals are to provide students with a thorough understanding of conditional statements, quantifiers, and logical equivalencies. Emphasis is placed on writing correct and useful negations of statements, especially those involving quantifiers. The logical equivalencies that are presented provide the logical basis for some of the standard proof techniques, such as proof by contrapositive, proof by contradiction, and proof using cases.

The standard methods for mathematical proofs are discussed in detail in Chapter 3. The mathematical content that is introduced to illustrate these proof methods is some elementary number theory, including congruence arithmetic. These concepts are used consistently throughout the text as a way to demonstrate ideas in direct proof, proof by contrapositive, proof by contradiction, proof by cases, and proofs using mathematical induction. This gives students a strong introduction to an important mathematical idea, while providing the instructor a consistent reference point and an example of how mathematical notation can greatly simplify a concept.

In Chapter 4, we take a break from introducing new proof techniques. Concepts of set theory are introduced, and the methods of proof studied in Chapter 3 are used to prove results about sets and operations on sets. The idea of an "element-chasing proof" is introduced in Section 4.2.

The three sections of Chapter 5 are devoted to proofs using mathematical induction. Again, the emphasis is not only on understanding mathematical induction but also on developing the ability to construct and write proofs that use mathematical induction.

The last four chapters are considered "mathematical content" chapters. Chapter 6 provides a thorough study of functions. The idea is to begin with a review of functions from previous courses so that students have a base from which to work. This notion of function is then extended to the general definition of function. Various proof techniques are employed in the study of injections, surjections, composition of functions, and inverses of functions.

Chapter 7 introduces the concepts of relations and equivalence relations. Section 7.4 is included to provide a link between the concept of an equivalence relation and the number theory that has been discussed throughout the text.

Chapter 8 continues the study of number theory. The highlights include problems dealing with greatest common divisors, prime numbers, the Fundamental Theorem of Arithmetic, and linear Diophantine equations.

Finally, Chapter 9 deals with further topics in set theory, focusing on cardinality, finite sets, countable sets, and uncountable sets.

A standard one-semester course in constructing and writing proofs should cover the first six chapters of the text and at least one of Chapter 7, Chapter 8, or Chapter 9. A class consisting of well-prepared and motivated students could cover two of the last three chapters. In addition, there are a few options that an instructor could choose to tailor the course to her or his needs. For example,

- Chapter 5 can be covered before Chapter 4 if it is desired to cover all methods of proof before beginning the "content" portion of the course. The only part of Chapter 5 that would need to be skipped is the material in Section 5.2 dealing with the cardinality of the power set. If desired, this material could be included when the power set is discussed in Chapter 4.

- Instructors who would like to cover topics in both Chapters 7 and 8 can omit a few selected sections from earlier chapters. Although it is an important and interesting section, Section 5.3 is not used in the remainder of the book. The same is true for Section 6.5. Finally, Section 3.5 can be skipped as long as the concept of a constructive proof is discussed during other parts of the course.

Supplementary Materials for the Instructor

The instructor's manual for this text includes suggestions on how to use the text, how to incorporate writing into the course, and how to use the preview activities and activities. The manual also includes solutions for all of the preview activities, activities, and exercises. In addition, for each section, there is a description of the purpose of each preview activity and how it is used in the corresponding section, and there are suggestions about how to use each activity in that section. The intention is to make it as easy as possible for the instructor to use the text in an active learning environment. These activities can also be used in a more traditional lecture-discussion course. In that case, some of the activities would be discussed in class or assigned as homework.

The instructor's manual is available by contacting the editorial offices at Prentice Hall in Upper Saddle River, NJ or by emailing a request to george_lobell@prenhall.com.

In addition, Adobe Acrobat (pdf) files are available to the instructor to assist the instructor in posting solutions to a course web page or distributing printed solutions to students. For each section, there is a file containing the solutions of the preview activities, and for each activity in the text, there is a file containing the solutions for that activity.

Instructors can contact the author through his email address for access to the files.

Acknowledgments

This text has benefited greatly from the comments of students who have used preliminary versions of the text and from ideas and insights of the reviewers of the text. I would like to thank the following reviewers:

> Frank Bäuerle, University of California, Santa Cruz; Michael C. Berg, Loyola Marymount University; Lifeng Ding, Georgia State University; Jeffrey Ehme, Spelman College; Christopher P. Grant, Brigham Young University; Joel Iiams, University of North Dakota; Robert Jajcay, Indiana State University; Iraj Kalantari, Western Illinois University; and William J. Keane, Boston College.

I would also like to express my sincere gratitude to my colleagues at Grand Valley State University who have used the text and have supported my work on this text. This includes Professors David Austin, Ed Aboufadel, Karen Brown, Will Dickinson, Philip Pratt, Steve Schlicker, Jody Sorensen, Clark Wells, and Pamela Wells. Special thanks go to Prof. Karen Novotny, who helped me immensely in writing the first draft of the text, Prof. Matt Boelkins, who carefully read the completed text and offered numerous helpful suggestions, and Prof. Larry King of the University of Michigan-Flint, who used an early version of the manuscript and made several useful suggestions for further development of the text.

Finally, a very special thank you goes to my wife, Karen, and my daughter, Laura, for their patience, understanding, and encouragement during the long time it took to develop this text.

Comments about the text and suggestions for improving it are welcome.

Ted Sundstrom
sundstrt@gvsu.edu

Mathematical Reasoning
Writing and Proof

Chapter 1

Introduction to Writing in Mathematics

1.1 Statements

Preview Activity 1 (Solving an Equation).
The following are some steps that can be used to begin solving the equation

$$x - 1 = \sqrt{x + 11},$$

where x represents a real number.

- Square both sides of the equation.

- Expand the left side of the equation.

- Write the resulting equation in standard quadratic form $\left(ax^2 + bx + c = 0\right)$.

- Solve the resulting quadratic equation.

- Check the solutions of the quadratic equation in the original equation.

Write a description of how to solve this equation. This description should be written for someone who already knows basic algebra and how to solve quadratic equations.

Preview Activity 2 (Statements).

In mathematics, a **statement** is a sentence that is either true or false but is not both true and false. A statement is sometimes called a proposition.

Which of the following sentences are statements? Do not worry about determining whether a statement is true or false. Just determine whether each sentence is a statement or not.

1. $3 + 4 = 8$.

2. $(x - 1) = \sqrt{x + 11}$.

3. $2x + 5y = 7$.

4. There are integers x and y such that $2x + 5y = 7$.

5. There are integers x and y such that $23x + 37y = 52$.

6. If x and y are odd integers, then $x \cdot y$ is an odd integer.

7. Given a line L and a point P not on that line, there is a unique line through P that does not intersect L.

8. $(a + b)^2 = a^2 + b^2$.

9. $(a + b)^2 = a^2 + b^2$ for all real numbers a and b.

10. If ABC is a right triangle with right angle at vertex B, and if D is the midpoint of the hypotenuse, then the line segment connecting vertex B to D is half the length of the hypotenuse.

11. If you pick N distinct points on the circumference of a circle and draw line segments connecting them all with each other, then the interior of the circle will be divided into 2^{N-1} portions.

12. There do not exist three real numbers x, y, and z such that $x^3 + y^3 = z^3$.

Preview Activity 3 (Conditional Statements).

A **conditional statement** is a statement that can be written in the form "If P then Q" where P and Q are statements. It seems reasonable that the truth value (true or false) of the conditional statement "If P then Q" depends on the truth values of P and Q. The statement "If P then Q" means that Q must be true whenever P is true. The statement P is called the **hypothesis** of the conditional statement, and the statement Q is called the **conclusion** of the conditional statement. We will now explore some examples.

1. "If it is raining, then Laura is at the theater." Under what conditions is this conditional statement false? For example,

   - Is it false if it is not raining and Laura is at the theatre?
   - Is it false if it is not raining and Laura is not at the theatre?
   - Is it false if it is raining and Laura is at the theatre?
   - Is it false if it is raining but Laura is not at the theatre?

2. "If x and y are odd integers, then $x \cdot y$ is an odd integer."

   (a) Notice that if $x = 7$ and $y = 2$, then $x \cdot y = 14$. So the statement that $x \cdot y$ is odd is false in this case. Does this mean that the given conditional statement is false?

   (b) Do you think this statement is true or false? Try and record at least five different examples where the hypothesis of this conditional statement is true.

3. "If n is a positive integer, then $(n^2 - n + 41)$ is a prime number." (Remember that a prime number is a positive integer greater than 1 whose only positive factors are 1 and itself.)

   Do you think this statement is true? Record your results for $n = 1$, $n = 2$, $n = 3$, $n = 4$, $n = 5$, and $n = 10$. Then record the result for at least four other values of n.

The purpose of Preview Activity 1 was to get you started writing in a mathematical setting. We will now introduce some basic writing guidelines for writing in mathematics.

1. **Know Your Audience**.

   Every writer should have a clear idea of the intended audience for a piece of writing. In that way, the writer can give the right amount of information at the proper level of sophistication to communicate effectively. This is especially true for mathematical writing. For example, if a mathematician is writing a solution to a textbook problem for a solutions manual for instructors, the writing would be brief with many details omitted. However, if the writing was for a students' solution manual, more details would be included. This is why the instructions for Preview Activity 1 stated that your descriptions should be written for someone who already knows basic algebra and how to solve quadratic equations.

2. Use the pronoun "we."

If you want to use a pronoun in mathematical writing, the usual convention is to use "we" instead of "I." The idea is that you and the reader are doing the mathematics together. For example, in Preview Activity 1, we could start the description of how to solve the equation $x - 1 = \sqrt{x + 11}$ as follows:

> We start by squaring both sides of the equation. This gives ...

3. Display important equations and mathematical expressions.

Most writing in mathematics will make extensive use of mathematical symbols, equations, and mathematical manipulations. It takes a great deal of attention and practice to develop skill at writing symbolic manipulations that communicate effectively with the reader. Do not write these equations and manipulations in one column with reasons given in another column. *These symbolic expressions and manipulations should be an integral part of the exposition, and important equations and manipulations should be displayed. This means that they should be centered with blank lines before and after the equation or manipulations.* An example of using symbolic displays follows:

In order to solve the equation

$$2x + 3 = (x - 1)^2$$

for x, we first expand the right side of the equation and then collect all terms on one side of the equation as follows:

$$2x + 3 = x^2 - 2x + 1$$
$$x^2 - 4x - 2 = 0.$$

We then solve the resulting quadratic equation using the quadratic formula. This gives

$$x = \frac{4 \pm \sqrt{4^2 - 4(1)(-2)}}{2(1)}$$

$$x = \frac{4 \pm \sqrt{24}}{2}.$$

Check both of these values in the original equation (with a calculator if needed), we see that

$$x = \frac{4 + \sqrt{24}}{2} \text{ and } x = \frac{4 - \sqrt{24}}{2}$$

are both solutions of the equation $2x + 3 = (x - 1)^2$.

4. **Use italics for variables when using a word processor.**

   When using a word processor to write mathematics, the word processor needs to be capable of producing the appropriate mathematical symbols and equations. The mathematics that is written with a word processor should look like typeset mathematics. This means that variables need to be italicized, boldface is used for vectors, and regular font is used for mathematical terms such as the names of the trigonometric functions and logarithmic functions.

Activity 1.1 (Rewriting a Solution of an Equation). Rewrite your description of how to solve the equation $(x - 1) = \sqrt{x + 11}$ following the writing guidelines stated in this section.

Statements

In Preview Activity 2, we tried to determine whether or not certain sentences were statements.

> **Definition.** A **statement** is a sentence that is either true or false. A statement is also called a **proposition**.

The key is that there must be no ambiguity. To be a statement, a sentence must be true or false, and it cannot be both. So a sentence such as "Tom is cute" is not a statement since whether it is true or not is a matter of opinion. There might be disagreement about whether or not Tom is cute. A second (more subtle) reason it is not a statement is that we may not be sure exactly who Tom is. We do not deal with these kinds of ambiguous sentences in mathematics.

Other sentences that seem more mathematical in nature often are not statements because we may not know precisely what a variable represents. Example (3) in Preview Activity 2 ($2x + 5y = 7$) is not a statement since we do not know what x and y represent. These are variables. If we substitute specific values for x and y (such as $x = 1$ and $y = 2$), then the resulting sentence would be a statement. However, Example (4) in Preview Activity 2

"There are integers x and y such that $2x + 5y = 7$"

is a statement since either there exist such integers or there do not exist such integers. Similarly, the sentence in Example (8) in Preview Activity 2, $(a + b)^2 = a^2 + b^2$, is not a statement, but the sentence in Example (9), $(a + b)^2 = a^2 + b^2$ for all real numbers a and b, is a statement.

How Do We Decide If a Statement Is True or False?

In mathematics, the ultimate method to establish that a statement is true is to write a mathematical proof. To establish that a statement is false, we often find a counterexample. (These ideas will be explored later in this chapter.) So mathematicians must be able to discover and construct proofs. In addition, once the discovery has been made, the mathematician must be able to communicate this discovery to others who speak the language of mathematics. We will be dealing with these ideas throughout the text.

For now, we want to focus on what happens when we start a proof. One thing that mathematicians often do is to make a conjecture beforehand as to whether the statement is true or false. This is done through exploration. The role of exploration in mathematics is often difficult because the goal is not to find a specific answer but simply to investigate. Following are some techniques of exploration that might be helpful.

Techniques of Exploration

- **Guesswork and Conjectures** Formulate and write down questions and conjectures. When we make a guess in mathematics, we usually call it a conjecture.

- **Examples** Constructing appropriate examples is extremely important. Exploration often requires looking at lots of examples. In this way, we can gather information that provides evidence that a statement is true, or we might find an example that shows the statement is false. This type of example is called a counterexample.

- **Use of Prior Knowledge** This also is very important. We cannot start from square one every time we explore a statement. We must make use of our acquired mathematical knowledge.

- **Cooperation and Brainstorming** Working together is often more fruitful than working alone. When we work with someone else, we can compare notes and articulate our ideas. Thinking out loud is often a useful brainstorming method that helps generate new ideas.

Activity 1.2 (Explorations). Use the techniques of exploration to investigate each of the following statements. Can you make a conjecture as to whether the statement is true or false? Can you determine whether it is true or false?

1. $(a + b)^2 = a^2 + b^2$, for all real numbers a and b.

2. There are integers x and y such that $2x + 5y = 9$.

3. If x and y are odd integers, then $x \cdot y$ is an odd integer.

Conditional Statements

Conditional statements are perhaps the most important type of statement in mathematics.

> **Definition. A conditional statement** A conditional statement is a statement that can be written in the form "If P then Q" where P and Q are statements. For this conditional statement, P is called the **hypothesis** and Q is called the **conclusion**.

Intuitively, "If P then Q" means that Q must be true whenever P is true. Because conditional statements are used so often, a symbolic shorthand notation is used to represent the conditional statement "If P then Q." We will use the notation $P \rightarrow Q$ to represent "If P then Q." It seems reasonable that the truth value (true or false) of the conditional statement $P \rightarrow Q$ depends on the truth values of P and Q. There are four cases to consider:

- P is true and Q is true.

- P is true and Q is false.

- P is false and Q is true.

- P is false and Q is false.

The conditional statement $P \rightarrow Q$ means that Q is true whenever P is true. It says nothing about the truth value of Q when P is false. Using this as a guide, we define the conditional statement $P \rightarrow Q$ to be false only when P is true and Q is false, that is, only when the hypothesis is true and the conclusion is false. In all other cases, $P \rightarrow Q$ is true. This can be summarized in a table that is called a **truth table** for the conditional statement $P \rightarrow Q$. (In Table 1.1, T stands for "true" and F stands for "false.")

P	Q	$P \rightarrow Q$
T	T	T
T	F	F
F	T	T
F	F	T

Table 1.1: Truth Table for $P \rightarrow Q$

The important thing to remember is that the conditional statement $P \rightarrow Q$ has its own truth value. It is either true or false (and not both). So consider the following conditional statement from Preview Activity 3:

If it is raining, then Laura is at the theater.

This statement is false only in the case when it is raining and Laura is not at the theater. In all other cases, it is true.

The fact that there is only one case when a conditional statement is false often provides a method to show that a given conditional statement is false. In Preview Activity 3, you were asked if you thought the following conditional statement was true or false.

If n is a positive integer, then $\left(n^2 - n + 41\right)$ is a prime number.

Perhaps for all of the values you tried for n, $\left(n^2 - n + 41\right)$ turned out to be a prime number. However, if we try $n = 41$, we get

$$n^2 - n + 41 = 41^2 - 41 + 41$$
$$n^2 - n + 41 = 41^2.$$

So in the case where $n = 41$, the hypothesis is true (41 is a positive integer) and the conclusion is false $\left(41^2 \text{ is not prime}\right)$. Therefore, the conditional statement

"If n is a positive integer, then $\left(n^2 - n + 41\right)$ is a prime number"

is false.

Exercises 1.1

1. Which of the following sentences are statements?

(a) $3^2 + 4^2 = 5^2$.

(b) $a^2 + b^2 = c^2$.

(c) If $x^2 = 4$, then $x = 2$.

(d) For each real number t, $\sin^2 t + \cos^2 t = 1$.

(e) $\sin x < \sin\left(\dfrac{\pi}{4}\right)$.

(f) If n is a prime number, then n^2 has three positive divisors.

(g) $1 + \tan^2 \theta = \sec^2 \theta$.

(h) Every rectangle is a parallelogram.

2. Identify the hypothesis and the conclusion for each of the following conditional statements.

(a) If n is a prime number, then n^2 has three positive divisors.

(b) If a is an irrational number and b is an irrational number, then $a \cdot b$ is an irrational number.

(c) If p is a prime number, then $p = 2$ or p is an odd number.

3. Determine whether each of the following conditional statements is true or false.

(a) If $8 < 5$, then $3 = 4$.   **(c)** If $8 < 5$, then $3 + 2 = 5$.

(b) If $5 < 8$, then $3 = 4$.   **(d)** If $5 < 8$, then $3 + 2 = 5$.

4. Determine the conditions under which each of the following conditional sentences will be a true statement.

   (a) If $a + 2 = 5$, then $8 < 5$.   **(b)** If $5 < 8$, then $a + 2 = 5$.

5. Let P be the statement "Student X passed every assignment in Calculus I," and let Q be the statement "Student X received a grade of C or better in Calculus I."

   (a) What does it mean for P to be true? What does it mean for Q to be true?

   (b) Suppose that Student X passed every assignment in Calculus I and received a grade of B$-$, and that the instructor made the statement $P \rightarrow Q$. Would you say that the instructor lied or told the truth?

   (c) Suppose that Student X passed every assignment in Calculus I and received a grade of C$-$, and that the instructor made the statement $P \rightarrow Q$. Would you say that the instructor lied or told the truth?

   (d) Now suppose that Student X did not pass two assignments in Calculus I and received a grade of D, and that the instructor made the statement $P \rightarrow Q$. Would you say that the instructor lied or told the truth?

   (e) How are Parts (5b), (5c), and (5d) related to the truth table for $P \rightarrow Q$?

6. Following is a statement of a theorem which can be proven using calculus or precalculus mathematics. For this theorem, a, b, and c are real numbers.

   Theorem If f is a quadratic function of the form $f(x) = ax^2 + bx + c$ and $a < 0$, then the function f has a maximum value when $x = \dfrac{-b}{2a}$.

   Using **only** this theorem, what can be concluded about each of the following functions?

(a) $g(x) = -8x^2 + 5x - 2$

(b) $h(x) = -\dfrac{1}{3}x^2 + 3x$

(c) $k(x) = 8x^2 - 5x - 7$

(d) $j(x) = -\dfrac{71}{99}x^2 + 210$

(e) $f(x) = -2x^4 - 3x + 7$

(f) $F(x) = -x^4 + x^2 + 9$

7. Following is a statement of a theorem which can be proven using the quadratic formula. For this theorem, a, b, and c are real numbers.

Theorem If f is a quadratic function of the form $f(x) = ax^2 + bx + c$ and $ac < 0$, then the function f has two x-intercepts.

Using **only** this theorem, what can be concluded about each of the following functions?

(a) $g(x) = -8x^2 + 5x - 2$

(b) $h(x) = -\dfrac{1}{3}x^2 + 3x$

(c) $k(x) = 8x^2 - 5x - 7$

(d) $j(x) = -\dfrac{71}{99}x^2 + 210$

(e) $f(x) = -2x^4 - 3x + 7$

(f) $F(x) = -x^4 + x^2 + 9$

1.2 Constructing Direct Proofs

Preview Activity 1 (Closure Properties of Number Systems).

The primary number system used in algebra and calculus is the **real number system**. We usually use the symbol $\mathbb{R}$ to stand for the set of all real numbers. The real numbers consist of the rational numbers and the irrational numbers. The **rational numbers** are those real numbers that can be written as a quotient of two integers (with a nonzero denominator), and the **irrational numbers** are those real numbers that cannot be written as a quotient of two integers. Some common irrational numbers are $\sqrt{2}$, π, and e. We usually use the symbol $\mathbb{Q}$ to represent the set of all rational numbers. (The letter $\mathbb{Q}$ is used because rational numbers are quotients of integers.) There is no standard symbol for the set of all irrational numbers.

One of the basic number systems that we will be working with is the set of **integers**. The integers consist of zero, the positive whole numbers, and the opposites of the positive whole numbers. If n is an integer, we can write $n = \frac{n}{1}$. So each integer is a rational number and hence also a real number.

We usually use the letter $\mathbb{Z}$ to stand for the set of integers. (The letter $\mathbb{Z}$ is from the German word, *Zahlen*, for numbers.) Two of the basic properties of the integers are that the set $\mathbb{Z}$ is **closed under addition** and the set $\mathbb{Z}$ is **closed under multiplication**. This means that

- If x and y are integers, then $x + y$ is an integer; and

- If x and y are integers, then $x \cdot y$ is an integer.

In addition to this, the set of all rational numbers is closed under addition and multiplication. That is:

- If x and y are rational numbers, then $x + y$ is an rational number; and

- If x and y are rational numbers, then $x \cdot y$ is an rational number.

Answer each of the following questions.

1. Is the set of integers closed under subtraction? Explain.

2. Is the set of integers closed under division? Explain.

3. Is the set of rational numbers closed under subtraction? Explain.

4. Is the set of rational numbers closed under division? Explain.

5. Is the set of nonzero rational numbers closed under division? Explain.

Preview Activity 2 (Definition of Even and Odd Integers).

Definitions play a very important role in mathematics. A direct proof of a proposition in mathematics is often a demonstration that the proposition follows logically from certain definitions (and sometimes, previously proven propositions). A **definition** is an agreement that a particular word or phrase will stand for some object, property, or other concept that we expect to refer to often. In many elementary proofs, the answer to the question, "How do we prove a certain proposition?", is often answered by means of a definition.

In Preview Activity 3 from Section 1.1, all of the examples you tried should have indicated that the following conditional statement is true:

If x and y are odd integers, then $x \cdot y$ is an odd integer.

In order to construct a mathematical proof of this conditional statement, we need a precise definition of the concept of an odd integer.

> **Definition.** An integer a is an **even integer** if there exists an integer n such that $a = 2n$. An integer a is an **odd integer** if there exists an integer n such that $a = 2n + 1$.

1. Use the definition of an even integer to explain why 8, -12, 24, and 0 are even integers.

2. Use the definition of an odd integer to explain why 7, -11, 51, 1, and -1 are odd integers.

Besides defining even and odd integers, another purpose of this activity is to indicate that mathematical definitions are not made randomly. In most cases, they are motivated by a mathematical concept that occurs frequently.

3. Are the definitions of even integers and odd integers consistent with your previous ideas about even and odd integers?

Preview Activity 3 (Thinking about a Proof).

In Activity 1.2 from Section 1.1, all of the examples you tried should have indicated that the following proposition is true.

Proposition: If x and y are odd integers, then $x \cdot y$ is an odd integer.

Think about how you might go about proving this proposition. A direct proof of a conditional statement is often a demonstration that the conclusion of the conditional statement follows logically from the hypothesis of the conditional statement. Definitions and previously proven propositions are frequently used to justify each step in the proof.

To help you get started in proving this proposition, try answering the following questions:

1. The proposition is a conditional statement. What is the hypothesis of this conditional statement? What is the conclusion of this conditional statement?

2. If $x = 2$ and $y = 3$, then $x \cdot y = 6$. Does this example prove that the proposition is false? Explain.

3. If $x = 5$ and $y = 3$, then $x \cdot y = 15$. Does this example prove that the proposition is true? Explain

4. What would we have to do to prove that this conditional statement is true?

5. To start a proof of this proposition, we will assume that the hypothesis of the conditional statement is true. So, what do we assume in this case?

6. We need to prove that if the hypothesis is true, then the conclusion is true. So, for this proposition, what do we need to prove?

7. How do we prove that an integer is an odd integer?

Properties of Number Systems

In Preview Activity 1, we introduced notations for the standard number systems we use in mathematics. We also discussed some closure properties of the standard number systems. For this text, it is assumed that the reader is familiar with these closure properties and the basic rules of algebra that apply to all real numbers. That is, it is assumed the reader is familiar with the properties of the real numbers shown in Table 1.2.

For all real numbers x, y, and z:	
Identity Properties	$x + 0 = x$ and $x \cdot 1 = x$
Inverse Properties	$x + (-x) = 0$ and if $x \neq 0$, then $x \cdot \dfrac{1}{x} = 1$.
Commutative Properties	$x + y = y + x$ and $x \cdot y = y \cdot x$
Associate Properties	$(x + y) + z = x + (y + z)$ and $(x \cdot y) \cdot z = x \cdot (y \cdot z)$
Distributive Properties	$x(y + z) = x \cdot y + x \cdot z$ and $(y + z)x = y \cdot x + z \cdot x$

Table 1.2: Properties of the Real Numbers

Constructing a Proof of a Conditional Statement

In order to prove that a conditional statement $P \rightarrow Q$ is true, we only need to prove that Q is true whenever P is true. This is because the conditional statement is true whenever the hypothesis is false. So in a direct proof of

$P \rightarrow Q$, we assume that P is true, and using this assumption, we proceed through a logical sequence of steps to arrive at the conclusion that Q is true.

Unfortunately, it is often not easy to discover how to start this logical sequence of steps or how to get to the conclusion that Q is true. We will attempt to describe a method of exploration that often can help in discovering the steps of a proof. This method will involve working forward from the hypothesis, P, and backward from the conclusion, Q. We will use a device called the **"know-show table"** to help organize our thoughts and the steps of the proof. This will be illustrated with the proposition from Preview Activity 3.

Proposition: *If x and y are odd integers, then $x \cdot y$ is an odd integer.*

The first step is to clearly identify the hypothesis, P, and the conclusion ,Q, of the conditional statement. The hypothesis consists of everything you are assuming, and the conclusion consists of everything you are trying to prove. In this case, we have the following:

P: x and y are odd integers.
Q: $x \cdot y$ is an odd integer.

We now treat P as what we know (we have assumed it to be true) and treat Q as what we want to show (that is, the goal). So we organize this by using P as the first step in the know portion of the table and Q as the last step in the show portion of the table. We will put the know portion of the table at the top and the show portion of the table at the bottom.

Step	Know	Reason
P	x and y are odd integers.	Hypothesis
$P1$		
$\vdots$	$\vdots$	$\vdots$
$Q1$		
Q	$x \cdot y$ is an odd integer.	?
Step	**Show**	**Reason**

We have not yet filled in the reason for the last step because we do not yet know how we will reach the goal. The idea now is to ask ourselves questions about what we know and what we are trying to prove. We usually start with the conclusion that we are trying to prove.

We will start with what is called a **backward question.** The basic form of the question is, "Under what conditions can we conclude that Q is true?"

How we ask the question is crucial since we must be able to answer it. We should first try to ask and answer the question in an abstract manner and then apply it to the particular form of statement Q.

In this case, we are trying to prove that some integer is an odd integer. So our backward question could be, "How do we prove that an integer is odd?" At this time, the only way we have of answering this question is to use the definition of an odd integer. So our answer could be, "We need to prove that there exists an integer q such that the integer equals $2q + 1$." We apply this answer to statement Q and insert it as the next to last line in the know-show table.

Step	Know	Reason
P	x and y are odd integers.	Hypothesis
$P1$		
$\vdots$	$\vdots$	$\vdots$
$Q1$	There exists an integer q such that $x \cdot y = 2q + 1$	
Q	$x \cdot y$ is an odd integer.	Definition of an odd integer.
Step	**Show**	**Reason**

The idea is to write the first step for the beginning of the proof (P) and the steps for the end of the proof (Q and $Q1$). We then try to fill in the steps for the middle of the proof, working backward from $Q1$ and working forward from P.

We now focus our effort on proving statement $Q1$ since we know that if we can prove $Q1$, then we can conclude that Q is true. We can ask a backward question about $Q1$ such as "How can we prove that there exists an integer q such that $x \cdot y = 2q + 1$?" We may not have a ready answer for this question. Instead, we look at the know portion of the table and try to connect the know portion to the show portion. To do this, we work forward from step P.

Working forward from P involves asking a **forward question.** The basic form of this type of question is, "What can we conclude from the fact that P is true?" In this case, we can use the definition of an odd integer to conclude that there exist integers m and n such that $x = 2m + 1$ and $y = 2n + 1$.

Important Note: When we claimed there exist integers m and n, we were very careful not to use the letter q (again) to denote these integers. If we had used q again, we would be claiming that the same integer that gives $x \cdot y = 2q + 1$ also gives $x = 2q + 1$. This is also why we used m and n for the integers x and y since there is no guarantee that x equals y. The basic rule of thumb is to use a different symbol for each new object we introduce in a proof. So our know-show table becomes:

Step	Know	Reason
P	x and y are odd integers.	Hypothesis
$P1$	There exist integers m and n such that $x = 2m + 1$ and $y = 2n + 1$.	Definition of an odd integer
$\vdots$	$\vdots$	$\vdots$
$Q1$	There exists an integer q such that $x \cdot y = 2q + 1$.	
Q	$x \cdot y$ is an odd integer.	Definition of an odd integer
Step	**Show**	**Reason**

We must always be looking for a way to link the two portions of the table so that we can proceed from P to Q. There are conclusions we can make from statement $P1$, but as we proceed, we must always keep in mind the form of statement $Q1$. The next forward question is, "What can I conclude about $x \cdot y$ from $P1$?" One way to answer this is to use our prior knowledge of algebra. That is, we can write $x \cdot y = (2m + 1)(2n + 1)$ and use algebra to try to arrive at step $Q1$.

Step	Know	Reason
P	x and y are odd integers.	Hypothesis
$P1$	There exist integers m and n such that $x = 2m + 1$ and $y = 2n + 1$.	Definition of an odd integer
$P2$	$x \cdot y = (2m + 1)(2n + 1)$	Substitution
$P3$	$x \cdot y = 4mn + 2m + 2n + 1$	Algebra
$\vdots$	$\vdots$	$\vdots$
$Q1$	There exists an integer q such that $x \cdot y = 2q + 1$.	
Q	$x \cdot y$ is an odd integer.	Definition of an odd integer
Step	**Show**	**Reason**

Now compare statement $P3$ to statement $Q1$. Sometimes the difficult part at this point is the realization that q stands for some integer and that we only have to show that $x \cdot y$ equals two times some integer plus one. Can we make that conclusion from $P3$? The answer is yes because we can factor a 2 from the first three terms on the right side of the equation. We can now complete the table showing the outline of the proof as follows:

Step	Know	Reason
P	x and y are odd integers.	Hypothesis
$P1$	There exist integers m and n such that $x = 2m + 1$ and $y = 2n + 1$.	Definition of an odd integer
$P2$	$x \cdot y = (2m + 1)(2n + 1)$	Substitution
$P3$	$x \cdot y = 4mn + 2m + 2n + 1$	Algebra
$P4$	$x \cdot y = 2(2mn + m + n) + 1$	Algebra
$P5$	$(2mn + m + n)$ is an integer.	Closure properties of the integers
$Q1$	There exists an integer q such that $x \cdot y = 2q + 1$.	
Q	$x \cdot y$ is an odd integer.	Definition of an odd integer

It is very important to realize that we have only constructed an outline of a proof. Mathematical proofs are not written in table form. They are written in narrative form using complete sentences and correct paragraph structure, and they follow certain conventions used in writing mathematics. In addition, most proofs are written only from the forward perspective. That is, although the use of the backward process was essential in discovering the proof, when we write the proof in narrative form, we use the forward process described in the preceding table. A completed proof follows.

Theorem 1.3. *If x and y are odd integers, then $x \cdot y$ is an odd integer.*

Proof. We assume that x and y are odd integers and will prove that $x \cdot y$ is an odd integer. Then there exist integers m and n such that

$$x = 2m + 1 \text{ and } y = 2n + 1.$$

Using algebra, we obtain

$$x \cdot y = (2m + 1)(2n + 1)$$
$$= 4mn + 2m + 2n + 1$$
$$= 2(2mn + m + n) + 1.$$

Since m and n are integers and the integers are closed under addition and multiplication, we conclude that $(2mn + m + n)$ is an integer. This means that $x \cdot y$ has been written in the form $(2q + 1)$ for some integer q, and hence, $x \cdot y$ is an odd integer. Consequently, it has been proven that if x and y are odd integers, then $x \cdot y$ is an odd integer. ■

Writing Guidelines

In this section, the emphasis is on constructing an outline of a proof using a know-show table. However, some proof writing will be done, and it is important to introduce some more writing guidelines. The preceding proof was written according to the guidelines introduced in Section 1.1 and the following guidelines for writing proofs:

1. **Begin with a carefully worded statement of the theorem or result to be proven.**

   The statement should be a simple declarative statement of the problem. Then skip a line and write "Proof" in italics or boldface font (when using a word processor). Begin the proof on the same line. Make sure that all paragraphs can be easily identified. Skipping a line between paragraphs or indenting each paragraph can accomplish this.

   Proof. We assume that . . .

2. **Begin the proof with a statement of your assumptions.**

   Follow the statement of your assumptions with a statement of what you will prove.

   Proof. We assume that x and y are odd integers and will prove that $x \cdot y$ is an odd integer.

3. **Use the pronoun "we."**

   If a pronoun is used in a proof, the usual convention is to use "we" instead of "I." The idea is that the author and the reader are proving the theorem together.

4. **Use italics for variables.**

   When using a word processor to write mathematics, the word processor needs to be capable of producing the appropriate mathematical symbols and equations. The mathematics that is written with a word processor should look like typeset mathematics. This means that variables need to be italicized, boldface is used for vectors, and regular font is used for mathematical terms such as the names of the trigonometric functions and logarithmic functions.

5. Use complete sentences and proper paragraph structure.

Good grammar is an important part of any writing. Therefore, conform to the accepted rules of grammar. Pay careful attention to the structure of sentences. Write proofs using **complete sentences** but avoid run-on sentences. Also, do not forget punctuation, and always use a spell checker when using a word processor.

6. Display important equations and mathematical expressions.

Equations and manipulations are often an integral part of the exposition. Do not write equations, algebraic manipulations, or formulas in one column with reasons given in another column (as is often done in geometry texts). Important equations and manipulations should be displayed. This means that they should be centered with blank lines before and after the equation or manipulations, and if one side of an equation does not change, it should not be repeated. For example,

Using algebra, we obtain

$$
\begin{aligned}
x \cdot y &= (2m + 1)(2n + 1) \\
&= 4mn + 2m + 2n + 1 \\
&= 2(2mn + m + n) + 1
\end{aligned}
$$

Since m and n are integers, we conclude that ...

7. Tell the reader when the proof has been completed.

Perhaps the best way to do this is to say outright that, "This completes the proof." Although it may seem repetitive, a good alternative is to finish a proof with a sentence that states precisely what has been proven. In any case, it is usually good practice to use some "end of proof symbol" such as ∎.

Some Comments about Constructing Proofs

1. When we constructed the know-show table prior to writing a proof for Theorem 1.3, we had only one answer for the backward question and one answer for the forward question. Often, there can be more than one answer for these questions. For example, consider the following statement:

$$\text{If } x \text{ is an odd integer, then } x^2 \text{ is an odd integer.}$$

The backward question for this could be, "How do I prove that an integer is an odd integer?" One way to answer this is to use the definition of an odd integer, but another way is to use the result of Theorem 1.3. That is, we can prove an integer is odd by proving that it is a product of two odd integers.

The difficulty then is deciding which answer to use. Sometimes we can tell by carefully watching the interplay between the forward process and the backward process. Other times, we may have to work with more than one possible answer.

2. Sometimes we can use previously proven results to answer a forward question or a backward question. This was the case in the example given in Comment (1) where Theorem 1.3 was used to answer a backward question.

3. Although we start with two separate processes (forward and backward), the key to constructing a proof is to find a way to link these two processes. This can be difficult. One way to proceed is to use the know portion of the table to motivate answers to backward questions and to use the show portion of the table to motivate answers to forward questions.

4. Answering a backward question can sometimes be tricky. If the goal is the statement Q, we must construct the know-show table so that if we know that $Q1$ is true, then we can conclude the Q is true. It is sometimes easy to answer this in a way that if it is known that Q is true, then we can conclude that $Q1$ is true. For example, suppose the goal is to prove

$$y^2 = 4$$

where y is a real number. A backward question could be, "How do we prove the square of a real number equals four?" One possible answer is to prove that the real number equals 2. Another way is to prove that the real number equals -2. This is an appropriate backward question, and these are appropriate answers.

However, if the goal is to prove

$$y = 2$$

where y is a real number, we could ask, "How do we prove a real number equals 2?" It is not appropriate to answer this question with "prove that the square of the real number equals 4." That is, we should not have the show portion of the table as follows:

$Q1$	$y^2 = 4$	
Q	$y = 2$	Square root of both sides
Step	**Show**	**Reason**

This is because if $y^2 = 4$, then it is not necessarily true that $y = 2$.

Activity 1.4 (Constructing a Know-Show Table). Construct a know-show table for the following statement:

If x is an odd integer, then x^2 is an odd integer.

Activity 1.5 (Constructing a Know-Show Table). Complete the know-show table for the following statement:

If x and y are real numbers such that $x^2 + y^2 = 13$ and $y^2 - x = 1$, then $|y| = 2$.

Step	**Know**	**Reason**
P	x and y are real numbers, $x^2 + y^2 = 13,$ and $y^2 - x = 1.$	Hypothesis
$P1$	$x = y^2 - 1$	Solve $y^2 - x = 1$ for x
$\vdots$	$\vdots$	$\vdots$
$Q1$	$y^2 = 4$	
Q	$|y| = 2$	$\sqrt{y^2} = |y|$ and $\sqrt{4} = 2$
Step	**Show**	**Reason**

Exercises 1.2

1. Construct a know-show table for each of the following statements:

    (a) If m is an even integer, then $m + 1$ is an odd integer.
    (b) If m is an odd integer, then $m + 1$ is an even integer.

2. Write a complete formal proof of one of the statements in Exercise (1).

3. Construct a know-show table for each of the following statements:

    (a) If x is an even integer and y is an even integer, then $x + y$ is an even integer.
    (b) If x is an even integer and y is an odd integer, then $x + y$ is an odd integer.
    (c) If x is an odd integer and y is an odd integer, then $x + y$ is an even integer.

4. Write a complete formal proof of one of the statements in Exercise (3).

5. Construct a know-show table for each of the following statements:

    (a) If x is an even integer and y is an integer, then $x \cdot y$ is an even integer.
    (b) If x is an even integer, then x^2 is an even integer.
    (c) If x is an odd integer, then x^2 is an odd integer.

6. Write a complete formal proof of one of the statements in Exercise (5).

7. In this section, it was noted that there is often more than one way to answer a backward question. For example, if the backward question is "How can we prove that two real numbers are equal?", one possible answer is to prove that their difference equals 0. Another possible answer is to prove that the first is less than or equal to the second and that the second is less than or equal to the first.

(a) Give at least one more answer to the backward question, "How can we prove that two real numbers are equal?"

(b) List as many answers as you can for the backward question, "How can we prove that a real number is equal to zero?"

(c) List as many answers as you can for the backward question, "How can we prove that two lines are parallel?"

(d) List as many answers as you can for the backward question, "How can we prove that a triangle is isosceles?"

8. Is the following statement true or false?

If a and b are nonnegative real numbers and $a+b = 0$, then $a = 0$.

Either give a counterexample to show that it is false or outline a proof by completing a know-show table.

9. An integer a is said to be **congruent to 1 modulo 3** if there exists an integer n such that $a = 3n + 1$. An integer a is said to be **congruent to 2 modulo 3** if there exists an integer m such that $a = 3m + 2$.

(a) Give examples of at least four different integers that are congruent to 1 modulo 3.

(b) Give examples of at least four different integers that are congruent to 2 modulo 3.

(c) By multiplying pairs of integers from the list in Exercise (9a), does it appear that the following statement is true or false?

If a is congruent to 1 modulo 3 and b is congruent to 1 modulo 3, then $a \cdot b$ is congruent to 1 modulo 3.

10. Write a proof for each of the following statements:

(a) If a is congruent to 1 modulo 3 and b is congruent to 1 modulo 3, then $a + b$ is congruent to 2 modulo 3.

(b) If a is congruent to 2 modulo 3 and b is congruent to 2 modulo 3, then $a + b$ is congruent to 1 modulo 3.

(c) If a is congruent to 1 modulo 3 and b is congruent to 2 modulo 3, then $a \cdot b$ is congruent to 2 modulo 3.

(d) If a is congruent to 2 modulo 3 and b is congruent to 2 modulo 3, then $a \cdot b$ is congruent to 1 modulo 3.

11. **(a)** Let a, b, and c be real numbers with $a \neq 0$. Use the quadratic formula to write the two solutions for the quadratic equation $ax^2 + bx + c = 0$.

(b) Prove that if a, b, and c are real numbers for which $a > 0$ and $c < 0$, then one solution of the quadratic equation

$$ax^2 + bx + c = 0$$

is a positive real number.

Chapter 2

Logical Reasoning

2.1 Predicates, Sets, and Quantifiers

Preview Activity 1 (Sentences That Are Not Statements).

1. Not all mathematical sentences are statements. For example, an equation such as

$$x^2 - 5 = 0$$

is not a statement.

In this sentence, the symbol x is a **variable**. It represents a number that may be chosen from some specified set of numbers. The sentence (equation) becomes true or false when a specific number is selected for x.

   (a) How many solutions does this equation have if x is restricted to being a real number?

   (b) How many solutions does this equation have if x is restricted to being an integer?

2. Compare the following two sentences:

   - $\sqrt{x}$ is a real number.
   - For each real number x, if $x \geq 0$, then $\sqrt{x}$ is a real number.

   What is the difference between these two sentences?

Preview Activity 2 (Variables).

For this course, we will consider a **set** to simply be some well-defined collection of objects. Some sets that we will frequently use are the usual number systems. In Preview Activity 1 from Section 1.2, we indicated that we would use the symbol $\mathbb{R}$ to stand for the set of all **real numbers**, the symbol $\mathbb{Q}$ to stand for the set of all **rational numbers**, and the symbol $\mathbb{Z}$ to stand for the set of all **integers**. We will also use the symbol $\mathbb{N}$ to stand for the set of all **natural numbers**. The natural numbers are the positive integers (that is, the positive whole numbers).

A **variable** is a symbol representing an unspecified object that can be chosen from some specified set of objects. This specified set of objects is agreed to in advance and is frequently called the **universal set**.

1. What real numbers will make the sentence "$\sqrt{x}$ is a real number" a true statement when substituted for x?

2. What real numbers will make the sentence "$\sin^2 x + \cos^2 x = 1$" a true statement when substituted for x?

3. What real numbers will make the sentence "$y^2 - 2y - 5 = 0$" a true statement when substituted for y?

4. What natural numbers will make the sentence "$\sqrt{n}$ is a natural number" a true statement when substituted for n?

5. What real numbers will make the sentence

$$\int_0^y t^2 dt > 9$$

a true statement when substituted for y?

Sets and Set Notation

The theory of sets is fundamental to mathematics in the sense that many areas of mathematics use set theory and its language and notation. This language and notation must be understood if we are to effectively communicate in mathematics. At this point, we will give a very brief introduction to some of the terminology used in set theory.

A **set** is a collection of objects that can be thought of as a single entity itself. For example, we can think of the set of all even integers. Even though we cannot write down all the integers that are in this set, it is still a perfectly well-defined set. By this we mean that if we are given a specific integer, we can tell whether or not it is in the set of all even integers. So for this course, a **set** is simply a well-defined collection of objects.

- If A is a set and y is one of the objects in the set A, we write $y \in A$ and read this as "y is an element of A" or "y is a member of A."

- If an object z is not an element in the set A, we write $z \notin A$ and read this as "z is not an element of A."

The most basic way of specifying the elements of a set is to simply list the elements of that set. This works well when the set contains only a small number of objects. The usual practice is to list these elements between braces. For example, if the set B consists of the integer solutions of the equation $x^2 = 9$, we would write

$$B = \{-3,\ 3\}.$$

If it is not possible to list all of the elements of a set, then we list several of them and then write a series of three dots ($\ldots$) to indicate that the pattern continues. For example, if E is the set of all even natural numbers, we could write

$$E = \{2,\ 4,\ 6,\ \ldots\}.$$

Listing the elements of a set inside braces is called the **roster method** of specifying the elements of the set.

Two sets, A and B, are considered to be **equal** when they have precisely the same elements. In this case, we write $A = B$. If the sets C and D are not equal, we write $C \neq D$. For example,

$$\{1,3,5\} = \{3,5,1\}$$
$$\{4,8,12\} = \{4,4,8,12,12\}$$
$$\{5,10\} = \{5,10,5\}$$
$$\{5,10\} \neq \{5,10,15\}.$$

In each of the first three examples, the two sets have exactly the same elements even though the elements may be repeated or written in a different order.

Variables and Predicates

As we have seen in Section 1.1 and in the Preview Activities for this section, not all mathematical sentences are statements. This is often true if the sentence contains a variable. The following terminology is useful in working with sentences and statements.

Definition. A **variable** is a symbol representing an unspecified object that can be chosen from a given set U. The set U is called the **universal set for the variable.** It is the set of specified objects from which objects may be chosen to substitute for the variable.

A **constant** is a specific member of the universal set.

A **predicate** is a sentence $P(x_1, x_2, \ldots, x_n)$ involving variables $x_1, x_2, \ldots, x_n$ with the property that when specific values from the universal set are assigned to $x_1, x_2, \ldots, x_n$, then the resulting sentence is either true or false. That is, the resulting sentence is a statement. A predicate is also called an **open sentence** or a **propositional function**.

Notation: One reason a predicate is sometimes called a propositional function is the fact that we use function notation $P(x_1, x_2, \ldots, x_n)$ for a predicate in n variables. When there is only one variable, such as x, we write $P(x)$, which is read "P of x." In this notation, x represents an arbitrary element of the universal set, and $P(x)$ represents a sentence. When we substitute a specific element of the universal set for x, the resulting sentence becomes a statement. This is illustrated in the next example.

Example 2.1. If the universal set is $\mathbb{R}$, then the sentence "$x^2 - 3x - 10 = 0$" is a predicate involving the one variable x.

- If we substitute $x = 2$, we obtain the false statement "$2^2 - 3 \cdot 2 - 10 = 0$."

- If we substitute $x = 5$, we obtain the true statement "$5^2 - 3 \cdot 5 - 10 = 0$."

In this example, we can let $P(x)$ be the predicate "$x^2 - 3x - 10 = 0$ " and then say that $P(2)$ is false and $P(5)$ is true.

Using similar notation, we can let $Q(x, y)$ be the predicate "$x + 2y = 7$." This predicate involves two variables. Then,

- $Q(1, 1)$ is false since "$1 + 2 \cdot 1 = 7$" is false; and

- $Q(3, 2)$ is true since "$3 + 2 \cdot 2 = 7$" is true.

Activity 2.2. Working with Predicates

1. Assume the universal set for all variables is $\mathbb{R}$ and let $P(x)$ be the predicate "$x^2 < 4$."

    (a) Find two values of x for which $P(x)$ is false.

    (b) Find two values of x for which $P(x)$ is true.

    (c) Describe the set of all x for which $P(x)$ is true.

2. Assume the universal set for all variables is $\mathbb{Z}$, and let $R(x, y, z)$ be the predicate "$x^2 + y^2 = z^2$."

    (a) Find two different examples for which $R(x, y, z)$ is false.

    (b) Find two different examples for which $R(x, y, z)$ is true.

3. Assume the universal set for the variable x is $\mathbb{R}$ and the universal set for the variable y is the set of all real numbers that are greater than or equal to 2. Let $Q(x, y)$ be the predicate "$y = x^2 + 2$."

    (a) Find two different examples for which $Q(x, y)$ is false.

    (b) Find two different examples for which $Q(x, y)$ is true.

    (c) Use a graph in the coordinate plane to describe the set of all ordered pairs (x, y) for which $Q(x, y)$ is true.

Without using the term, Example 2.1 and Activity 2.2 (and Preview Activity 2) dealt with a concept called the truth set of a predicate.

> **Definition.** The **truth set of a predicate with one variable** is the collection of objects in the universal set that can be substituted for the variable to make the predicate a true statement.

For example,

- If the universal set is $\mathbb{R}$, then the truth set of "$x^2 - 3x - 10 = 0$" is $\{-2, 5\}$.

- If the universal set is $\mathbb{N}$, then the truth set of "$\sqrt{n} \in \mathbb{N}$" is $\{1, 4, 9, 16, \ldots\}$.

Set Builder Notation

Sometimes it is not possible to list all the elements of a set. For example, if the universal set is $\mathbb{R}$, we cannot list all the elements of the truth set of "$x^2 < 4$." In this case, it is sometimes convenient to use the so-called **set builder notation** in which the set is defined by stating a rule which all elements of the set must satisfy. If $P(x)$ is a predicate in the variable x, then the notation

$$\{x \in U \mid P(x)\}$$

stands for the set of all elements x in the universal set U for which $P(x)$ is true. If it is clear what set is being used for the universal set, this notation is sometimes shortened to $\{x \mid P(x)\}$. This is usually read as "the set of all x such that $P(x)$." The vertical bar stands for the phrase "such that." Some writers will use a colon (:) instead of the vertical bar.

For a nonmathematical example, P could be the property that a college student is a mathematics major. Then $\{x \mid P(x)\}$ denotes the set of all college students who are mathematics majors. This would usually be written as

$$\{x \mid x \text{ is a college student who is a mathematics major}\}.$$

Example 2.3. Assume the universal set is $\mathbb{R}$ and $P(x)$ is "$x^2 < 4$." The truth set of $P(x)$ can be written as

$$\left\{x \in \mathbb{R} \mid x^2 < 4\right\}.$$

However, if we solve the inequality $x^2 < 4$, we obtain $-2 < x < 2$. So we could also write the truth set as

$$\{x \in \mathbb{R} \mid -2 < x < 2\}.$$

Activity 2.4 (Working with Truth Sets). Let $P(x)$ be the predicate "$x^2 \leq 9$."

1. If the universal set is $\mathbb{R}$, write the truth set of $P(x)$ using set builder notation.

2. If the universal set is $\mathbb{Z}$, then what is the truth set of $P(x)$? You should be able to use the roster method to list all the elements of this truth set.

3. Are the truth sets in parts (1) and (2) equal?

The purpose of this activity is to show that the truth set of a predicate depends on the predicate and on the universal set.

An Introduction to Quantifiers

We have seen that one way to create a statement from a predicate is to substitute a specific element from the appropriate universal set for each variable in the predicate. Another way is to make some claim about the truth set of the predicate. This is often done by using a quantifier. For example, if the universal set is $\mathbb{R}$, then is the following sentence a statement or a predicate?

For each real number x, $x^2 > 0$.

The phrase "For each real number x" is said to quantify the variable that follows it in the sense that the sentence is claiming that something is true for all real numbers. This sentence is a statement (which happens to be false).

> **Definition.** The phrase "for every" (or its equivalents) is called a **universal quantifier**. The phrase "there exists" (or its equivalents) is called an **existential quantifier**. The symbol $\forall$ is used to denote a universal quantifier, and the symbol $\exists$ is used to denote an existential quantifier.

For example, the statement "For each real number x, $x^2 > 0$" could be written in symbolic form as follows:

$$(\forall x \in \mathbb{R}) \left(x^2 > 0 \right).$$

The followiing is an example of a statement involving an existential quantifier.

There exists a rational number x such that $x^2 - 3x - 7 = 0$.

This could be written in symbolic form as

$$(\exists x \in \mathbb{Q}) \left(x^2 - 3x - 7 = 0 \right).$$

This statement is false because there are no rational numbers that are solutions of the quadratic equation $x^2 - 3x - 7 = 0$.

The following table summarizes the facts about the two types of quantifiers.

A statement involving	Often has the form	The statement is true provided that
a universal quantifier: $(\forall x, P(x))$	"For every x, $P(x)$," where $P(x)$ is a predicate.	Every value of x in the universal set makes $P(x)$ true.
an existential quantifier: $(\exists x, P(x))$	"There exists an x such that $P(x)$," where $P(x)$ is a predicate.	There is at least one value of x in the universal set that makes $P(x)$ true.

In effect, the table indicates that the universally quantified statement is true provided that the truth set of the predicate equals the universal set, and the existentially quantified statement is true provided that the truth set of the predicate contains at least one element.

The Empty Set

When a set contains no elements, we say that the set is the **empty set**. For example, if the universal set is the set of rational numbers $\mathbb{Q}$, then the truth set of the predicate $x^2 - 3x - 7 = 0$ is the empty set. This means that the statement $(\exists x \in \mathbb{Q})\left(x^2 - 3x - 7 = 0\right)$ is false. In mathematics, the empty set is usually designated by the symbol $\emptyset$. We usually read the symbol $\emptyset$ as "the empty set" or "the null set." (The symbol $\emptyset$ is actually the last letter in the Danish-Norwegian alphabet.)

Forms of Quantified Statements in English

There are many ways to write statements involving quantifiers in English. In some cases, the quantifiers are not apparent, and this often happens with conditional statements. The following examples will illustrate these points. Each example contains a quantified statement written in symbolic form followed by several ways to write the statement in English.

1. $(\forall x \in \mathbb{R})\left(x^2 > 0\right)$.

   - For any real number x, $x^2 > 0$.
   - For each real number x, $x^2 > 0$.
   - The square of every real number is greater than 0.
   - The square of a real number is greater than 0.
   - If $x \in \mathbb{R}$, then $x^2 > 0$.

In the second to the last example, the quantifier is not stated explicitly. Care must be taken when reading this because it really does say the same thing as the previous examples.

The last example illustrates the fact that conditional statements often contain a "hidden" universal quantifier.

If the universal set is $\mathbb{R}$, then the truth set of the predicate $x^2 > 0$ is the set of all nonzero real numbers. That is, the truth set is

$$\{x \in \mathbb{R} \mid x \neq 0\}.$$

So the preceding statements are false. For the conditional statement, the example using $x = 0$ produces a true hypothesis and a false conclusion. This is a **counterexample** that shows that the statement with a universal quantifier is false.

2. $(\exists x \in \mathbb{R}) \left(x^2 = 5\right)$.

- There exists a real number x such that $x^2 = 5$.
- $x^2 = 5$ for some real number x.
- There is a real number whose square equals 5.

The second example is usually not used since it is not considered good writing practice to start a sentence with a mathematical symbol.

If the universal set is $\mathbb{R}$, then the truth set of the predicate "$x^2 = 5$" is $\left\{-\sqrt{5}, \sqrt{5}\right\}$. So these are all true statements.

When the Truth Set Is the Universal Set

The truth set of a predicate can be the universal set. For example, if the universal set is the set of real numbers $\mathbb{R}$, then the truth set of the predicate

$$x + 0 = x$$

is $\mathbb{R}$. Notice that the sentence "$x + 0 = x$" has not been quantified and a particular element of the universal set has not been substituted for the variable x. Even though the truth set for this sentence is the universal set, we will adopt the convention that in this case, unless the quantifier is explicitly stated, we will consider the sentence to be a predicate or open sentence. So, with this convention, if the universal set is $\mathbb{R}$, then

- $x + 0 = x$ is a predicate;

- $(\forall x \in \mathbb{R})\,(x + 0 = x)$ is a statement;

- $(\exists x \in \mathbb{R})\,(x + 0 = x)$ is a statement.

Exercises 2.1

1. Use the roster method to specify the elements in each of the following sets:

   (a) $\{x \in \mathbb{R} \mid 2x^2 + 3x - 2 = 0\}$ **(d)** $\{x \in \mathbb{N} \mid x^2 < 25\}$

   (b) $\{x \in \mathbb{Z} \mid 2x^2 + 3x - 2 = 0\}$ **(e)** $\{y \in \mathbb{Q} \mid |y - 2| = 2.5\}$

   (c) $\{x \in \mathbb{Z} \mid x^2 < 25\}$ **(f)** $\{y \in \mathbb{Z} \mid |y - 2| = 2.5\}$

2. Which of the following sets are equal to the set
   $$\left\{x \in \mathbb{R} \mid x\,(x + 2)^2 \left(x - \tfrac{3}{2}\right) = 0\right\}?$$

   (a) $\{-2, 0, 3\}$ **(c)** $\{-2, -2, 0, \tfrac{3}{2}\}$

   (b) $\{\tfrac{3}{2}, -2, 0\}$ **(d)** $\{-2, 3\}$

3. Use set builder notation to specify the following sets:

   (a) The set of all integers greater than or equal to 5

   (b) The set of all even integers

   (c) The set of all positive rational numbers

   (d) The set of all real numbers greater than 1 and less than 7

   (e) The set of all real numbers whose square is greater than 10

4. Each of the following sentences is a predicate or a statement. Assume that the universal set for each variable in these sentences is the set of all real numbers. If a sentence is a predicate, determine its truth set. If a sentence is a statement, determine whether it is true or false.

   (a) $\forall a \in \mathbb{R},\ a + 0 = a$.

   (b) $3x - 5 = 9$.

   (c) $\sqrt{x} \in \mathbb{R}$.

(d) $\forall x \in \mathbb{R}$, $\sin(2x) = 2(\sin x)(\cos x)$.

(e) $\sin(2x) = 2(\sin x)(\cos x)$.

(f) $\exists x \in \mathbb{R}$ such that $x^2 + 1 = 0$.

(g) $\forall x \in \mathbb{R}$, $x^3 \geq x^2$.

(h) $x^2 + 1 = 0$.

(i) If $x \geq 1$, then $x^2 \geq 1$.

(j) $\forall x \in \mathbb{R}, \exists y \in \mathbb{R}$ such that $x + y = 0$.

(k) $\exists y \in \mathbb{R}$ such that $\forall x \in \mathbb{R}, x + y = 0$.

(l) $\sqrt{x} \in \mathbb{Z}$.

5. Assume that the universal set is $\mathbb{Z}$. Consider the following sentence:

$$(\exists t \in \mathbb{Z})(t \cdot x = 20)$$

(a) Explain why this sentence is not a statement.

(b) If 5 is substituted for x, is the resulting sentence a statement? If it is a statement, is the statement true or false?

(c) If 8 is substituted for x, is the resulting sentence a statement? If it is a statement, is the statement true or false?

(d) If -2 is substituted for x, is the resulting sentence a statement? If it is a statement, is the statement true or false?

(e) What is the truth set of the predicate $(\exists t \in \mathbb{Z})(t \cdot x = 20)$?

2.2   Statements and Logical Operators

Preview Activity 1 (Truth Values of Statements).
Use the following statements for all of this Preview Activity:

- P is the statement "It is raining."

- Q is the statement "Daisy is playing golf."

In each of the following four questions, a truth value will be assigned to statements P and Q. For example, in Question (1), we will assume that each statement is true. In Question (2), we will assume that P is true and Q is false.

1. Assume that it is raining (P is true) and Daisy is playing golf (Q is true). Do you think the following statements are true or false?

    (a) (P and Q)   It is raining and Daisy is playing golf.

    (b) (P or Q)   It is raining or Daisy is playing golf.

    (c) (If P then Q)   If it is raining, then Daisy is playing golf.

    (d) (P or not Q)   It is raining or Daisy is not playing golf.

    (e) (P and not Q)   It is raining and Daisy is not playing golf.

2. Assume that it is raining (P is true) and Daisy is not playing golf (Q is false). Do you think the following statements are true or false?

    (a) (P and Q)   It is raining and Daisy is playing golf.

    (b) (P or Q)   It is raining or Daisy is playing golf.

    (c) (If P then Q)   If it is raining, then Daisy is playing golf.

    (d) (P or not Q)   It is raining or Daisy is not playing golf.

    (e) (P and not Q)   It is raining and Daisy is not playing golf.

3. Assume that it is not raining (P is false) and Daisy is playing golf (Q is true). Do you think the following statements are true or false?

    (a) (P and Q)   It is raining and Daisy is playing golf.

    (b) (P or Q)   It is raining or Daisy is playing golf.

    (c) (If P then Q)   If it is raining, then Daisy is playing golf.

    (d) (P or not Q)   It is raining or Daisy is not playing golf.

    (e) (P and not Q)   It is raining and Daisy is not playing golf.

4. Assume that it is not raining (P is false) and Daisy is not playing golf (Q is false). Do you think the following statements are true or false?

    (a) (P and Q)   It is raining and Daisy is playing golf.

    (b) (P or Q)   It is raining or Daisy is playing golf.

    (c) (If P then Q)   If it is raining, then Daisy is playing golf.

    (d) (P or not Q)   It is raining or Daisy is not playing golf.

    (e) (P and not Q)   It is raining and Daisy is not playing golf.

Preview Activity 2 (Truth Tables).

In Section 1.1, we introduced the concept of a truth table for the conditional statement $P \to Q$. See Table 1.1 on page 8.

The symbol $\to$ is called a **logical operator** and by using it, we can form a new statement from existing statements. In this case, we form the conditional statement "If P then Q", or symbolically, $P \to Q$. Some other logical operators use the words "and," "or," and "not." Given existing statements P and Q, we can form the new statements "not P," "P and Q," and "P or Q."

Your task is to complete the following truth tables as you think they should be completed.

P	not P
T	
F	

P	Q	P and Q
T	T	
T	F	
F	T	
F	F	

P	Q	P or Q
T	T	
T	F	
F	T	
F	F	

P	Q	$Q \to P$
T	T	
T	F	
F	T	
F	F	

P	Q	not Q	not P	not $Q \to$ not P
T	T			
T	F			
F	T			
F	F			

As we have seen in the Preview Activities for this section, it is possible to form new statements from existing statements by connecting the statements with words such as "and" and "or" or by negating the statement. A **logical operator** (or **connective**) on mathematical statements is a word or combination of words that combines one or more mathematical statements to make a new mathematical statement. We often refer to a statement as a **compound statement** if it contains one or more operators. Because some operators are used so frequently in logic and mathematics, we give them names and use special symbols to represent them. The following table gives the names and symbolic forms for the resulting compound statements:

Operator	Common Connective	Usual Form	Symbolic Form
Conjunction	and	P and Q	$P \wedge Q$
Disjunction	or	P or Q	$P \vee Q$
Negation	not	not P	$\neg P$
Conditional	if - then	If P then Q	$P \rightarrow Q$

In Preview Activity 2, we completed truth tables for these compound statements. Following are the standard truth tables used in mathematics and logic.

P	$\neg P$
T	F
F	T

P	Q	$P \wedge Q$
T	T	T
T	F	F
F	T	F
F	F	F

P	Q	$P \vee Q$
T	T	T
T	F	T
F	T	T
F	F	F

P	Q	$P \rightarrow Q$
T	T	T
T	F	F
F	T	T
F	F	T

It is important to understand the use of the operator "or." In mathematics, we use the **"inclusive or"** unless stated otherwise. This means that $P \vee Q$ is true when both P and Q are true and also when only one of them is true. That is, $P \vee Q$ is true when at least one of P or Q is true, or $P \vee Q$ is false only when both P and Q are false.

A different use of the word "or" is the "exclusive or." For the exclusive or, the resulting statement is false when both statements are true. That is, "P exclusive or Q" is true only when exactly one of P or Q is true. In everyday life, we often use the exclusive or. When someone says, "At the intersection, turn left or go straight," this person is using the exclusive or.

Rather than memorizing the truth tables, for many people it is easier to remember the rules summarized in the following table:

Operator	Symbolic Form	Summary of Truth Values
Conjunction	$P \wedge Q$	True only when both P and Q are true.
Disjunction	$P \vee Q$	False only when both P and Q are false.
Negation	$\neg P$	Opposite truth value of P.
Conditional	$P \rightarrow Q$	False only when P is true and Q is false.

Conditional Statements

Conditional statements are extremely important in mathematics because almost all mathematical theorems are (or can be) stated in the form of a conditional statement. That is, they are almost always in the following form:

   If "certain conditions are met," then "something happens."

It is imperative that all students studying mathematics thoroughly understand the meaning of a conditional statement and the truth table for a conditional statement.

   Another problem that can cause some difficulty for people who do not frequently work with conditional statements is that in the English language, there are other ways for expressing the conditional statement $P \rightarrow Q$ other than "If P, then Q." Some of these will be listed after the next activity.

Activity 2.5 (The "Only If" Statement). Recall that a quadrilateral is a four-sided polygon. Let S represent the following true conditional statement:

   If a quadrilateral is a square, then it is a rectangle.

Now let T represent the following statement:

   A quadrilateral is a square only if it is a rectangle.

   1. Is T a true statement? Explain your reasoning.

   2. Let P represent "The quadrilateral is a square," and let Q represent "The quadrilateral is a rectangle." Using P and Q and logical operators, write symbolic expressions for statements S and T.

Other Forms of the Conditional Statement

Following are some common ways to express the conditional statement $P \rightarrow Q$ in the English language:

- If P, then Q.

- P implies Q.

- P only if Q.

- Q if P.

- Q is necessary for P. (This basically means that if P is true, then Q is necessarily true.)

- P is sufficient for Q. (This basically means that if you want Q to be true, it is sufficient to show that P is true.)

- If P is true, then Q is also true.

- Whenever P is true, Q must also be true.

- Q is true whenever P is true.

In all of these cases, P is the **hypothesis** of the conditional statement and Q is the **conclusion** of the conditional statement.

Activity 2.6 (Working with Conditional Statements). Complete the following table:

English Form	Hypothesis	Conclusion	Symbolic Form
If P, then Q.	P	Q	$P \rightarrow Q$
Q only if P.	Q	P	$Q \rightarrow P$
P is necessary for Q.			
P is sufficient for Q.			
Q is necessary for P.			
P implies Q.			
P only if Q.			
P if Q.			
If Q then P.			
If $\neg Q$, then $\neg P$.			
If P, then $Q \wedge R$.			
If $P \vee Q$, then R.			

Constructing Truth Tables

We can use the information contained in the truth tables for the basic connectives to construct truth tables for more complex statements. To illustrate this, we will construct a truth table for $(P \wedge \neg Q) \rightarrow R$. The first step is to determine the number of rows needed.

- For a truth table with two different simple statements, four rows are needed since there are four different combinations of truth values for the two statements. We should be consistent with how we set up the rows. The way we will do it in this text is to label the rows for the first statement with (T, T, F, F) and the rows for the second statement with (T, F, T, F). All truth tables in the text have this scheme.

- For a truth table with three different simple statements, eight rows are needed since there are eight different combinations of truth values for the three statements. Our standard scheme for this type of truth table is shown in Table 2.1.

The next step is to determine the columns to be used. One way to do this is to work backward from the form of the given statement. In this example, $(P \wedge \neg Q) \rightarrow R$, the last step is to deal with the conditional operator ($\rightarrow$). To do this, we need to know the truth values of $(P \wedge \neg Q)$ and R. To determine the truth values for $(P \wedge \neg Q)$, we need to apply the rules for the conjunction operator ($\wedge$) and we need to know the truth values for P and $\neg Q$.

Table 2.1 is a completed truth table for $(P \wedge \neg Q) \rightarrow R$ with the step numbers indicated at the bottom of each column. The step numbers correspond to the order in which the columns were completed.

- When completing the column for $P \wedge \neg Q$, remember that the only time the conjunction is true is when both P and $\neg Q$ are true.

- When completing the column for $(P \wedge \neg Q) \rightarrow R$, remember that the only time the conditional statement is false is when the hypothesis $(P \wedge \neg Q)$ is true and the conclusion, R, is false.

The last column entered is the truth table for the statement $(P \wedge \neg Q) \rightarrow R$ using the set up in the first three columns, which show the eight different truth value combinations for the statements P, Q, and R.

The Biconditional Statement

Some mathematical results are stated in the form "P if and only if Q" or "P is necessary and sufficient for Q." For example,

P	Q	R	$\neg Q$	$P \wedge \neg Q$	$(P \wedge \neg Q) \rightarrow R$
T	T	T	F	F	T
T	T	F	F	F	T
T	F	T	T	T	T
T	F	F	T	T	F
F	T	T	F	F	T
F	T	F	F	F	T
F	F	T	T	F	T
F	F	F	T	F	T
Step No.	1	1	1	2	3	4

Table 2.1: Truth Table for $(P \wedge \neg Q) \rightarrow R$

A triangle is equilateral if and only if its three interior angles are congruent.

The symbolic form for the biconditional statement "P if and only if Q" is $P \leftrightarrow Q$. In order to determine a truth table for a biconditional statement, it is instructive to look carefully at the form of the phrase "P if and only if Q." The word "and" suggests that this statement is a conjunction. Actually it is a conjunction of the following two statements:

$$P \text{ if } Q \qquad Q \rightarrow P$$
$$P \text{ only if } Q \qquad P \rightarrow Q$$

The symbolic form of this conjunction is $[(Q \rightarrow P) \wedge (P \rightarrow Q)]$.

Activity 2.7 (The Truth Table for the Biconditional Statement).

Complete a truth table for $[(Q \rightarrow P) \wedge (P \rightarrow Q)]$. Use the following columns: P, Q, $Q \rightarrow P$, $P \rightarrow Q$, and $[(Q \rightarrow P) \wedge (P \rightarrow Q)]$.

The last column of this table will be the truth table for $P \leftrightarrow Q$.

Other Forms of the Biconditional Statement

As with the conditional statement, there are some common ways to express the biconditional statement, $P \leftrightarrow Q$, in the English language. Some of these are as follows:

- P if and only if Q.

- P implies Q and Q implies P.

- P is necessary and sufficient for Q.

- P is equivalent to Q.

Exercises 2.2

1. Suppose that Daisy says, "If it does not rain, then I will play golf." Later in the day you come to know that it did rain but Daisy still played golf. Was Daisy's statement true or false? Support your conclusion.

2. Suppose that P and Q are statements for which $P \to Q$ is true and for which $\neg Q$ is true.

   (a) What conclusion (if any) can be made about the truth values of P and Q?

   (b) What conclusion (if any) can be made about the truth value of $P \wedge Q$?

   (c) What conclusion (if any) can be made about the truth value of $P \vee Q$?

3. Suppose that P and Q are statements for which $P \to Q$ is false.

   (a) What conclusion (if any) can be made about the truth value of $\neg P \to Q$?

   (b) What conclusion (if any) can be made about the truth value of $Q \to P$?

   (c) What conclusion (if any) can be made about the truth value of $P \vee Q$?

4. Construct a truth table for each of the following statements:

   (a) $P \wedge \neg Q$ (c) $\neg P \wedge \neg Q$

   (b) $\neg(P \wedge Q)$ (d) $\neg P \vee \neg Q$

   Do any of these statements have the same truth table?

5. Construct a truth table for each of the following statements:

(a) $P \rightarrow Q$ (c) $\neg P \rightarrow \neg Q$

(b) $Q \rightarrow P$ (d) $\neg Q \rightarrow \neg P$

Do any of these statements have the same truth table?

6. Construct truth tables for $P \wedge (Q \vee R)$ and $(P \wedge Q) \vee (P \wedge R)$. What do you observe?

7. Suppose each of the following statements is true.

- Laura is in the seventh grade.

- Laura got an A on the mathematics test or Sarah got an A on the mathematics test.

- If Sarah got an A on the mathematics test, then Laura is not in the seventh grade.

If possible, determine the truth value of each of the following statements. Carefully explain your reasoning.

(a) Laura got an A on the mathematics test.

(b) Sarah got an A on the mathematics test.

(c) Either Laura or Sarah did not get an A on the mathematics test.

8. Let P stand for "the integer x is even," and let Q stand for "x^2 is even." Express the conditional statement $P \rightarrow Q$ in English using

(a) The "if then" form of the conditional statement

(b) The word "implies"

(c) The "only if" form of the conditional statement

(d) The phrase "is necessary for"

(e) The phrase "is sufficient for"

9. Repeat Exercise (8) for the conditional statement $Q \rightarrow P$.

2.3   Logically Equivalent Statements

Preview Activity 1 (Logically Equivalent Statements).

> **Definition.** Two expressions are **logically equivalent** provided that they have the same truth value for all possible combinations of truth values for all variables appearing in the two expressions. In this case, we write $X \equiv Y$ and say that X and Y are logically equivalent.

1. Complete truth tables for $\neg(P \wedge Q)$ and $\neg P \vee \neg Q$.

2. Are the expressions $\neg(P \wedge Q)$ and $\neg P \vee \neg Q$ logically equivalent?

3. Suppose that the statement "I will play golf and I will mow the lawn" is false. Then its negation is true. Write the negation of this statement in the form of a disjunction. Does this make sense?

Preview Activity 2 (Converse and Contrapositive).

> **Definition.** If P and Q are statements, then
>
> - The **converse** of the conditional statement $P \to Q$ is the conditional statement $Q \to P$.
>
> - The **contrapositive** of the conditional statement $P \to Q$ is the conditional statement $\neg Q \to \neg P$.

1. For the following, the universal set is the set of all real numbers $\mathbb{R}$. So, x represents a real number. Label each of the following statements as true or false.

    (a) If $x = 3$, then $x^2 = 9$.
    (b) If $x^2 = 9$, then $x = 3$.
    (c) If $x^2 \neq 9$, then $x \neq 3$.
    (d) If $x \neq 3$, then $x^2 \neq 9$.

2. Which statement in the list of conditional statements in part (1) is the converse of Statement (1a)? Which is the contrapositive of Statement (1a)?

3. Complete the following truth table to show that:

- $P \to Q$ is logically equivalent to its contrapositive $\neg Q \to \neg P$.
- $P \to Q$ is not logically equivalent to its converse $Q \to P$.

P	Q	$P \to Q$	$Q \to P$	$\neg Q$	$\neg P$	$\neg Q \to \neg P$
T	T					
T	F					
F	T					
F	F					

Preview Activity 3 (Conditional Statements).

Sometimes, we actually use logical reasoning in our everyday living! Consider the following two statements.

> Statement 1   If you do not clean your room, then you cannot watch TV.
>
> Statement 2   You clean your room or you cannot watch TV.

(Perhaps you can imagine a parent making these statements.)

1. Do these two statements mean the same thing? Explain.

2. Let P be "you do not clean your room," and let Q be "you cannot watch TV." Use these to translate Statement 1 and Statement 2 into symbolic forms.

3. Construct a truth table for each of the expressions you determined in part (2). Are the expressions logically equivalent?

4. Assume that Statement 1 and Statement 2 are false. In this case, what is the truth value of P and what is the truth value of Q? Now, write a true statement in symbolic form that is a conjunction and involves P and Q.

5. Write a truth table for the (conjunction) statement in part (4) and compare it to a truth table for $\neg (P \to Q)$. What do you observe?

In Preview Activity 1, we introduced the concept of logically equivalent expressions. When two expressions X and Y are logically equivalent, we write $X \equiv Y$ and say that X and Y are logically equivalent.

The following theorem gives two important logical equivalencies. They are sometimes referred to as **De Morgan's Laws**.

Theorem 2.8 (De Morgan's Laws).

- *The statement* $\neg(P \wedge Q)$ *is logically equivalent to* $\neg P \vee \neg Q$. *This can be written as* $\neg(P \wedge Q) \equiv \neg P \vee \neg Q$.

- *The statement* $\neg(P \vee Q)$ *is logically equivalent to* $\neg P \wedge \neg Q$. *This can be written as* $\neg(P \vee Q) \equiv \neg P \wedge \neg Q$.

The first equivalency in Theorem 2.8 was established in Preview Activity 1. The following truth table establishes the second equivalency.

P	Q	$P \vee Q$	$\neg(P \vee Q)$	$\neg P$	$\neg Q$	$\neg P \wedge \neg Q$
T	T	T	F	F	F	F
T	F	T	F	F	T	F
F	T	T	F	T	F	F
F	F	F	T	T	T	T

It is possible to develop and state several different logical equivalencies at this time. However, we will restrict ourselves to what we consider to be some of the most important ones. Since many mathematical statements are written in the form of conditional statements, logical equivalencies related to conditional statements are quite important.

Logical Equivalencies Related to Conditional Statements

1. The conditional statement $P \to Q$ is logically equivalent to its contrapositive $\neg Q \to \neg P$.

2. The conditional statement $P \to Q$ is logically equivalent to $\neg P \vee Q$.

3. The statement $\neg(P \to Q)$ is logically equivalent to $P \wedge \neg Q$.

The first logical equivalency was established in Preview Activity 2, and the second and third were established in Preview Activity 3. The truth table that establishes the third logical equivalency is repeated here.

P	Q	$P \to Q$	$\neg(P \to Q)$	$\neg Q$	$P \wedge \neg Q$
T	T	T	F	F	F
T	F	F	T	T	T
F	T	T	F	F	F
F	F	T	F	T	F

Example 2.9 (A Conditional Statement as a Disjunction).

The logical equivalency $(P \to Q) \equiv (\neg P \vee Q)$ was illustrated in Preview Activity 3. To see this, let P stand for " You do not clean your room" and let Q stand for "You cannot watch TV." Then the conditional statement $(P \to Q)$ is

> If you do not clean your room, then you cannot watch TV.

This is logically equivalent to the statement $(\neg P \vee Q)$, which can be written as follows:

> You clean your room or you cannot watch TV.

Example 2.10 (The Negation of a Conditional Statement).

The logical equivalency $\neg (P \to Q) \equiv P \wedge \neg Q$ is interesting because it shows us that **the negation of a conditional statement is not another conditional statement.** The negation of a conditional statement can be written in the form of a conjunction. So what does it mean to say that the conditional statement in Example 2.9 is false? To answer this, we can use the logical equivalency $\neg (P \to Q) \equiv P \wedge \neg Q$. The idea is that if $P \to Q$ is false, then its negation must be true. So the negation of the conditional statement

> If you do not clean your room, then you cannot watch TV,

can be written as

> You do not clean your room and you can watch TV.

For another example, consider the following conditional statement:

> If $-5 < -3$, then $(-5)^2 < (-3)^2$.

This conditional statement is false since its hypothesis is true and its conclusion is false. Consequently, its negation must be true. Its negation is not a conditional statement. The negation can be written in the form of a conjunction by using the logical equivalency $\neg (P \to Q) \equiv P \wedge \neg Q$. So, the negation can be written as follows:

$$-5 < -3 \text{ and } \neg \left((-5)^2 < (-3)^2 \right).$$

However, the second part of this conjunction can be written in a simpler manner by noting that "not less than" means the same thing as "greater than or equal to." So we use this to write the negation of the original conditional statement as follows:

$$-5 < -3 \text{ and } (-5)^2 \geqslant (-3)^2.$$

This conjunction is true since each of the individual statements in the conjunction are true.

Other Methods of Establishing Logical Equivalencies

Truth tables are one way to establish logical equivalencies. Following are two other methods for establishing the logical equivalence

$$\neg(P \to Q) \equiv (P \wedge \neg Q).$$

Method 1: Consider Cases in Which Each Expression Is True

A conditional statement is false only when the hypothesis is true and the conclusion is false. So $P \to Q$ is false only when P is true and Q is false. Consequently, $\neg(P \to Q)$ is true only when P is true and Q is false.

A conjunction of two statements is true only when both statements are true. So $(P \wedge \neg Q)$ is true only when P is true and $\neg Q$ is true. That is, $(P \wedge \neg Q)$ is true only when P is true and Q is false.

Comparing the last sentence in the two preceding paragraphs, we see that $\neg(P \to Q)$ is logically equivalent to $(P \wedge \neg Q)$.

Method 2: Use Previously Established Logical Equivalencies

Sometimes a logical equivalence can be proven by a sequence of previously established logical equivalencies. For example,

- $P \to Q$ is logically equivalent to $\neg P \vee Q$. So

- $\neg(P \to Q)$ is logically equivalent to $\neg(\neg P \vee Q)$. Hence, by one of De Morgan's Laws (Theorem 2.8),

- $\neg(P \to Q)$ is logically equivalent to $\neg(\neg P) \wedge \neg Q$. This means that

- $\neg(P \to Q)$ is logically equivalent to $P \wedge \neg Q$.

The last step used the fact that $\neg(\neg P)$ is logically equivalent to P.

When proving theorems in mathematics, it is often important to be able to decide if two expressions are logically equivalent. Sometimes when we are attempting to prove a theorem, we may be unsuccessful in developing a proof for the original statement of the theorem. However, in some cases, it is possible to prove an equivalent statement. Knowing that the statements are equivalent tells us that if we prove one, then we have also proven the other. In fact, once we know the truth value of a statement, then we know the truth value of any other logically equivalent statement.

This is illustrated in the next activity.

Activity 2.11 (Working with a Logical Equivalency). Suppose we are trying to prove the following for integers x and y:

If $x \cdot y$ is even, then x is even or y is even.

1. Show that we can write this statement in the following symbolic form:

$$P \to (Q \vee R).$$

2. Write the contrapositive of $P \to (Q \vee R)$. Then use one of De Morgan's Laws (Theorem 2.8) to rewrite the hypothesis of this conditional statement.

3. Use the result from part (2) to explain why the given statement is logically equivalent to the following statement:

If x is odd and y is odd, then $x \cdot y$ is odd.

The two statements in this activity are logically equivalent. We now have the choice of proving either of these statements. If we prove one, we prove the other, or if we show one is false, the other is also false.

The second statement is Theorem 1.3, which was proven in Section 1.2.

Activity 2.12 (Working with a Logical Equivalency).

In Section 2.2, we constructed a truth table for $(P \wedge \neg Q) \to R$. We will use that truth table here.

P	Q	R	$\neg Q$	$P \wedge \neg Q$	$(P \wedge \neg Q) \to R$
T	T	T	F	F	T
T	T	F	F	F	T
T	F	T	T	T	T
T	F	F	T	T	F
F	T	T	F	F	T
F	T	F	F	F	T
F	F	T	T	F	T
F	F	F	T	F	T

Now complete a truth table for $P \rightarrow (Q \vee R)$.

1. What conclusion can be made from these two truth tables?

2. Let a and b be integers. Suppose we are trying to prove the following:

- If 3 is a factor of $a \cdot b$, then 3 is a factor of a or 3 is a factor of b.

Explain why we will have proven this statement if we prove the following result:

- If 3 is a factor of $a \cdot b$ and 3 is not a factor of a, then 3 is a factor of b.

In Activities 2.11 and 2.12, we worked with the following two logical equivalencies:

- $P \rightarrow (Q \vee R)$ is logically equivalent to $(\neg Q \wedge \neg R) \rightarrow \neg P$.

- $P \rightarrow (Q \vee R)$ is logically equivalent to $(P \wedge \neg Q) \rightarrow R$.

As we will see later, it is often difficult to construct a direct proof for a conditional statement of the form $P \rightarrow (Q \vee R)$. These two logical equivalencies give us other ways to attempt to prove a statement of the form $P \rightarrow (Q \vee R)$.

- The first logical equivalency is basically the contrapositive of $P \rightarrow (Q \vee R)$.

- The advantage of the equivalent form, $(P \wedge \neg Q) \rightarrow R$, is that we have an additional assumption, $\neg Q$, in the hypothesis. This just gives us more information with which to work.

The following theorem states some of the most frequently used logical equivalencies used when writing mathematical proofs. We have already established many of these equivalencies. Others will be established in the exercises.

Theorem 2.13 (Important Logical Equivalencies).

De Morgan's Laws

$$\neg (P \wedge Q) \equiv \neg P \vee \neg Q$$
$$\neg (P \vee Q) \equiv \neg P \wedge \neg Q$$

Conditional Statements

$$P \rightarrow Q \equiv \neg Q \rightarrow \neg P \ \ (contrapositive)$$
$$P \rightarrow Q \equiv \neg P \vee Q$$
$$\neg (P \rightarrow Q) \equiv P \wedge \neg Q$$

Biconditional Statement	$(P \leftrightarrow Q) \equiv (P \to Q) \wedge (Q \to P)$
Double Negation	$\neg(\neg P) \equiv P$
Distributive Laws	$P \vee (Q \wedge R) \equiv (P \vee Q) \wedge (P \vee R)$
	$P \wedge (Q \vee R) \equiv (P \wedge Q) \vee (P \wedge R)$
Conditionals with	$P \to (Q \vee R) \equiv (P \wedge \neg Q) \to R$
Disjunctions	$(P \vee Q) \to R \equiv (P \to R) \wedge (Q \to R)$

Exercises 2.3

1. Write each of the following conditional statements as a disjunction.

   (a) If $a = 5$, then $a^2 = 25$.

   (b) If it is not raining, then Laura is playing golf.

   (c) If $a \neq b$, then $a^4 \neq b^4$.

   (d) If a is an odd integer, then $3a$ is an odd integer.

2. Write the negation of each of the conditional statements in Exercise (1) as a conjunction.

3. Write a useful negation of each of the following statements. Do not leave a negation as a prefix of a statement. For example, we would write the negation of "I will play golf and I will mow the lawn" as "I will not play golf or I will not mow the lawn."

   (a) We will win the first game and we will win the second game.

   (b) They will lose the first game or they will lose the second game.

   (c) If you mow the lawn, then I will pay you $20.

   (d) If we do not win the first game, then we will not play a second game.

   (e) I will wash the car or I will mow the lawn.

   (f) If you graduate from college, then you will get a job or you will go to graduate school.

   (g) If I play tennis, then I will wash the car or I will do the dishes.

   (h) If you clean your room or do the dishes, then you can go to see a movie.

(i) It is warm outside and if it does not rain, then I will play golf.

4. Use truth tables to establish each of the following logical equivalencies dealing with biconditional statements:

(a) $(P \leftrightarrow Q) \equiv (P \rightarrow Q) \wedge (Q \rightarrow P)$

(b) $(P \leftrightarrow Q) \equiv (Q \leftrightarrow P)$

(c) $(P \leftrightarrow Q) \equiv (\neg P \leftrightarrow \neg Q)$

5. Use truth tables to establish the following logical equivalencies known as the distributive laws.

(a) $P \vee (Q \wedge R) \equiv (P \vee Q) \wedge (P \vee R)$

(b) $P \wedge (Q \vee R) \equiv (P \wedge Q) \vee (P \wedge R)$

6. Prove the logical equivalency, $[(P \vee Q) \rightarrow R] \equiv (P \rightarrow R) \wedge (Q \rightarrow R)$.

7. Prove each of the following logical equivalencies about **conditionals with conjunctions:**

(a) $[(P \wedge Q) \rightarrow R] \equiv (P \rightarrow R) \vee (Q \rightarrow R)$

(b) $[P \rightarrow (Q \wedge R)] \equiv (P \rightarrow Q) \wedge (P \rightarrow R)$

8. Prove the logical equivalency $[\neg P \rightarrow (Q \wedge \neg Q)] \equiv P$.

9. Let a be a real number and let f be a real-valued function defined on an interval containing $x = a$. Consider the following conditional statement:

If f is differentiable at $x = a$, then f is continuous at $x = a$.

Which of the following statements have the same meaning as this conditional statement and which ones are negations of this conditional statement?

Note: This is not asking which statements are true and which are false. It is asking which statements are logically equivalent to the given statement. It might be helpful to let P represent the hypothesis of the given statement, Q represent the conclusion, and then determine a symbolic representation for each statement.

(a) If f is continuous at $x = a$, then f is differentiable at $x = a$.

(b) If f is not differentiable at $x = a$, then f is not continuous at $x = a$.

(c) If f is not continuous at $x = a$, then f is not differentiable at $x = a$.

(d) f is not differentiable at $x = a$, or f is continuous at $x = a$.

(e) f is not continuous at $x = a$, or f is differentiable at $x = a$.

(f) f is differentiable at $x = a$, and f is not continuous at $x = a$.

10. Let a, b, and c be integers. Consider the following conditional statement:

> If a divides bc, then a divides b or a divides c.

Which of the following statements have the same meaning as this conditional statement and which ones are negations of this conditional statement? (See the note in the instructions for Exercise (9)).

(a) If a divides b or a divides c, then a divides bc.

(b) If a does not divide b or a does not divide c, then a does not divide bc.

(c) a divides bc, a does not divide b, and a does not divide c.

(d) If a does not divide b and a does not divide c, then a does not divide bc.

(e) a does not divide bc or a divides b or a divides c.

(f) If a divides bc and a does not divide c, then a divides b.

(g) If a divides bc or a does not divide b, then a divides c.

2.4 Quantifiers and Negations

Preview Activity 1 (Quantifiers).

The following table from Section 2.1 summarizes the facts about the two types of quantifiers that we use in mathematics:

A statement involving	Often has the form	The statement is true provided that
a universal quantifier: $(\forall x, P(x))$	"For every x, $P(x)$," where $P(x)$ is a predicate.	Every value of x in the universal set makes $P(x)$ true.
an existential quantifier: $(\exists x, P(x))$	"There exists an x such that $P(x)$," where $P(x)$ is a predicate.	There is at least one value of x in the universal set that makes $P(x)$ true.

1. Consider the following statement written in symbolic form:
   $(\forall x \in \mathbb{Z}) \, (x \text{ is even})$.

   (a) Write this statement as an English sentence.

   (b) Is the statement true or false? Why?

   (c) How would you write the negation of this statement as an English sentence?

   (d) How would you write the negation of this statement symbolically (using a quantifier)?

2. Consider the following statement written in symbolic form:
   $(\exists x \in \mathbb{Z}) \, (x \text{ is a multiple of } 5)$.

   (a) Write this statement as an English sentence.

   (b) Is the statement true or false? Why?

   (c) How would you write the negation of this statement as an English sentence?

   (d) How would you write the negation of this statement symbolically (using a quantifier)?

Preview Activity 2 (Statements with Two Quantifiers).
   This activity is similar to Exercise (5) in Section 2.1. One important difference is that the universal set has been changed. Assume the universal set for the variables is the set of all real numbers $\mathbb{R}$. Consider the following sentence:

$$(\exists x \in \mathbb{R}) \, (x \cdot y = 100).$$

1. Explain why this sentence is not a statement.

2. If 5 is substituted for y, is the resulting sentence a statement? If it is a statement, is it true or false?

3. If -3 is substituted for y, is the resulting sentence a statement? If it is a statement, is it true or false?

4. If π is substituted for y, is the resulting sentence a statement? If it is a statement, is it true or false?

5. What is the truth set of the predicate $(\exists x \in \mathbb{R}) \, (x \cdot y = 100)$?

6. The following sentence, written in symbolic form, is a statement.

$$(\forall y \in \mathbb{R}) \left[(\exists x \in \mathbb{R}) \left(x \cdot y = 100 \right) \right].$$

(a) Is this statement true or false? Explain.

(b) Write this statement in the form of an English sentence.

7. The following sentence, written in symbolic form, is a statement.

$$(\exists x \in \mathbb{R}) \left[(\forall y \in \mathbb{R}) \left(x \cdot y = 100 \right) \right]$$

(a) Is this statement true or false? Explain.

(b) Write this statement in the form of an English sentence.

We introduced the concepts of predicates and quantifiers in Section 2.1. Following are some of the relevant definitions.

> **Definition.** A **variable** is a symbol representing an unspecified object that can be chosen from a given set U. The set U is called the **universal set for the variable.** It is the set of specified objects from which objects may be chosen to substitute for the variable.
>
> A **constant** is a specific member of the universal set.
>
> A **predicate** is a sentence $P(x_1, x_2, \ldots, x_n)$ involving variables $x_1, x_2, \ldots, x_n$ with the property that when specific values from the universal set are assigned to $x_1, x_2, \ldots, x_n$, then the resulting sentence is either true or false. That is, the resulting sentence is a statement. A predicate is also called an **open sentence** or a **propositional function.**

> **Definition.** The **truth set of a predicate** is the collection of objects in the universal set that can be substituted to make the propositional function a true statement.

The table in Preview Activity 1 summarizes the facts about the two types of quantifiers we use in mathematics.

Negations of Quantified Statements

In Preview Activity 1, we tried to write a negation of some quantified statements. This is a very important mathematical activity. As we will see in future sections, it is sometimes just as important to be able to describe when some object does not satisfy a certain property as it is to describe when the object satisfies the property. Our next task is to learn how to write negations of quantified statements in a useful English form.

We first look at the negation of a statement involving a universal quantifier. The general form for such a statement can be written as

$$(\forall x)\,(P\,(x))\,,$$

where $P\,(x)$ is a predicate. When we write

$$\neg\,(\forall x)\,[P\,(x)]\,,$$

we are asserting that the statement $(\forall x)\,[P\,(x)]$ is false. This is equivalent to saying that the truth set of the predicate $P\,(x)$ is not the universal set. That is, there exists an element x in the universal set U such that $P\,(x)$ is false. This in turn means that there exists an element x in U such that $\neg P\,(x)$ is true. This is equivalent to saying that $(\exists x)\,[\neg P\,(x)]$ is true. This explains why the following result is true:

$$\neg\,(\forall x)\,[P\,(x)] \equiv (\exists x)\,[\neg P\,(x)]\,.$$

Similarly, when we write

$$\neg\,(\exists x)\,[P\,(x)]\,,$$

we are asserting that the statement $(\exists x)\,[P\,(x)]$ is false. This is equivalent to saying that the truth set of the predicate $P\,(x)$ is empty set. That is, there is no element x in the universal set U such that $P\,(x)$ is true. This in turn means that for each element x in U, $\neg P\,(x)$ is true. This is equivalent to saying that $(\forall x)\,[\neg P\,(x)]$ is true. This explains why the following result is true:

$$\neg\,(\exists x)\,[P\,(x)] \equiv (\forall x)\,[\neg P\,(x)]\,.$$

We summarize these results in the following theorem.

Theorem 2.14. *For any predicate $P\,(x)$,*

$$\neg\,(\forall x)\,[P\,(x)] \equiv (\exists x)\,[\neg P\,(x)]\,,\ and$$
$$\neg\,(\exists x)\,[P\,(x)] \equiv (\forall x)\,[\neg P\,(x)]\,.$$

Example 2.15. Consider the following statement: $(\forall x \in \mathbb{R}) \left(x^3 \geq x^2 \right)$.

We can write this statement as an English sentence in several ways. Following are two different ways to do so.

- For each real number x, $x^3 \geq x^2$.

- If x is a real number, then x^3 is greater than or equal to x^2.

The second statement shows that in a conditional statement, there is often a hidden universal quantifier. This statement is false since there are real numbers x for which x^3 is not greater than or equal to x^2. For example, we could use $x = -1$ or $x = \frac{1}{2}$.

Since the phrase "is not greater than or equal to" means the same thing as "is less than," we usually say that there are real numbers x for which $x^3 < x^2$. This means that the negation must be true. We can form the negation as follows:

$$\neg \left(\forall x \in \mathbb{R} \right) \left(x^3 \geq x^2 \right) \equiv \left(\exists x \in \mathbb{R} \right) \neg \left(x^3 \geq x^2 \right).$$

In most cases, we want to write this negation in a way that does not use the negation symbol. In this case, we can now write the predicate $\neg \left(x^3 \geq x^2 \right)$ as $\left(x^3 < x^2 \right)$. (That is, the negation of "is greater than or equal to" is "is less than.") So, we obtain the following:

$$\neg \left(\forall x \in \mathbb{R} \right) \left(x^3 \geq x^2 \right) \equiv \left(\exists x \in \mathbb{R} \right) \left(x^3 < x^2 \right).$$

The statement $\left(\exists x \in \mathbb{R} \right) \left(x^3 < x^2 \right)$ could be written in English as follows:

- There exists a real number x such that $x^3 < x^2$.

- There exists an x such that x is a real number and $x^3 < x^2$.

Counterexamples and Negations of Conditional Statements

The number $x = -1$ in the previous example was used to show that $(\forall x \in \mathbb{R}) \left(x^3 \geq x^2 \right)$ is false. This is called a **counterexample** to the statement. In general, a **counterexample** to a statement of the form $(\forall x) \left[P \left(x \right) \right]$ is an object a in the universal set U so that $P \left(a \right)$ is false. It is an example that proves that $(\forall x) \left[P \left(x \right) \right]$ is a false statement, and hence its negation, $(\exists x) \left[\neg P \left(x \right) \right]$, is a true statement.

Also, in the preceding example, we wrote the universally quantified statement as a conditional statement. The number $x = -1$ is a counterexample for the statement

If x is a real number, then x^3 is greater than or equal to x^2.

It is an example that makes the hypothesis of the conditional statement true and the conclusion false. Remember that a conditional statement often contains a "hidden" universal quantifier. Also, recall that in Section 2.3 we saw that the negation of the conditional statement "If P then Q" is the statement "P and not Q." Symbolically, this can be written as follows:

$$\neg (P \to Q) \equiv P \wedge \neg Q.$$

So when we specifically include the universal quantifier, the symbolic form of the negation of a conditional statement is

$$\neg (\forall x) [P(x) \to Q(x)] \equiv (\exists x) \neg [P(x) \to Q(x)]$$
$$\equiv (\exists x) [P(x) \wedge \neg Q(x)].$$

That is,

$$\neg (\forall x) [P(x) \to Q(x)] \equiv (\exists x) [P(x) \wedge \neg Q(x)].$$

Activity 2.16 (Negating Quantified Statements). For each of the following statements:

- Write the statement in the form of an English sentence that does not use the symbols for quantifiers.

- Write the negation of the statement in a symbolic form that does not use the negation symbol.

- Write the negation of the statement in the form of an English sentence that does not use the symbols for quantifiers.

1. $\forall a \in \mathbb{R}$, $a + 0 = a$.

2. $\forall x \in \mathbb{R}$, $\sin(2x) = 2(\sin x)(\cos x)$.

3. $\forall x \in \mathbb{R}$, $\tan^2 x + 1 = \sec^2 x$.

4. $\exists x \in \mathbb{Q}$ such that $x^2 - 3x - 7 = 0$.

5. $\exists x \in \mathbb{R}$ such that $x^2 + 1 = 0$.

Activity 2.17 (Perfect Squares). Definitions of terms in mathematics often involve quantifiers. Most often, these definitions are given in a form that does not use the symbols for quantifiers. Not only is it important to know a definition, it is also important to be able to write a negation of the definition. This will be illustrated with the definition of what it means to say that a natural number is a perfect square.

Recall that the natural numbers, denoted by $\mathbb{N}$, consist of the positive whole numbers. That is, $\mathbb{N} = \{1, 2, 3, \ldots\}$.

Definition. A natural number n is a **perfect square** provided that there exists a natural number k such that $n = k^2$.

This definition can be written in symbolic form using appropriate quantifiers as follows:

A natural number n is a **perfect square** provided $(\exists k \in \mathbb{N})\left(n = k^2\right)$.

1. Give several examples of natural numbers that are perfect squares.

2. Give several examples of natural numbers that are not perfect squares.

3. Use the symbolic form of the definition of a perfect square to complete the following sentence: "A natural number n is not a perfect square provided that ..."

4. Without using the symbols for quantifiers, complete the following sentence: "A natural number n is not a perfect square ..."

Activity 2.17 illustrates a good method for trying to understand a new definition. Most textbooks will simply define a concept and leave it to the reader to do the steps outlined in Activity 2.17. Frequently, it is not sufficient to just read a definition and expect to understand the new term. We must provide examples that satisfy the definition, examples that do not satisfy the definition, and we must be able to write a coherent negation of the definition

---

Activity 2.18 (Prime Numbers and Composite Numbers).

Definition. A natural number p is a **prime number** provided that it is greater than 1 and the only natural numbers that are factors of p are 1 and p. A natural number other than 1 that is not a prime number is a **composite number**. The number 1 is neither prime nor composite.

For example, 2, 3, 5, and 7 are prime numbers. Also, 4 is a composite number since $4 = 2 \cdot 2$; 10 is a composite number since $10 = 2 \cdot 5$; and 60 is a composite number since $60 = 4 \cdot 15$.

1. Give examples of four natural numbers other than 2, 3, 5, and 7 that are prime numbers.

2. Explain why a natural number p that is greater than 1 is a prime number provided that

   For all $d \in \mathbb{N}$, if d is a factor of p, then $d = 1$ or $d = p$.

3. Give examples of four natural numbers that are composite numbers and explain why they are composite numbers.

4. Write a useful description of what it means to say that a natural number is a composite number (other than saying that it is not prime).

Statements with More than One Quantifier

When a predicate contains more than one variable, each variable must be quantified to create a statement that is either true or false. For example, assume the universal set is the set of integers, $\mathbb{Z}$, and let $P(x, y)$ be the predicate, "$x + y = 0$." We can create a statement from this predicate in several ways.

1. $(\forall x \in \mathbb{Z}) \, (\forall y \in \mathbb{Z}) \, (x + y = 0)$

   We could read this as, "For all integers x and y, $x + y = 0$." This is a false statement since it is possible to find two integers whose sum is not zero $(2 + 3 \neq 0)$.

2. $(\forall x \in \mathbb{Z}) \, (\exists y \in \mathbb{Z}) \, (x + y = 0)$

   We could read this as, "For every integer x, there exists an integer y such that $x + y = 0$." This is a true statement.

3. $(\exists x \in \mathbb{Z}) \, (\forall y \in \mathbb{Z}) \, (x + y = 0)$

   We could read this as, "There exists an integer x such that for each integer y, $x + y = 0$." This is a false statement since there is no integer whose sum with each integer is zero.

4. $(\exists x \in \mathbb{Z})\,(\exists y \in \mathbb{Z})\,(x + y = 0)$

We could read this as, "There exist integers x and y such that $x + y = 0$." This is a true statement. For example, $2 + (-2) = 0$.

When we negate a statement with more than one quantifier, we consider each quantifier in turn and apply the appropriate part of Theorem 2.14. As an example, we will negate Statement (3) from the preceding list. The statement is

$$(\exists x \in \mathbb{Z})\,(\forall y \in \mathbb{Z})\,(x + y = 0)\,.$$

We first treat this as a statement in the following form: $(\exists x \in \mathbb{Z})\,(P\,(x))$ where $P\,(x)$ is the predicate $(\forall y \in \mathbb{Z})\,(x + y = 0)$. Using Theorem 2.14, we have

$$\neg\,(\exists x \in \mathbb{Z})\,(P\,(x)) \equiv (\forall x \in \mathbb{Z})\,(\neg P\,(x))\,.$$

Using Theorem 2.14 again, we obtain the following:

$$
\begin{aligned}
\neg P\,(x) &\equiv \neg\,(\forall y \in \mathbb{Z})\,(x + y = 0)\\
&\equiv (\exists y \in \mathbb{Z})\,\neg\,(x + y = 0)\\
&\equiv (\exists y \in \mathbb{Z})\,(x + y \neq 0)\,.
\end{aligned}
$$

Combining these two results, we obtain

$$\neg\,(\exists x \in \mathbb{Z})\,(\forall y \in \mathbb{Z})\,(x + y = 0) \equiv (\forall x \in \mathbb{Z})\,(\exists y \in \mathbb{Z})\,(x + y \neq 0)\,.$$

This process can be written as follows:

$$
\begin{aligned}
\neg\,(\exists x \in \mathbb{Z})\,(\forall y \in \mathbb{Z})\,(x + y = 0) &\equiv (\forall x \in \mathbb{Z})\,[\neg\,(\forall y \in \mathbb{Z})\,(x + y = 0)]\\
&\equiv (\forall x \in \mathbb{Z})\,[(\exists y \in \mathbb{Z})\,\neg\,(x + y = 0)]\\
&\equiv (\forall x \in \mathbb{Z})\,(\exists y \in \mathbb{Z})\,(x + y \neq 0)\,.
\end{aligned}
$$

The results are summarized in the following table.

	Symbolic Form	**English Form**
Statement	$(\exists x \in \mathbb{Z})\,(\forall y \in \mathbb{Z})\,(x + y = 0)$	There exists an integer x such that for each integer y, $x + y = 0$.
Negation	$(\forall x \in \mathbb{Z})\,(\exists y \in \mathbb{Z})\,(x + y \neq 0)$	For each integer x, there exists an integer y such that $x + y \neq 0$.

Since the given statement is false, its negation is true.

We can construct a similar table for each of the four statements. The next table shows Statement (2) and its negation.

	Symbolic Form	**English Form**
Statement	$(\forall x \in \mathbb{Z})\,(\exists y \in \mathbb{Z})\,(x + y = 0)$	For every integer x, there exists an integer y such that $x + y = 0$.
Negation	$(\exists x \in \mathbb{Z})\,(\forall y \in \mathbb{Z})\,(x + y \neq 0)$	There exists an integer x such that for every integer y, $x + y \neq 0$.

Since the given statement is true, its negation is false.

Activity 2.19 (Upper Bounds for Subsets of $\mathbb{R}$). Let A be a subset of the real numbers. A number b is called an **upper bound** for the set A provided that for each element x in A, $x \leq b$.

1. Write this definition in symbolic form by completing the following:

   Let A be a subset of the real numbers. A number b is called an upper bound for the set A provided that ...

2. Give examples of three different upper bounds for the set $A = \{x \in \mathbb{R} \mid 1 \leq x \leq 3\}$.

3. Does the set $B = \{x \in \mathbb{R} \mid x > 0\}$ have an upper bound? Explain.

4. Give examples of three different real numbers that are not upper bounds for the set $A = \{x \in \mathbb{R} \mid 1 \leq x \leq 3\}$.

5. Complete the following in symbolic form: "Let A be a subset of the real numbers. A number b is not an upper bound for the set A provided that ..."

6. Without using the symbols for quantifiers, complete the following sentence: "Let A be a subset of the real numbers. A number b is not an upper bound for the set A provided that ..."

7. Are your examples in part (4) consistent with your work in part (6)? Explain.

Activity 2.20 (Least Upper Bound for a Subset of $\mathbb{R}$). In Activity 2.19, we introduced the definition of an upper bound for a subset of the real numbers. Assume that we know this definition and that we know what it means to say that a number is not an upper bound for a subset of the real numbers.

Let A be a subset of $\mathbb{R}$. A real number α is the **least upper bound** for A provided that α is an upper bound for A, and if β is an upper bound for A, then $\alpha \leq \beta$.

Note: The symbol α is the lowercase Greek letter alpha, and the symbol β is the lowercase Greek letter beta.

If we define $P(x)$ to be "x is an upper bound for A," then we can write the definition for least upper bound as follows:

A real number α is the **least upper bound** for A provided that $P(\alpha) \wedge [(\forall \beta \in \mathbb{R})(P(\beta) \rightarrow (\alpha \leq \beta))]$.

1. Why is a universal quantifier used for the real number β?

2. How do we negate a conjunction?

3. Complete the following sentence in symbolic form: "A real number α is not the least upper bound for A provided that ...".

4. Complete the following sentence as an English sentence: "A real number α is not the least upper bound for A provided that ...".

Exercises 2.4

1. Explain why each of the following statements is false.

   (a) $(\forall x \in \mathbb{R})(x^2 > 0)$.

   (b) $(\forall a \in \mathbb{Z})\left(\sqrt{a^2} = a\right)$.

   (c) $(\forall x \in \mathbb{R})(\tan^2 x + 1 = \sec^2 x)$.

   (d) $\exists x \in \mathbb{Q}$ such that $x^2 - 3x - 7 = 0$.

   (e) $(\forall m \in \mathbb{Z})(m^2 \text{ is even})$.

   (f) $\exists x \in \mathbb{R}$ such that $x^2 + 1 = 0$.

2. For each of the following statements

   - Write the statement as an English sentence that does not use the symbols for quantifiers.

   - Write the negation of the statement in symbolic form in which the negation symbol is not used.

- Write a useful negation of the statement in an English sentence that does not use the symbols for quantifiers.

(a) $(\exists x \in \mathbb{Q})\,(x > \sqrt{2})$.

(b) $(\forall x \in \mathbb{Q})\,(x^2 - 2 \neq 0)$.

(c) $(\forall x \in \mathbb{Z})\,(x$ is even or x is odd$)$.

(d) $(\exists x \in \mathbb{Q})\,(\sqrt{2} < x < \sqrt{3})$. <u>Note</u>: The sentence "$\sqrt{2} < x < \sqrt{3}$" is actually a conjuction. It means $\sqrt{2} < x$ and $x < \sqrt{3}$.

(e) $(\forall x \in \mathbb{Z})\,\big($If x^2 is odd, then x is odd$\big)$.

(f) $(\forall n \in \mathbb{N})\,[$ If n is a perfect square, then $(2^n - 1)$ is not a prime number$]$.

(g) $(\forall n \in \mathbb{N})\,(n^2 - n + 41$ is a prime number$)$.

(h) $(\exists x \in \mathbb{R})\,(\cos(2x) = 2\,(\cos x))$.

3. Assume the universal set for each variable is the set of integers.

Write each of the following statements as an English sentence that does not use the symbols for quantifiers.

(a) $(\exists m)\,(\exists n)\,(m > n)$

(b) $(\exists m)\,(\forall n)\,(m > n)$

(c) $(\forall m)\,(\exists n)\,(m > n)$

(d) $(\forall m)\,(\forall n)\,(m > n)$

(e) $(\exists n)\,(\forall m)\,(m^2 > n)$

(f) $(\forall n)\,(\exists m)\,(m^2 > n)$

4. Write the negation of each statement in Exercise (3) in symbolic form and as an English sentence that does not use the symbols for quantifiers.

5. In calculus, we define a function f with domain $\mathbb{R}$ to be **strictly increasing** provided that for all real numbers x and y, $f(x) < f(y)$ whenever $x < y$. Complete each of the following sentences using the appropriate symbols for quantifiers:

(a) A function f with domain $\mathbb{R}$ is strictly increasing provided that

...

(b) A function f with domain $\mathbb{R}$ is not strictly increasing provided that ...

Complete the following sentence in English without using symbols for quantifiers:

(c) A function f with domain $\mathbb{R}$ is not strictly increasing provided that ...

6. In calculus, we define a function f to be **continuous** at a real number a provided that for every $\varepsilon > 0$, there exists a $\delta > 0$ such that if $|x - a| < \delta$, then $|f(x) - f(a)| < \varepsilon$.

<u>Note</u>: The symbol ε is the lowercase Greek letter epsilon, and the symbol δ is the lowercase Greek letter delta.

Complete each of the following sentences using the appropriate symbols for quantifiers:

(a) A function f is continuous at the real number a provided that ...

(b) A function f is not continuous at the real number a provided that ...

Complete the following sentence in English without using symbols for quantifiers:

(c) A function f is not continuous at the real number a provided that ...

Chapter 3

Constructing and Writing Proofs in Mathematics

3.1 Direct Proofs

Preview Activity 1 (Definition of Divides, Divisor, Multiple).

> **Definition.** An integer m **divides** an integer n provided that there is an integer q such that $n = m \cdot q$. We also say that m is a **divisor** of n, m is a **factor** of n, and n is a **multiple** of m.

This definition can be written in symbolic form using appropriate quantifiers as follows:

An integer m **divides** an integer n provided that $(\exists q \in \mathbb{Z}) \, (n = m \cdot q)$.

1. Give three different examples of three integers where the first integer divides the second integer and the second integer divides the third integer.

2. In your examples in Part (1), is there any relationship between the first and the third integer? Explain, and formulate a conjecture. Write your conjecture in the form of a conditional statement.

3. Give several examples of two integers where the first integer does not divide the second integer.

4. According to the definition of "divides," does the integer 0 divide the integer 10? That is, is 0 a divisor of 10? Explain.

5. According to the definition of "divides," does the integer 10 divide the integer 0? That is, is 10 a divisor of 0? Explain.

6. Use the definition of "divides" to complete the following sentence in symbolic form: "The integer m does not divide the integer n means that ...."

7. Use the definition of "divides" to complete the following sentence without using the symbols for quantifiers: "The integer m does not divide the integer n ...."

Preview Activity 2 (A Proposition about Even Integers).

1. Multiply several pairs of integers where at least one of the integers is an even integer.

2. Do your examples in Part (1) support the following proposition or prove that the proposition is false? Explain.

   Proposition: If m is an even integer and n is an integer, then $m \cdot n$ is an even integer.

3. Write this conditional statement in the proposition in symbolic form using the symbols for logical operators.

4. What is the hypothesis of the conditional statement? What is the conclusion of the conditional statement?

5. What is the definition of an even integer? (See Preview Activity 2 in Section 1.2.)

6. Construct a know-show table for this proposition. See Section 1.2 to review this concept.

Preview Activity 3 (Calendars and Clocks).

1. Suppose that it is currently Tuesday.

   (a) What day will it be 3 days from now?

   (b) What day will it be 10 days from now?

(c) What day will it be 17 days from now? What day will it be 24 days from now?

(d) Find several other natural numbers x such that it will be Friday x days from now.

(e) Create a list (in increasing order) of the numbers $3, 10, 17, 24$, and the numbers you generated in Part (1d). Pick any two numbers from this list and subtract one from the other. Repeat this several times.

(f) What do the numbers you obtained in Part (1e) have in common?

2. Suppose that we are using a twelve-hour clock with no distinction between A.M. and P.M. Also, suppose that the current time is 5:00.

(a) What time will it be 4 hours from now?

(b) What time will it be 16 hours from now? What time will it be 28 hours from now?

(c) Find several other natural numbers x such that it will be 9:00 x hours from now.

(d) Create a list (in increasing order) of the numbers $4, 16, 28$, and the numbers you generated in Part (2c). Pick any two numbers from this list and subtract one from the other. Repeat this several times.

(e) What do the numbers you obtained in Part (2d) have in common?

3. This is a continuation of Part (1). Suppose that it is currently Tuesday.

(a) What day was it 4 days ago?

(b) What day was it 11 days ago? What day was it 18 days ago?

(c) Find several other natural numbers x such that it was Friday x days ago.

(d) Create a list (in increasing order) consisting of the numbers $-18, -11, -4$, the opposites of the numbers you generated in Part (3c) and the positive numbers in the list from Part (1e). Pick any two numbers from this list and subtract one from the other. Repeat this several times.

(e) What do the numbers you obtained in Part (3d) have in common?

Some Mathematical Terminology

In Section 1.2, we introduced the idea of a direct proof. Since then, we have used some common terminology in mathematics without much explanation. Before we proceed further, we will discuss some frequently used mathematical terms.

A **proof** in mathematics is a convincing argument that some mathematical statement is true. A proof should contain enough mathematical detail to be convincing to the person(s) to whom the proof is addressed. A proof must use correct, logical reasoning and be based on previously established results. These previous results can be axioms, definitions, or previously proven theorems. These terms are discussed below.

Surprising to some is the fact that in mathematics, there are always **undefined terms**. This is because if we tried to define everything, then we would end up going in circles. Simply put, we must start somewhere. For example, in Euclidean geometry, the terms "point," "line," and "contains" are undefined terms. In this text, we are using our number systems such as the natural numbers and integers as undefined terms. We often assume that these undefined objects satisfy certain properties. These assumed relationships are accepted as true without proof and are called axioms (or postulates). An **axiom** is a mathematical statement that is accepted without proof. Euclidean geometry starts with undefined terms and a set of postulates and axioms. For example, the following statement is an axiom of Euclidean geometry:

> *Given any two distinct points, there is exactly one line that contains these two points.*

The properties we listed in Preview Activity 1 from Section 1.2 are being used as axioms in this text.

A **definition** is simply an agreement as to the meaning of a particular term. For example, in this text, we have defined the terms "even integer" and "odd integer." Definitions are not made at random, but rather, a definition is usually made because a certain property is observed to occur frequently. As a result, it becomes convenient to give this property its own special name. Definitions that have been made can be used in developing mathematical proofs. In fact, most proofs require the use of some definitions.

In dealing with mathematical statements, we frequently use the terms "conjecture," "theorem," "proposition," "lemma," and "corollary." A **conjecture** is a statement that we believe is plausible. That is, we think it is true, but we have not yet developed a proof that it is true. A **theorem** is a mathematical statement for which we have a proof. A term that is often considered to be synonymous with theorem is **proposition**.

Often the proof of a theorem can be quite long. In this case, it is often easier to communicate the proof in smaller "pieces." These supporting pieces are often called lemmas. A **lemma** is a true mathematical statement that was proven mainly to help in the proof of some theorem. Once a given theorem has been proven, it is often the case that other propositions follow immediately from the fact that the theorem is true. These are called corollaries of the theorem. The term **corollary** is used to refer to a theorem that is easily proven once some other theorem has been proven.

Constructing Mathematical Proofs

To create a proof of a theorem, we must use correct logical reasoning and mathematical statements that we already accept as true. These statements include axioms, definitions, theorems, lemmas, and corollaries.

In Section 1.2, we introduced the use of a **know-show table** to help us organize our work when we are attempting to prove a statement. We also introduced some guidelines for writing mathematical proofs once we have created the proof. These guidelines should be reviewed before proceeding.

Please remember that when we start the process of writing a proof, we are essentially "reporting the news." That is, we have already discovered the proof (discovered the news), and now we need to report it. This reporting often does not describe the process of discovering the news (the investigative portion of the process).

The first step is to develop a conjecture. This is often done after working within certain objects for some time. For example, after dealing with the definition of one integer dividing another integer in Preview Activity 1, several examples where an integer a divided an integer b and the integer b divided the integer c were investigated. In all cases, it should have been observed that a also divided c. So we can form the following conjecture.

Conjecture: *Let a, b, and c be integers. If a divides b and b divides c, then a divides c.*

To try to prove this conjecture, we will, of course, have to use the definitions in Preview Activity 1. They are repeated here along with some new notation.

> **Definition.** An integer m **divides** an integer n provided that there is an integer q such that $n = m \cdot q$. We also say that m is a **divisor** of n, m is a **factor** of n, and n is a **multiple** of m.

Important Comment about Notation When an integer m divides an integer n, we frequently use the notation $m \mid n$. Be careful with this notation. It does not represent the rational number $\frac{m}{n}$. The notation $m \mid n$ represents a relationship between the integers m and n and is simply a shorthand for "m divides n."

Back to the conjecture. Before trying to prove it, we should make sure we have explored some examples. This simply means to construct some specific examples where the integers a, b, and c satisfy the hypothesis of the conjecture in order to see if they also satisfy the conclusion. If we happen to find an example of three integers that satisfy the hypothesis but make the conclusion false, then we would have found a counterexample for the conjecture. We could then conclude the conjecture is false. This will not happen for the current conjecture.

One example for this conjecture is $a = 3, b = 12, c = 48$. Notice that $3 \mid 12$ and $12 \mid 48$, and we observe that $3 \mid 48$. In particular, if we use the definition of divides, we see that

$$12 = 3 \cdot 4 \text{ and that } 48 = 12 \cdot 4.$$

Now, substitute the right side of the first equation for 12 in the second equation. This gives

$$48 = (3 \cdot 4) \cdot 4$$
$$48 = 3 \cdot (4 \cdot 4).$$

This last equation shows that 3 divides 48. While in this case the examples may seem trivial, this is not always the case. Exploring examples can sometimes lead to a counterexample for the conjecture, and other times examples can suggest a method of proof. In this case, the main step was to substitute the expression $3 \cdot 4$ for 12 from one equation into the other equation.

We will now start a know-show table for this conjecture.

Step	Know	Reason
P	$a \mid b$ and $b \mid c$	Hypothesis
$P1$		
$\vdots$	$\vdots$	$\vdots$
$Q1$		
Q	$a \mid c$	
Step	**Show**	**Reason**

The backward question we ask is, "How can we prove that a divides c?" One answer is to use the definition and show that there exists an integer q such that $c = a \cdot q$.

The forward question we ask is, "What can we conclude from the facts that $a \mid b$ and $b \mid c$?" Again, using the definition, we know that there exist integers s and t such that $b = a \cdot s$ and $c = b \cdot t$. So we now have the following know-show table.

Step	Know	Reason
P	$a \mid b$ and $b \mid c$	Hypothesis
$P1$	$(\exists s \in \mathbb{Z})\,(b = a \cdot s)$ $(\exists t \in \mathbb{Z})\,(c = b \cdot t)$	Definition of "divides"
$\vdots$	$\vdots$	$\vdots$
$Q1$	$(\exists q \in \mathbb{Z})\,(c = a \cdot q)$	
Q	$a \mid c$	Definition of "divides"
Step	**Show**	**Reason**

The key now is to determine how to get from $P1$ to $Q1$. We might get some motivation from the numerical example we explored. Using the equation $b = a \cdot s$, we can substitute $a \cdot s$ for b in the second equation, $c = b \cdot t$. This gives

$$c = b \cdot t$$
$$= (a \cdot s) \cdot t$$
$$= a\,(s \cdot t).$$

The last step used the associative property of multiplication. (See Preview Activity 1 from Section 1.2.) The last equation shows that c is equal to a times some integer. (This is because $s \cdot t$ is an integer by the closure property for integers.) So although we did not use the letter q, we have arrived at step $Q1$. The completed know-show table follows.

Step	Know	Reason
P	$a \mid b$ and $b \mid c$	Hypothesis
$P1$	$(\exists s \in \mathbb{Z})\,(b = a \cdot s)$ $(\exists t \in \mathbb{Z})\,(c = b \cdot t)$	Definition of "divides"
$P2$	$c = (a \cdot s) \cdot t$	Substituting for b
$P3$	$c = a \cdot (s \cdot t)$	Associative property of multiplication
$Q1$	$(\exists q \in \mathbb{Z})\,(c = a \cdot q)$	Step $P3$ and closure property of the integers
Q	$a \mid c$	Definition of "divides"

We can now report the news by writing a formal proof.

Theorem 3.1. *Let a, b, and c be integers. If a divides b and b divides c, then a divides c.*

Proof. We assume that a, b, and c are integers, and we further assume that a divides b and that b divides c. We will prove that a divides c.

Since a divides b, there exists an integer s such that

$$b = a \cdot s. \tag{1}$$

Similarly, because b divides c, there exists an integer t such that

$$c = b \cdot t. \tag{2}$$

We can now substitute the expression for b from Equation (1) into Equation (2). This gives

$$c = (a \cdot s) \cdot t. \tag{3}$$

Using the associate property for multiplication, we can rearrange the right side of Equation (3) to obtain

$$c = a \cdot (s \cdot t). \tag{4}$$

Because both s and t are integers, and since the integers are closed under multiplication, we know that $s \cdot t \in \mathbb{Z}$. Therefore, Equation (4) proves that a divides c. Consequently, we have proven that whenever a, b, and c are integers such that a divides b and b divides c, then a divides c. ∎

Writing Guidelines for Equation Numbers

We have attempted to write the proof for Theorem 3.1 according to the guidelines introduced in Section 1.2. A new element that appeared in this proof was the use of equation numbers. Following are some guidelines that can be used for **equation numbers**.

If it is necessary to refer to an equation later in a proof, that equation should be centered and displayed. It should then be given it a number. The number for the equation should be written in parentheses on the same line as the equation at the right-hand margin.

Example 3.2.

Since x is an odd integer, there exists an integer n such that

$$x = 2n + 1. \tag{1}$$

Later in the proof, there may be a line such as

Then, using the result in Equation (1), we obtain ....

Please note that we should only number those equations we will be referring to later in the proof. Also, note that the word "Equation" begins with a capital "E" when we are referring to an equation by number.

Activity 3.3 (A Property of Divisors).   In this activity, the universal set for each variable is the set of integers.

1. Give at least four different examples of integers a, b, and c such that a divides b and a divides c.

2. For each example in Part (1), calculate the sum $b + c$. Do you notice any relationship between the integer a and the sum $b+c$? If so, describe the relationship.

3. Based upon your work in Parts (1) and (2) formulate a conjecture concerning the relationship between the integer a and the sum $b + c$. Your conjecture should start with the sentence, "Let a, b, and c be integers.", and should be completed with a conditional statement.

4. Construct a Know-Show-Table for a proof of the conjecture in Part (3).

Congruence

What mathematicians call congruence is a concept used to describe cycles in the world of the integers. For example, the day of the week is a cyclic phenomenon in that the day of the week repeats every seven days. The time of the day is a cyclic phenomenon because it repeats every 12 hours if we use a 12-hour clock or every 24 hours if we use a 24-hour clock. We explored these two cyclic phenomena in Preview Activity 3.

Similar to what we saw in Preview Activity 3, if it is currently Monday, then it will be Wednesday 2 days from now, 9 days from now, 16 days from now, 23 days from now, and so on. In addition, it was Wednesday 5 days ago, 12 days ago, 19 days ago, and so on. Using negative numbers for time in the past, we generate the following list of numbers:

$$\ldots, -19, -12, -5, 2, 9, 16, 23, \ldots.$$

Since days of the week have a 7-day cycle, if we subtract any number in the list above from any other number in that list, we will obtain a multiple of 7. For example

$$16 - 2 = 14 = 7 \cdot 2$$
$$(-5) - (9) = -14 = 7 \cdot (-2)$$
$$16 - (-12) = 28 = 7 \cdot 4$$

Using the concept of congruence, we would say that all the numbers in this list are congruent modulo 7. We start with the following definition that defines when two numbers are congruent modulo some natural number n.

Definition. Let $n \in \mathbb{N}$. If a and b are integers, then we say that **a is congruent to b modulo n** provided that n divides $a - b$. A standard notation for this is $a \equiv b \pmod{n}$. This is read as "a is congruent to b modulo n" or "a is congruent to b mod n ."

Notice that $n \mid (a - b)$ if and only if there exists an integer k such that $a - b = nk$. So we can write

$$a \equiv b \pmod{n} \text{ means } (\exists k \in \mathbb{Z}) (a - b = nk) \text{, or}$$
$$a \equiv b \pmod{n} \text{ means } (\exists k \in \mathbb{Z}) (a = b + nk).$$

We will study the concept of congruence modulo n in much more detail later in the text. For now, we will work with the definition of congruence modulo n in the context of proofs.

Activity 3.4 (Congruence Modulo 6).

1. Find several integers that are congruent to 5 modulo 6 and then square each of these integers.

2. For each integer m from Part (1), determine an integer k so that $0 \le k < 6$ and $m^2 \equiv k$ (mod 6). What do you observe?

3. Complete the following know-show table for the conjecture that if $m \equiv 5$ (mod 6), then $m^2 \equiv 1$ (mod 6).

Step	Know	Reason
P	$m \equiv 5$ (mod 6)	Hypothesis
$P1$	$6 \mid (m - 5)$	Definition of "congruence modulo 6"
$P2$	$(\exists k \in \mathbb{Z}) \, (m - 5 = 6k)$	Definition of "divides"
$P3$		Algebra
$\vdots$	$\vdots$	$\vdots$
$Q1$	$6 \mid (m^2 - 1)$	
Q	$m^2 \equiv 1$ (mod 6)	Definition of "congruence modulo 6"
Step	**Show**	**Reason**

We will complete this section with some standard textbook proofs of two of the basic properties of congruence modulo n. Three properties will be listed; the proof of one of them will be left as an exercise. Please remember that textbook proofs are usually written in final form of reporting the news. Before reading these proofs, it might be instructive to first try to construct a know-show table for each proof.

Theorem 3.5 (Properties of Congruence Modulo n). *Let $n \in \mathbb{N}$, and let a, b, and c be integers.*

1. *For every integer a, $a \equiv a$ (mod n).*

   *This is called the **reflexive property** of congruence modulo n.*

2. *If $a \equiv b$ (mod n), then $b \equiv a$ (mod n).*

   *This is called the **symmetric property** of congruence modulo n.*

3. *If $a \equiv b$ (mod n) and $b \equiv c$ (mod n), then $a \equiv c$ (mod n).*

  *This is called the **transitive property** of congruence modulo n.*

Proof of the reflexive property: Let $n \in \mathbb{N}$, and let $a \in \mathbb{Z}$. We will show that $a \equiv a$ (mod n). Notice that

$$a - a = 0 = n \cdot 0.$$

This proves that n divides $(a - a)$ and hence, by the definition of congruence modulo n, we have proven that $a \equiv a$ (mod n). ■

The proof of the symmetric property is included in the exercises.

Proof of the transitive property: Let $n \in \mathbb{N}$, and let a, b, and c be integers. We assume that $a \equiv b$ (mod n) and that $b \equiv c$ (mod n). We will use the definition of congruence modulo n to prove that $a \equiv c$ (mod n). Since $a \equiv b$ (mod n) and $b \equiv c$ (mod n), we know that $n \mid (a - b)$ and $n \mid (b - c)$. Hence, there exist integers k and q such that

$$a - b = nk$$
$$b - c = nq$$

By adding these two equations, we obtain

$$(a - b) + (b - c) = nk + nq.$$

If we now simplify the left side of this last equation and factor the right side, we get

$$a - c = n(k + q).$$

By the closure property of the integers, $(k + q) \in \mathbb{Z}$, and so this equation proves that $n \mid (a - c)$ and hence that $a \equiv c$ (mod n). This completes the proof of the transitive property of congruence modulo n. ■

Exercises 3.1

For Exercises (1) through (7), you are asked to determine if statements are true or false. This means that if a statement is true, then write a formal proof of that statement, and if it is false, then provide a counterexample that shows it is false.

1. Are the following propositions true or false?

    (a) Let a, b, and c be integers. If $a \mid b$ and $a \mid c$, then $a \mid (b + c)$.
    (b) Let a, b, and c be integers. If $a \mid b$ and $a \mid c$, then $a \mid (b - c)$.

2. Is the following statement true or false?
    If n is an odd integer, then n^3 is an odd integer.

3. Is the following proposition true or false?
    Let a, b, and c be integers. If $a \mid b$, then $a \mid (bc)$.

4. Is the following proposition true or false?
    Let a, b, and c be integers. If $a \mid (bc)$, then $a \mid b$ or $a \mid c$.

5. Are the following propositions true or false?

    (a) For every integer n, $4n^2 + 7n + 6$ is an odd integer.
    (b) For every odd integer n, $4n^2 + 7n + 6$ is an odd integer.

6. (a) If x and y are integers and $xy = 1$, explain why $x = 1$ or $x = -1$.
    (b) Is the following proposition true or false?
        Let a and b be nonzero integers. If $a \mid b$ and $b \mid a$, then $a = \pm b$.

7. Is the following proposition true or false?
    Let a, b, and d be integers. If d divides both $a - b$ and $a + b$, then d divides a.

8. Let $n \in \mathbb{N}$. Prove the symmetric property of congruence modulo n.

9. Let a be an integer and let $n \in \mathbb{N}$.

    (a) Prove that if $a \equiv 0 \pmod{n}$, then $n \mid a$.
    (b) Prove that if $n \mid a$, then $a \equiv 0 \pmod{n}$.

10. Prove the following proposition:
    Let a be an integer. If there exists an integer n such that $a \mid (4n + 3)$ and $a \mid (2n + 1)$, then $a = 1$ or $a = -1$.

    Hint: Use the fact that the only divisors of 1 are 1 and -1.

11. Let n be a natural number and let a, b, c, and d be integers. Prove the following propositions.

(**a**) If $a \equiv b \pmod{n}$ and $c \equiv d \pmod{n}$, then
$(a + c) \equiv (b + d) \pmod{n}$.

(**b**) If $a \equiv b \pmod{n}$ and $c \equiv d \pmod{n}$, then $ac \equiv bd \pmod{n}$.

3.2 More Methods of Proof

Preview Activity 1 (Attempting a Proof).

The following statement was proven in Exercise (5c) in Section 1.2.

If n is an odd integer, then n^2 is an odd integer.

It is also a direct consequence of Theorem 1.3 in Section 1.2.

If x and y are odd integers, then $x \cdot y$ is an odd integer.

Now consider the following proposition:

Let n be an integer. If n^2 is an odd integer, then n is an odd integer.

1. After examining several examples, decide whether you think this proposition is true or false.

2. Try completing the following know-show table for a direct proof of this proposition.

Step	**Know**	**Reason**
P	n^2 is an odd integer.	Hypothesis
$P1$		
$\vdots$	$\vdots$	$\vdots$
$Q1$	There exists an integer q such that $n = 2q + 1$	
Q	n is an odd integer.	Definition of "odd integer"
Step	**Show**	**Reason**

The question is, "Can we perform algebraic manipulations to get from the 'know' portion of the table to the 'show' portion of the table?" Be careful with this! Remember that we are working with integers and we want to make sure that we can end up with an integer q as stated in Step $Q1$.

Preview Activity 2 (The Contrapositive).

1. Complete the following truth table. What does this truth table prove?

P	Q	$\neg Q$	$\neg P$	$P \to Q$	$\neg Q \to \neg P$
T	T				
T	F				
F	T				
F	F				

Consider the following proposition again:

Let n be an integer. If n^2 is an odd integer, then n is an odd integer.

2. Write the contrapositive of this conditional statement in this proposition. Remember that "not odd" means "even."

3. Complete a know-show table for the contrapositive statement from Part (2).

4. By completing the proof in Part (3), have you proven the proposition from Part (1)? That is, have you proven that if n^2 is an odd integer, then n is an odd integer? Explain.

Preview Activity 3 (A Biconditional Statement).

1. Recall that $P \leftrightarrow Q \equiv (P \to Q) \wedge (Q \to P)$.

Suppose that we want to prove a biconditional statement of the form $P \leftrightarrow Q$. Explain a method for completing this proof based on this logical equivalency.

2. Let n be an integer. Assume that we have completed the proofs of the following two statements:

- If n is an odd integer, then n^2 is an odd integer.

- If n^2 is an odd integer, then n is an odd integer.

(See Preview Activity 1 and Preview Activity 2.) Have we completed the proof of the following proposition?

The integer n is an odd if and only if n^2 is an odd integer.

Explain.

Review of Direct Proofs

In Sections 1.2 and 3.1, we studied what is called a direct proof of a mathematical statement. Most of the statements we prove in mathematics are conditional statements that can be written in the form $P \to Q$.

A direct proof of a statement of the form $P \to Q$ is based on the definition that a conditional statement can only be false when the hypothesis, P, is true and the conclusion, Q, is false. Thus, if the conclusion is true whenever the hypothesis is true, then the conditional statement must be true. So, in a direct proof,

- We start by assuming that P is true.

- From this assumption, we logically deduce that Q is true.

We have used the so-called forward and backward method with a know-show table to discover how to logically deduce Q from the assumption that P is true.

Proof Using the Contrapositive

As we saw in Preview Activity 1, it is sometimes difficult to construct a direct proof of a conditional statement. This is one reason we studied logical equivalencies in Section 2.3. Knowing that two expressions are logically equivalent tells us that if we prove one, then we have also proven the other. In fact, once we know the truth value of a statement, then we know the truth value of any other statement that is logically equivalent to it.

One of the most useful logical equivalencies in this regard is the fact that a conditional statement $P \to Q$ is logically equivalent to its contrapositive, $\neg Q \to \neg P$. (We established this in Section 2.3.) This means that if we prove the contrapositive of the conditional statement, then we have proven the conditional statement. The following are some important points to remember.

- A conditional statement is logically equivalent to its contrapositive. If the conditional statement is, "If P then Q," then the contrapositive is "If not Q then not P."

- Thus, $P \rightarrow Q$ can be proven to be true by proving that its contrapositive $\neg Q \rightarrow \neg P$ is true.

- Use a direct proof to prove that $\neg Q \rightarrow \neg P$ is true.

- Caution: One difficulty with this type of proof is in the formation of correct negations. (We need to be very careful doing this.)

- We might consider using a proof by contrapositive when the statements P and Q are stated as negations.

Writing Guidelines

One of the basic rules of writing mathematical proofs is to keep the reader informed. So when we prove a result using the contrapositive, we indicate this within the first few lines of the proof.

Examples:

- We will prove this theorem by proving its contrapositive.

- We will prove the contrapositive of this statement.

In addition, make sure the reader knows the status of every assertion that you make. That is, make sure you state whether an assertion is an assumption of the theorem, a previously proven result, a well-known result, or something from the reader's mathematical background.

Following is a completed proof of a theorem similar to the statement from Preview Activity 1 from this section.

Theorem 3.6. *Let n be an integer. If n^2 is an even integer, then n is an even integer.*

Proof. We will prove this result by proving the contrapositive of the statement, which is

If n is an odd integer, then n^2 is an odd integer.

So assume that n is an odd integer. We will show that n^2 is an odd integer. By definition, there exists an integer k such that

$$n = 2k + 1.$$

Squaring both sides of this equation yields

$$n^2 = 4k^2 + 4k + 1.$$

However, we can rewrite this equation in the form

$$n^2 = 2\left(2k^2 + 2k\right) + 1.$$

Using the closure properties of the integers, we conclude that $2k^2 + 2k$ is an integer. Hence, the last equation proves that n^2 is an odd integer. Thus, we have proven the contrapositive of the theorem, and consequently, we have proven that if n^2 is an even integer, then n is an even integer. ■

Proofs of Biconditional Statements

In Preview Activity 3, we used the following logical equivalency:

$$(P \leftrightarrow Q) \equiv (P \rightarrow Q) \wedge (Q \rightarrow P).$$

This logical equivalency suggests one method for proving a biconditional statement written in the form "P if and only if Q." This method is to construct separate proofs of the two conditional statements $P \rightarrow Q$ and $Q \rightarrow P$. For example, since we have now proven that

- If n is an even integer, then n^2 is an even integer; and that

- If n^2 is an even integer, then n is an even integer.

we can state the following theorem.

Theorem 3.7. *The integer n is an even integer if and only if n^2 is an even integer.*

Writing Guidelines

When proving a biconditional statement using the logical equivalency $(P \leftrightarrow Q) \equiv (P \rightarrow Q) \wedge (Q \rightarrow P)$, we actually need to prove two conditional statements. The proof of each conditional statement can be considered as one of two parts of the proof of the biconditional statement. Make sure that the start and end of each of these parts is clearly indicated.

Proposition 3.8. *Let $x \in \mathbb{R}$. The real number x equals 2 if and only if $x^3 - 2x^2 + x = 2$.*

Proof. We will prove this biconditional statement by proving the following two conditional statements:

- If x equals 2 , then $x^3 - 2x^2 + x = 2$.

- If $x^3 - 2x^2 + x = 2$, then x equals 2.

For the first part, we assume $x = 2$ and prove that $x^3 - 2x^2 + x = 2$. We can do this by substituting $x = 2$ into the expression $x^3 - 2x^2 + x$. This gives

$$x^3 - 2x^2 + x = 2^3 - 2\left(2^2\right) + 2$$
$$= 8 - 8 + 2$$
$$= 2.$$

This completes the first part of the proof.

For the second part, we assume that $x^3 - 2x^2 + x = 2$ and from this assumption, we will prove that $x = 2$. We will do this by solving this equation for x. To do so, we first rewrite the equation $x^3 - 2x^2 + x = 2$ by subtracting 2 from both sides:

$$x^3 - 2x^2 + x - 2 = 0.$$

We can now factor the left side of this equation by factoring an x from the first two terms and then factoring $(x - 2)$ from the resulting two terms. This is shown below.

$$x^3 - 2x^2 + x - 2 = 0$$
$$x^2 (x - 2) + (x - 2) = 0$$
$$(x - 2)\left(x^2 + 1\right) = 0$$

Now, in the real numbers, if a product of two factors is equal to zero, then one of the factors must be zero. So this last equation implies that

$$x - 2 = 0 \text{ or } x^2 + 1 = 0.$$

The equation $x^2 + 1 = 0$ has no real number solution. So since x is a real number, the only possibility is that $x - 2 = 0$. From this we can conclude that x must be equal to 2.

Since we have now proven both conditional statements, we have proven that $x = 2$ if and only if $x^3 - 2x^2 + x = 2$. ∎

Using Other Logical Equivalencies

As was noted in Section 2.3, there are several different logical equivalencies. Fortunately, there are only a small number that we often use when trying to write proofs, and many of these are listed in Theorem 2.13 at the end of Section 2.3. We will illustrate the use of one of these logical equivalencies with the following proposition:

Let a and b be real numbers. If $a \neq 0$ and $b \neq 0$, then $ab \neq 0$.

First, notice that the hypothesis and the conclusion of the conditional statement are stated in the form of negations. This suggests that we consider the contrapositive. Care must be taken when we negate the hypothesis since it is a conjunction. We use one of De Morgan's Laws as follows:

$$\neg\,(a \neq 0 \;\wedge\; b \neq 0) \equiv (a = 0) \vee (b = 0)\,.$$

So the contrapositive is

Let a and b be real numbers. If $ab = 0$, then $a = 0$ or $b = 0$.

The contrapositive is a conditional statement in the form $P \to (Q \vee R)$.

The difficulty is that there is not much we can do with the hypothesis $(ab = 0)$ since we know nothing else about the real numbers a and b. However, if we knew that a was not equal to zero, then we could multiply both sides of the equation $ab = 0$ by $1/a$. This suggests that we consider using the following logical equivalency from Theorem 2.13:

$$P \to (Q \vee R) \equiv (P \wedge \neg Q) \to R.$$

This logical equivalency makes sense because if we are trying to prove $Q \vee R$, we only need to prove that at least one of Q or R is true. So the idea is to prove that if Q is false, then R must be true. In the following completed proof, we will make use of this logical equivalency in this informal manner rather than explicitly stating the logical equivalency. (However, it is perfectly acceptable to make explicit use of the logical equivalency.)

Proposition 3.9. *Let a and b be real numbers. If $a \neq 0$ and $b \neq 0$, then $ab \neq 0$.*

Proof. We will prove the contrapositive of this proposition. So let a and b be real numbers. We will prove that

If $ab = 0$, then $a = 0$ or $b = 0$.

To begin, assume that $ab = 0$. We need to prove that $a = 0$ or $b = 0$. If $a = 0$, then the conclusion is true. So we assume that $a \neq 0$. We can then multiply both sides of the equation $ab = 0$ by $1/a$. This gives

$$\frac{1}{a}(ab) = \frac{1}{a} \cdot 0.$$

We now use the associative property on the left side of this equation and simplify both sides of the equation to obtain

$$\left(\frac{1}{a} \cdot a\right) b = 0$$
$$1 \cdot b = 0$$
$$b = 0.$$

Therefore, either $a = 0$ or $b = 0$. This completes the proof of the contrapositive and hence, we have proven the proposition. ∎

Results about Even and Odd Integers and Divisors

We conclude this section with a summary of the results we have obtained about even and odd integers and about divisors. This will give us a handy reference for use in later sections.

Theorem 3.10 (Properties of Even and Odd Integers).

1. *(Exercise (1), Section 1.2)*
   If m is an even integer, then m + 1 is an odd integer.
   If m is an odd integer, then m + 1 is an even integer.

2. *(Exercise (3), Section 1.2)*
   If x is an even integer and y is an even integer, then x + y is an even integer.
   If x is an even integer and y is an odd integer, then x + y is an odd integer.
   If x is an odd integer and y is an odd integer, then x + y is an even integer.

3. *(Exercise (5), Section 1.2)*
   If x is an even integer and y is an integer, then x · y is an even integer.
   (Theorem 1.3)
   If x is an odd integer and y is an odd integer, then x · y is an odd integer.

4. (*Theorem 3.6*)
The integer n is an even integer if and only if n^2 is an even integer.
(*Preview Activity 2*)
The integer n is an odd integer if and only if n^2 is an odd integer.

Theorem 3.11 (Properties of Divisors).

Let a, b, and c be integers.

1. (*Theorem 3.1*)
   If $a \mid b$ and $b \mid c$, then $a \mid c$.

2. (*Exercise (1), Section 3.1*)
   If $a \mid b$ and $a \mid c$, then $a \mid (b + c)$.
   If $a \mid b$ and $a \mid c$, then $a \mid (b - c)$.

3. (*Exercise (3), Section 3.1*)
   If $a \mid b$, then $a \mid (bc)$.

4. (*Exercise (6), Section 3.1*)
   If $a \mid b$ and $b \mid a$, then $a = \pm b$.

Activity 3.12 (Using a Logical Equivalency).

Consider the following proposition:

> Let a and b be integers. If 3 does not divide a and 3 does not divide b, then 3 does not divide the product $a \cdot b$.

1. Notice that the hypothesis of the proposition is stated as a conjunction of two negations ("3 does not divide a and 3 does not divide b"). Also, the conclusion is stated as the negation of a sentence. ("3 does not divide the product $a \cdot b$.") This often indicates that we should consider using a proof of the contrapositive. If we use the symbolic form $(\neg Q \wedge \neg R) \rightarrow \neg P$ as a model for this proposition, what is P, what is Q, and what is R?

2. Write a symbolic form for the contrapositive of $(\neg Q \wedge \neg R) \rightarrow \neg P$.

3. Write the contrapositive of the proposition as a conditional statement in English.

We do not yet have all the tools needed to prove the proposition or its contrapositive. However, later in the text, we will learn that the following proposition is true.

Proposition X. Let a be an integer. If 3 does not divide a, then there exist integers x and y such that $3x + ay = 1$.

4. **(a)** Find integers x and y guaranteed by Proposition X when $a = 5$.

   (b) Find integers x and y guaranteed by Proposition X when $a = 2$.

   (c) Find integers x and y guaranteed by Proposition X when $a = -2$.

5. Assume that Proposition X is true and use it to help construct a proof of the contrapositive of the given proposition. In doing so, you will most likely have to use the following logical equivalency:
   $$P \to (Q \vee R) \equiv (P \wedge \neg Q) \to R.$$

Exercises 3.2

1. Let n be an integer. Prove each of the following:

   (a) If n is even, then n^3 is even.

   (b) If n^3 is even, then n is even.

   (c) The integer n is even if and only if n^3 is an even integer.

   (d) The integer n is odd if and only if n^3 is an odd integer.

2. In Section 3.1, we defined congruence modulo n where n is a natural number. If a and b are integers, we will use the notation $a \not\equiv b \pmod{n}$ to mean that a is not congruent to b modulo n.

   (a) Write the contrapositive of the conditional statement in the following proposition:

   > Let a and b be integers. If $a \not\equiv 0 \pmod 6$ and $b \not\equiv 0 \pmod 6$, then $ab \not\equiv 0 \pmod 6$.

   (b) Is this proposition true or false? Explain.

3. Let a and b be positive real numbers.

   (a) Write the contrapositive of the following statement:

   $$\text{If } \sqrt{ab} \neq \frac{a+b}{2}, \text{ then } a \neq b.$$

(b) Is this statement true or false? Prove the statement if it is true or provide a counterexample if it is false.

4. Let a be an integer. Are the following statements true or false? Explain. This means to prove the statement if it is true or provide a counterexample if it is false.

(a) If $a \equiv 2 \pmod 5$, then $a^2 \equiv 4 \pmod 5$.

(b) If $a^2 \equiv 4 \pmod 5$, then $a \equiv 2 \pmod 5$.

(c) $a \equiv 2 \pmod 5$ if and only if $a^2 \equiv 4 \pmod 5$.

5. For a right triangle, suppose that the hypotenuse has length c feet and the lengths of the sides are a feet and b feet.

(a) What is a formula for the area of this right triangle?

(b) What can be concluded from the Pythagorean Theorem for right triangles?

(c) What is an isosceles triangle?

(d) Prove that the right triangle described above is an isosceles triangle if and only if the area of the right triangle is $\frac{1}{4}c^2$.

6. A real number x is defined to be a **rational number** provided

$$\text{there exist integers } m \text{ and } n \text{ with } n \neq 0 \text{ such that } x = \frac{m}{n}.$$

A real number that is not a rational number is called an **irrational number.**

It is known that if x is a positive rational number, then there exist positive integers m and n with $n \neq 0$ such that $x = \frac{m}{n}$.

Is the following proposition true or false? Explain.

Let x be a positive real number. If x is irrational, then $\sqrt{x}$ is irrational.

7. Is the following proposition true or false? Justify your conclusion.

Let $n \in \mathbb{Z}$. The integer n is even if and only if 4 divides n^2.

8. Let a and b be natural numbers such that $a^2 = b^3$. Prove each of the propositions in Parts (8a) through (8d). (The results of Exercise (1) and Theorem 3.7 may be helpful.)

(a) If a is even, then 4 divides a.

(b) If 4 divides a, then 4 divides b.

(c) If 4 divides b, then 8 divides a.

(d) If a is even, then 8 divides a.

(e) Give an example of natural numbers a and b such that a is even and $a^2 = b^3$, but b is not divisible by 8.

9. Prove the following proposition:

> Let a and b be integers with $a \neq 0$. If a does not divide b, then the equation $ax^3 + bx + (b + a) = 0$ does not have a solution that is an integer.

Hint: It may be necessary to factor a sum cubes. Recall that

$$u^3 + v^3 = (u + v)\left(u^2 - uv + v^2\right).$$

3.3   Proof by Contradiction

Preview Activity 1 (Proof by Contradiction).

A **tautology** is a propositional expression that yields a true statement regardless of what statements replace its variables. A **contradiction** is a propositional expression that yields a false statement regardless of what statements replace its variables. That is, a tautology is necessarily true in all circumstances, and a contradiction is necessarily false in all circumstances.

1. Use a truth table to explain why $(P \vee \neg P)$ is a tautology.

2. Use a truth table to explain why $(P \wedge \neg P)$ is a contradiction.

Another method of proof that is frequently used to prove a given proposition is the **contradiction method of proof.** This method is based on the fact that a statement Q can only be true or false (and not both). The idea is to prove that the statement Q is true by showing that it cannot be false. This is done by assuming that Q is false and proving that this leads to a contradiction. (Usually the contradiction has the form $(R \wedge \neg R)$ where R is some statement.) This means that the assumption that the statement Q is false is incorrect and hence Q cannot be false. Since it cannot be false, then Q must be true.

The logical basis for the contradiction method of proof is the tautology

$$[\neg Q \rightarrow (R \wedge \neg R)] \rightarrow Q.$$

3. Use a truth table to show that $[\neg Q \to (R \wedge \neg R)] \to Q$ is a tautology.

4. Explain why this tautology shows that if the assumption that Q is false leads to a contradiction, then you have proven that Q is true.

Preview Activity 2 (Proof by Contradiction [continued]).

The idea of a proof by contradiction is to assume that the statement you want to prove is false and reach a contradiction based on this assumption. When we try to prove the conditional statement, "If P then Q" using a proof by contradiction, we must assume that $P \to Q$ is false and show that this leads to a contradiction.

Recall that

$$P \to Q \quad \text{is logically equivalent to} \quad \neg P \vee Q,$$

and that

$$\neg (P \to Q) \quad \text{is logically equivalent to} \quad P \wedge \neg Q.$$

The preceding logical equivalencies show that when you assume that $P \to Q$ is false, you are assuming that P is true and Q is false. If you can prove that this leads to a contradiction, then you have shown that $\neg (P \to Q)$ is false and hence that $P \to Q$ is true.

1. Give an example to show that the following statement is false.

For all real numbers x and y, if $x \neq y$, then $\dfrac{x}{y} + \dfrac{y}{x} > 2$.

2. Instead of working with the statement in (1), we will work with a related statement that is obtained by adding conditions to the hypothesis.

For all real numbers x and y, if

if $x \neq y$, $x > 0$, and $y > 0$, then $\dfrac{x}{y} + \dfrac{y}{x} > 2$.

What assumptions need to be made for a proof by contradiction for this statement? Carefully write down all assumptions made at the beginning of a proof by contradiction.

Preview Activity 3 (Rational Numbers).

In Exercise (6) from Section 3.2, we defined a rational number as follows:

Definition. A real number r is a **rational number** provided that there exist integers m and n , with $n \neq 0$, such that $r = \dfrac{m}{n}$. A real number r is an **irrational number** if it is not a rational number.

---

1. Give examples of at least five different rational numbers.

2. Are integers rational numbers? Explain.

3. The real numbers $\dfrac{2}{3}, \dfrac{4}{6}, \dfrac{-15}{12}$, and $\dfrac{20}{-16}$ are rational numbers. Are any of these rational numbers equal?

4. The real numbers $\dfrac{-5}{4}, \dfrac{-10}{8}, \dfrac{18}{27}$, and $\dfrac{-8}{-12}$ are rational numbers. Are any of these rational numbers equal?

5. What does it mean to say that a real number r is an irrational number? Explain by writing a precise negation of the definition of a rational number.

Questions (3) and (4) were included to illustrate the fact that a rational number can be written as a fraction in "lowest terms" with a positive denominator. This means that any rational number can be written as a quotient $\dfrac{m}{n}$, where m and n are integers, $n > 0$, and m and n have no common factor greater than 1. This fact will be used in a proof by contradiction later in this section.

When mathematicians want to prove that a certain statement is true, they frequently assume that the negation of the statement is true and then prove that this assumption leads to a contradiction. In effect, this shows that the statement cannot be false and hence must be true. The fact that this is a valid method of proof was established in Preview Activity 1.

When we try to prove the conditional statement "If P then Q" using a proof by contradiction, we must assume that $P \rightarrow Q$ is false and show that this leads to a contradiction. Two important logical equivalencies for this process are

$$P \rightarrow Q \qquad \text{is logically equivalent to} \quad \neg P \vee Q.$$

$$\neg (P \rightarrow Q) \quad \text{is logically equivalent to} \quad P \wedge \neg Q.$$

The preceding logical equivalencies show that when we assume that $P \to Q$ is false, we are assuming that P is true and Q is false. If we can prove that this leads to a contradiction, then we have shown that $\neg(P \to Q)$ is false and hence that $P \to Q$ is true.

In the following activity, we will outline a proof by contradiction for the statement from Preview Activity 2.

Activity 3.13 (Outlining a Proof by Contradiction).
Consider the following statement:

For all real numbers x and y, if $x \neq y$, $x > 0$, and $y > 0$, then $\frac{x}{y} + \frac{y}{x} > 2$.

To start a proof by contradiction, we assume that x and y are real numbers. In addition, we assume that $x \neq y$, $x > 0$, and $y > 0$ and that
$$\frac{x}{y} + \frac{y}{x} \leq 2.$$
Because this is proof by contradiction, we will only work with the know column of a know-show table since the goal is simply to obtain some contradiction. Using our assumptions, we try to perform some algebraic operations on the inequality

$$\frac{x}{y} + \frac{y}{x} \leq 2$$

until we obtain a contradiction. This can be done by completing the following table:

Know	Reason
$x, y \in \mathbb{R}$, $x \neq y$, $x > 0$, and $y > 0$	Hypothesis
$\dfrac{x}{y} + \dfrac{y}{x} \leq 2$	Negation of the conclusion.
	Multiply both sides of the inequality by the positive number $x \cdot y$.
	Subtract $2xy$ from both sides of the inequality.
	Factor the left side of the inequality.

Now try to explain why this leads to a contradiction. A completed proof of this statement is given after the following writing guidelines.

Writing Guidelines: Keep the Reader Informed

A very important piece of information about a proof is the method of proof to be used. So when we are going to prove a result using the contrapositive or a proof by contradiction, we indicate this within the first few lines of the proof.

- We will prove this result by proving the contrapositive of the statement.

- We will prove this statement using a proof by contradiction.

- We assume to the contrary that . . .

Proposition 3.14. *For all real numbers x and y, if $x \neq y$, $x > 0$, and $y > 0$, then $\dfrac{x}{y} + \dfrac{y}{x} > 2$.*

Proof. This proposition will be proven using a proof by contradiction. So we assume that x and y are real numbers. In addition, we assume that $x \neq y$, $x > 0$, and $y > 0$ and that

$$\frac{x}{y} + \frac{y}{x} \leq 2.$$

Since x and y are positive, we can multiply both sides of this inequality by xy to obtain

$$x^2 + y^2 \leq 2xy.$$

We will obtain a contradiction using the assumption that $x \neq y$. Using basic algebra, it is seen that

$$x^2 - 2xy + y^2 \leq 0$$
$$(x - y)^2 \leq 0. \tag{1}$$

However, one of the assumptions we made was that $x \neq y$. This implies that $x - y \neq 0$ and hence that

$$(x - y)^2 > 0. \tag{2}$$

Inequalities (1) and (2) cannot both be true, and so we have obtained a contradiction. Thus, our given statement cannot be false and we have proven that if $x \neq y$, $x > 0$, and $y > 0$, then $\frac{x}{y} + \frac{y}{x} > 2$. ∎

Important Note

A proof by contradiction is often used to prove a conditional statement $P \rightarrow Q$ when a direct proof has not been found and it is relatively easy to form the negation of the conclusion. The advantage of a proof by contradiction is that we have an additional assumption with which to work (since we assume not only P but also $\neg Q$). The disadvantage is that there is no well defined goal to work toward. The goal is simply to obtain some contradiction. There usually is no way of telling beforehand what that contradiction will be, so we have to stay alert for a possible absurdity. Thus, when we set up a know-show table for a proof by contradiction, we really only work with the know portion of the table. This was illustrated in Activity 3.13.

Preliminary Comments about Theorem 3.15

The proof that the square root of 2 is an irrational number is one of the classic proofs by contradiction. The proposition can be stated as follows:

If r is a real number such that $r^2 = 2$, then r is an irrational number.

This is stated in the form of a conditional statement, but it basically means that $\sqrt{2}$ is irrational (and that $-\sqrt{2}$ is irrational). That is, $\sqrt{2}$ cannot be written as a quotient of integers with the denominator not equal to zero.

The proof of this proposition uses the fact that a rational number can be written as a fraction "in lowest terms." (This was discussed in Preview Activity 3.) This means that any rational number can be written as a quotient $\dfrac{m}{n}$, where m and n are integers, $n > 0$, and m and n have no common factor greater than 1.

In the following proof, notice that it is assumed that r is a real number, $r^2 = 2$, and r is not irrational (that is, r is rational). That is, we are assuming $(P \wedge \neg Q)$ given that the statement is of the form $(P \rightarrow Q)$.

Theorem 3.15. *If r is a real number such that $r^2 = 2$, then r is an irrational number.*

Proof. We will use a proof by contradiction. So we assume that the statement of the theorem is false. That is, we assume that

r is a real number, $r^2 = 2$, and that r is a rational number.

Since r is a rational number, there exist integers m and n with $n > 0$ such that

$$r = \frac{m}{n}$$

and m and n have no common factor greater than 1. We will obtain a contradiction by showing that m and n must both be even.

Squaring both sides of this equation and using the fact that $r^2 = 2$, we obtain

$$2 = \frac{m^2}{n^2}$$
$$m^2 = 2n^2. \tag{1}$$

Equation (1) means that m^2 must be even. Now, since m^2 is even, we can conclude that m must be even, by Theorem 3.10. This means that there is an integer p such that

$$m = 2p.$$

We now substitute this into Equation (1). This gives

$$(2p)^2 = 2n^2$$
$$4p^2 = 2n^2. \tag{2}$$

But Equation (2) implies that $2p^2 = n^2$. Consequently, n^2 is even and thus n must be even, again by Theorem 3.10.

We have now established that both m and n are even. This means that 2 is a common factor of m and n, which contradicts the assumption that m and n have no common factor greater than 1. Consequently, the statement of the theorem cannot be false, and we have proven that if r is a real number such that $r^2 = 2$, then r is an irrational number. ∎

Activity 3.16 (Exploration and a Proof by Contradiction).
Consider the following proposition:

Let $n \in \mathbb{Z}$. If $n \equiv 2 \pmod 4$, then $n \not\equiv 3 \pmod 6$.

1. Determine at least five different integers that are congruent to 2 modulo 4.

2. Determine at least five different integers that are congruent to 3 modulo 6.

3. Are there any integers that are in both of the lists in Parts (1) and (2)?

4. For this proposition, why does it seem reasonable to try a proof by contradiction?

5. For this proposition, clearly state the assumptions that need to be made at the beginning of a proof by contradiction.

6. Use a proof by contradiction to prove this proposition.

Activity 3.17 (A Proof by Contradiction).
Consider the following proposition:

> Let a, b, and c be integers. If 3 divides a, 3 divides b, and $c \equiv 1$ (mod 3), then the equation
>
> $$ax + by = c$$
>
> has no solution in which both x and y are integers.

Complete the following proof of this proposition:

Proof. A proof by contradiction will be used. So we assume that the statement is false. That is, we assume that a, b, and c are integers, that 3 divides both a and b, that $c \equiv 1$ (mod 3), and that the equation

$$ax + by = c$$

has a solution in which both x and y are integers.
  So let m and n be integers such that

$$am + bn = c.$$

[Hint: Now use the facts that 3 divides a, 3 divides b, and $c \equiv 1$ (mod 3).]

A Comparison of Direct Proofs, Proofs Using the Contrapositive, and Proofs by Contradiction

The following tables provide descriptions of three of the most common methods of proving a conditional statement.

Direct Proof of $P \to Q$	
When Is It Indicated?	**Description of the Process**
This type of proof is often used when the hypothesis and the conclusion are both stated in a "positive" manner. That is, no negations are evident in the hypothesis and conclusion.	Assume that P is true and use this to conclude that Q is true. That is, we use the forward-backward method and work forward from P and backward from Q.
Why the Process Makes Sense	
We know that the conditional statement $P \to Q$ is automatically true when the hypothesis is false. Therefore, because our goal is to prove that $P \to Q$ is true, there is nothing to do in the case that P is false. Consequently, we may assume that P is true. Then, in order for $P \to Q$ to be true, the conclusion Q must also be true. (When P is true, but Q is false, $P \to Q$ is false.) Thus, we must use our assumption that P is true to show that Q is also true.	

Proof of $P \to Q$ Using the Contrapositive	
When Is It Indicated?	**Description of the Process**
This type of proof is often used when both the hypothesis and the conclusion are stated in the form of negations. This often works well if the conclusion contains the operator "or"; that is, if the conclusion is in the form of a disjunction. In this case, the negation will be a conjunction.	We prove the logically equivalent statement $\neg Q \to \neg P$. The forward-backward method is used to prove $\neg Q \to \neg P$. That is, we work forward from $\neg Q$ and backward from $\neg P$.
Why the Process Makes Sense	
When we prove $\neg Q \to \neg P$, we are also proving $P \to Q$ because these two statements are logically equivalent. When we prove the contrapositive of $P \to Q$, we are doing a direct proof of $\neg Q \to \neg P$. So we assume $\neg Q$ because, when doing a direct proof, we assume the hypothesis, and $\neg Q$ is the hypothesis of the contrapositive. We must show $\neg P$ because it is the conclusion of the contrapositive.	

Proof of $P \rightarrow Q$ Using a Proof by Contradiction	
When Is It Indicated?	**Description of the Process**
This type of proof is often used when the conclusion is stated in the form of a negation, but the hypothesis is not. This often works well if the conclusion contains the operator "or"; that is, if the conclusion is in the form of a disjunction. In this case, the negation will be a conjunction.	Assume P and $\neg Q$ and work forward from these two assumptions until a contradiction is obtained.

Why the Process Makes Sense

The statement $P \rightarrow Q$ is either true or false. In a proof by contradiction, we show that it is true by eliminating the only other possibility (that it is false). We show that $P \rightarrow Q$ cannot be false by assuming it is false and reaching a contradiction. Since we assume that $P \rightarrow Q$ is false, and the only way for a conditional statement to be false is for its hypothesis to be true and its conclusion to be false, we assume that P is true and that Q is false (or equivalently, that $\neg Q$ is true.) When we reach a contradiction, we know that our original assumption that $P \rightarrow Q$ is false is incorrect. Hence, $P \rightarrow Q$ cannot be false, and so it must be true.

Please keep in mind that these tables summarize three of the most common methods of proving a conditional statement of the form $P \rightarrow Q$. As was indicated in Section 3.2, other methods can sometimes be used. Quite often, these other methods involve the use of a logical equivalency. For example, in order to prove a statement of the form

$$P \rightarrow (Q \vee R),$$

it is sometimes possible to use the logical equivalency

$$[P \rightarrow (Q \vee R)] \equiv [(P \wedge \neg Q) \rightarrow R].$$

We would then prove the statement

$$(P \wedge \neg Q) \rightarrow R.$$

Because of the logical equivalency, by proving one statement, we have also proven the other statement.

Activity 3.18 (Exploring a Quadratic Equation).
 Consider the following proposition:

 If m and n are integers and n is odd, then the equation

$$x^2 + 2mx + 2n = 0$$

 has no integer solution for x.

1. What are the solutions of the equation when $m = 1$ and $n = -1$? That is, what are the solutions of the equation $x^2 + 2x - 2 = 0$?

2. What are the solutions of the equation when $m = 2$ and $n = 1$? That is, what are the solutions of the equation $x^2 + 4x + 2 = 0$?

3. What are the solutions of the equation when $m = 1$ and $n = 3$? That is, what are the solutions of the equation $x^2 + 2x + 6 = 0$?

4. Solve the resulting quadratic equation for at least one more example using values of m and n that satisfy the hypothesis of the proposition.

5. For this proposition, why does it seem reasonable to try a proof by contradiction?

6. For this proposition, clearly state the assumptions that need to be made at the beginning of a proof by contradiction.

7. Use a proof by contradiction to prove this proposition.

Exercises 3.3

1. This exercise is intended to provide another rationale as to why a proof by contradiction works.

   Suppose that we are trying to prove that a statement P is true. Instead of proving this statement, assume that we prove that the conditional statement "If $\neg P$, then C" is true, where C is some contradiction. Recall that a contradiction is a statement that is always false.

   (a) In symbols, write a statement that is a disjunction and that is logically equivalent to $\neg P \rightarrow C$.

(b) Since we have proven that $\neg P \to C$ is true, then the disjunction in Exercise (1a) must also be true. Use this to explain why the statement P must be true.

(c) Now explain why P must be true if we prove that the negation of P implies a contradiction.

2. Consider the following statement:

If r is a real number such that $r^2 = 18$, then r is irrational.

(a) If you were setting up a proof by contradiction for this statement, what would you assume? Carefully write down all conditions that you would assume.

(b) Complete a proof by contradiction for this statement. <u>Hint</u>: Do not attempt to mimic the proof that the square root of 2 is irrational. You should still use the definition of a rational number but then use the fact that $\sqrt{18} = \sqrt{9 \cdot 2} = \sqrt{9}\sqrt{2} = 3\sqrt{2}$.

3. Prove the following proposition:

For all real numbers x and y, if x is rational and y is irrational, then $x + y$ is irrational.

4. Are the following statements true or false? Justify each conclusion.

(a) For any positive real number x, if x is irrational, then x^2 is irrational.

(b) For any positive real number x, if x is irrational, then $\sqrt{x}$ is irrational.

5. Are the following statements true or false? Justify your conclusions.

(a) For every pair of real numbers x and y, if $x + y$ is irrational, then x is irrational and y is irrational.

(b) For every pair of real numbers x and y, if $x + y$ is irrational, then x is irrational or y is irrational.

6. (a) Is the following statement true or false?

For all real numbers x and y, if $x > 0$ and $y > 0$, then $\sqrt{x + y} = \sqrt{x} + \sqrt{y}$.

(b) Prove that if x and y are positive real numbers, then $\sqrt{x + y} \le \sqrt{x} + \sqrt{y}$

7. Is the following statement true or false? Justify your conclusion.

> For each integer n that is greater than 1, if a is the smallest positive factor of n, then a is prime.

See Activity 2.18 in Section 2.4 for the definition of a prime number and the definition of a composite number.

8. (a) Is the base 2 logarithm of 32, $\log_2 32$, a rational number or an irrational number? Justify your conclusion.

  (b) Is the base 2 logarithm of 3, $\log_2 3$, a rational number or an irrational number? Justify your conclusion.

9. Is the real number $\sqrt{2} + \sqrt{5}$ a rational number or an irrational number? Justify your conclusion.

10. Prove the following proposition:

> For any real number θ, if $0 < \theta < \frac{\pi}{2}$, then $(\sin\theta + \cos\theta) > 1$.

11. Prove the following proposition:

> If n is an integer greater than 2, then for all integers m, n does not divide m or $n + m \neq nm$.

12. Is the following proposition true or false? Justify your conclusion.

> Let a and b be real numbers. If $a > 0$ and $b > 0$, then

$$\frac{2}{a} + \frac{2}{b} \neq \frac{4}{a+b}.$$

13. (a) Verify that if $a = 20$, $b = 21$, and $c = 29$, then $a^2 + b^2 = c^2$. We say that the integers 20, 21, and 29 for a **Pythagorean triple**.

  (b) Determine two other Pythagorean triples. That is, find integers a, b, and c such that $a^2 + b^2 = c^2$.

  (c) Is the following proposition true or false? Justify your conclusion. Let a, b, and c be integers. If $a^2 + b^2 = c^2$, then a is even or b is even.

14. Consider the following proposition: There are no integers a and b such that $b^2 = 4a + 2$.

  (a) Rewrite this statement in an equivalent form using a universal quantifier by completing the following:

For all integers a and b, $\cdots$.

(b) Prove the statement in Part (a).

15. A **magic square** is a square array of natural numbers whose rows, columns, and diagonals all sum to the same number. For example, the following is a 3 by 3 magic square since the sum of 3 numbers in each row is equal to 15, the sum of the 3 numbers in each column is equal to 15, and the sum of the 3 numbers in each diagonal is equal to 15.

8	3	4
1	5	9
6	7	2

Prove that the following 4 by 4 square cannot be completed to form a magic square.

	1		2
3	4	5	
6	7		8
9		10	

Hint: Assign each of the six blank cells in the square a name. One possibility is to use a, b, c, d, e, and f.

3.4   Using Cases in Proofs

Preview Activity 1 (Quotients and Remainders).

1. Determine several integers q and r so that $27 = 4 \cdot q + r$. For example, we could write

$$27 = 4 \cdot 2 + 19 \text{ and } 27 = 4\,(9) + (-9).$$

Include at least five other examples where r is positive and at least two other examples where r is negative.

2. What is the smallest positive value for r that you obtained in your examples from Part (1)?

3. Division is not considered an operation on the set of integers since the quotient of two integers need not be an integer. However, we have all divided one integer by another and obtained a quotient and a remainder. For example, if we divide 113 by 5, we obtain a quotient of 22 and a remainder of 3. We can write this as follows:

$$\frac{113}{5} = 22 + \frac{3}{5}.$$

What is the resulting equation if we multiply both sides of the equation above by 5 ?

4. What are the quotient and the remainder when we divide 27 by 4? How is this related to your answer for Part (2)?

Preview Activity 2 (A Logical Equivalency).

1. Complete a truth table to show that $(P \vee Q) \rightarrow R$ is logically equivalent to $(P \rightarrow R) \wedge (Q \rightarrow R)$.

2. Suppose that you are trying to prove a statement that is written in the form $(P \vee Q) \rightarrow R$. Explain why you can complete this proof by writing separate and independent proofs of $P \rightarrow R$ and $Q \rightarrow R$.

3. Explain why the statement

   If n is an integer, then $n^2 + n$ is an even integer

   is logically equivalent to

   If n is an even integer, then $n^2 + n$ is an even integer and if n is an odd integer, then $n^2 + n$ is an even integer.

Preview Activity 3 (A Property of the Integers).

1. Complete the proof for the following proposition:

   Proposition 1: If n is an even integer, then $n^2 + n$ is an even integer.

   Proof. Let n be an even integer. Then there exists an integer m such that

   $$n = 2m.$$

   Substituting this into the expression $n^2 + n$ yields ...

2. Construct a proof for the following proposition:

   Proposition 2: If n is an odd integer, then $n^2 + n$ is an even integer.

3. Assume that you have completed the proofs of Proposition 1 and Proposition 2. Do these two proofs constitute a proof of the following proposition?

   Proposition 3: If n is an integer, then $n^2 + n$ is an even integer.

Preview Activity 1 was an introduction to a mathematical result known as the Division Algorithm. One of the purposes of this activity was to illustrate that we have already worked with this result, perhaps without knowing its name. For example, when we divide 337 by 6, we often write

$$\frac{337}{6} = 56 + \frac{1}{6}.$$

When we multiply both sides of this equation by 6, we get

$$337 = 6 \cdot 56 + 1.$$

When we are working within the system of integers, the second equation is preferred over the first since the second one uses only integers and the operations of addition and multiplication, and the integers are closed under addition and multiplication. Following is a complete statement of the Division Algorithm.

The Division Algorithm

Let a and b be integers with $b > 0$. Then there exist unique integers q and r such that
$$a = bq + r \text{ and } 0 \le r < b.$$

Some Comments about the Division Algorithm

1. The Division Algorithm can be proven, but we have not yet studied the methods that are usually used to do so. In this text, we will treat the Division Algorithm as an axiom of the integers. Preview Activity 1 provides some rationale that this is a reasonable axiom.

2. The statement of the Division Algorithm contains the new phrase, "there exist unique integers q and r such that ...". This means that there is only one pair of integers q and r that satisfy both the conditions $a = bq + r$ and $0 \le r < b$.

   As we saw in Preview Activity 1, there are several different ways to write the integer a in the form $a = bq + r$. However, there is only one way to do this and satisfy the additional condition that $0 \le r < b$.

3. In light of the previous comment, when we speak of **the quotient** and **the remainder** when we "divide an integer a by the positive integer b," we will always mean the quotient (q) and the remainder (r) guaranteed by the Division Algorithm. So the remainder r is the least nonnegative integer such that there exists an integer (quotient) q with $a = bq + r$.

4. If $a < 0$, then we must be careful when writing the result of the Division Algorithm. For example, if $a = -17$ and $b = 5$, we often write $\frac{-17}{5} = -3.4$. But this means

$$\frac{-17}{5} = -\left(3 + \frac{4}{10}\right) = -3 - \frac{2}{5}.$$

If we multiply both sides of this equation by 5, we obtain

$$-17 = 5\left(-3\right) + \left(-2\right).$$

This is not the result guaranteed by the Division Algorithm. If we write the result of "dividing -17 by 5" guaranteed by the Division Algorithm, we get

$$-17 = 5\left(-4\right) + 3.$$

5. One way to look at the Division Algorithm is that the integer a is either going to be a multiple of b, or it will lie between two multiples of b. Say that a is not a multiple of b and that it lies between the multiples $b \cdot q$ and $b\left(q + 1\right)$. On the number line, this would look something like the following:

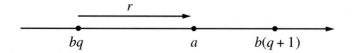

If r represents the distance from $b \cdot q$ to a, then

$$r = a - b \cdot q, \text{ or}$$
$$a = b \cdot q + r$$

From the diagram, also notice that r is less than the distance between $b \cdot q$ and $b\left(q + 1\right)$. Algebraically, this distance is

$$b\left(q + 1\right) - b \cdot q = b \cdot q + b - b \cdot q$$
$$= b.$$

Thus, in the case where a is not a multiple of b, we get $0 < r < b$.

6. We have been implicitly using the fact that an integer cannot be both even and odd. There are several ways to understand this fact, but one way is through the Division Algorithm. When we classify an integer as even or odd, we are doing so on the basis of the remainder (according to the Division Algorithm) when the integer is "divided" by 2. If $a \in \mathbb{Z}$, then by the Division Algorithm, there exist unique integers q and r such that

$$a = 2q + r \text{ and } 0 \leq r < 2.$$

This means that the remainder, r, can only be zero or one (and not both). When $r = 0$, the integer is even, and when $r = 1$, the integer is odd.

Using Cases

When we are trying to prove a proposition or a theorem, we often run into the problem that there does not seem to be enough information to proceed. In this situation, we will sometimes use cases to provide additional assumptions for the forward process of the proof. When this is done, the original proposition is divided into a number of separate cases that are proven independently of each other. The cases must be chosen so that they exhaust all possibilities for the hypothesis of the original proposition.

This is what we did in Preview Activities 2 and 3. The proposition was as follows:

If n is an integer, then $n^2 + n$ is an even integer.

Although the hypothesis is not divided into cases as it is stated, we know that an integer must be even or it must be odd. Thus, we can divide the hypothesis into the following two cases:

- The integer n is an even integer;

- The integer n is an odd integer.

We then constructed separate proofs of the following:

- If n is an even integer, then $n^2 + n$ is an even integer.

- If n is an odd integer, then $n^2 + n$ is an even integer.

Technically, this method of case analysis is justified by the logical equivalency $(P \lor Q) \rightarrow R \equiv (P \rightarrow R) \land (Q \rightarrow R)$, which was established in Preview Activity 2.

Some Common Situations to Use Cases

When the hypothesis is, "n is an integer":
   Case 1: n is an even integer.
   Case 2: n is an odd integer.

When the hypothesis is, "x is a real number":
   Case 1: $x > 0$
   Case 2: $x = 0$
   Case 3: $x < 0$
   Often, Cases (1) and (2) can be combined.

When the hypothesis is, "x is a real number":
   Case 1: x is rational.
   Case 2: x is irrational.

Using Cases Determined by the Division Algorithm

The Division Algorithm can sometimes be used to construct a proof by cases of a statement that is true for all integers. We did this when we divided the integers into the even integers and the odd integers.

The Division Algorithm states that when we divide an integer a by 2, there are two possible remainders: 0 and 1. The remainder is 0 if and only if a is even, and the remainder is 1 if and only if a is odd.

Sometimes it is more useful to divide the integer a by an integer other than 2. For example, if a is divided by 3, there are three possible remainders: 0, 1, and 2. If a is divided by 4, there are four possible remainders: 0, 1, 2, and 3. The remainders form the basis for the cases.

For example, if the hypothesis of a proposition is that "n is an integer," then we can use the Division Algorithm to claim that there are unique integers q and r such that

$$n = 3q + r \text{ and } 0 \le r < 3.$$

We can then divide the problem into the following three cases: (1) $r = 0$; (2) $r = 1$; and (3) $r = 2$.

Activity 3.19 (A Proof Using the Division Algorithm).

Complete the details for Case (2) and Case (3) in the proof of the following proposition:

Proposition 3.20. *If n is an integer, then 3 divides $n^3 - n$.*

Proof. Let n be an integer. We will show that 3 divides $n^3 - n$ be examining the three cases for the remainder when n is divided by 3. By the Division Algorithm, there exist unique integers q and r such that

$$n = 3q + r, \text{ and } 0 \leq r < 3.$$

This means that we can consider the following three cases: (1) $r = 0$; (2) $r = 1$; and (3) $r = 2$.

Case 1: $r = 0$. In this case, $n = 3q$. By substituting this into the expression $n^3 - n$, we get

$$n^3 - n = (3q)^3 - (3q)$$
$$= 27q^3 - 3q$$
$$= 3\left(9q^3 - q\right).$$

Since $\left(9q^3 - q\right)$ is an integer, this proves that $3 \mid \left(n^3 - n\right)$.

Case 2: $r = 1$. In this case, $n = 3q + 1$, and when we substitute this into $\left(n^3 - n\right)$, we obtain

$$n^3 - n = (3q + 1)^3 - (3q + 1)$$
$$= \cdots .$$

Case 3: $r = 2$. In this case, $n = 3q + 2$, and when we substitute this into $\left(n^3 - n\right)$, we obtain

$$n^3 - n = (3q + 2)^3 - (3q + 2)$$
$$= \cdots .$$

So, in all three cases, we have proven that 3 divides $n^3 - n$. Hence, we may conclude that if n is an integer, then 3 divides $n^3 - n$. ■

Another Look at Congruence

Let $n \in \mathbb{N}$. Recall that if a and b are integers, then we say that a is congruent to b modulo n provided that n divides $a - b$, and we write $a \equiv b \pmod{n}$. For example, each integer in the following list is congruent to 2 modulo 7.

$$\ldots, -19, -12, -5, 2, 9, 16, 23, \ldots$$

If we divide each integer in the list by 7 and write the result according to the Division Algorithm, we will get a remainder of 2. For example,

$$2 = 7 \cdot 0 + 2 \qquad\qquad -5 = 7(-1) + 2$$
$$9 = 7 \cdot 1 + 2 \qquad\qquad -12 = 7(-2) + 2$$
$$16 = 7 \cdot 2 + 2 \qquad\qquad -19 = 7(-3) + 2$$
$$23 = 7 \cdot 3 + 2$$

Is this a coincidence or is this always true? Let's look at the general case. For this, let n be a natural number and let $a \in \mathbb{Z}$. By the Division Algorithm, there exist unique integers q and r such that

$$a = nq + r \text{ and } 0 \le r < n.$$

By subtracting r from both sides of the equation $a = nq + r$, we obtain

$$a - r = nq.$$

But this implies that $n \mid (a - r)$ and hence that $a \equiv r \pmod{n}$. We have proven the following result.

Theorem 3.21. *Let $n \in \mathbb{N}$ and let $a \in \mathbb{Z}$. If $a = nq + r$ and $0 \le r < n$ for some integers q and r, then $a \equiv r \pmod{n}$.*

This theorem says that an integer is congruent (mod n) to its remainder when it is divided by n. Since this remainder is unique and since the only possible remainders for division by n are $0, 1, 2, \ldots, n - 1$, we can state the following result.

Corollary 3.22. *If $n \in \mathbb{N}$, then each integer is congruent, modulo n, to precisely one of the integers $0, 1, 2, \ldots, n - 1$.*

Corollary 3.22 can be used as the basis for a proof by cases. If $n \in \mathbb{N}$ and $a \in \mathbb{Z}$, then we can consider n cases for a. The integer a could be congruent to $0, 1, 2, \ldots,$ or $n - 1$ modulo n. This method will be illustrated in Proposition 3.25.

Activity 3.23 (Working with Congruences). Following are two lists. List 1 contains some integers that are congruent to 1 modulo 3, and List 2 contains some integers that are congruent to 2 modulo 3.

List 1: $-8, -5, -2, 1, 4, 7, 10$ List 2: $-7, -4, -1, 2, 5, 8, 11$

1. Square each number in List 1. For each integer a in this list, determine r so that $a^2 \equiv r \pmod 3$ and $0 \le r < 3$.

2. Square each number in List 2. For each integer a in this list, determine r so that $a^2 \equiv r \pmod 3$ and $0 \le r < 3$.

3. Now multiply an integer from List 1 with an integer from List 2. Repeat this several times. For each a from List 1 and each b from List 2, determine r so that $ab \equiv r \pmod 3$ and $0 \le r < 3$.

These activities should have provided examples of the general result stated in Exercise (11b) from Section 3.1. This exercise was to prove the first two parts of the following theorem:

Theorem 3.24. *Let n be a natural number and let $a, b, c,$ and d be integers. If $a \equiv b \pmod n$ and $c \equiv d \pmod n$, then*

1. $(a + c) \equiv (b + d) \pmod n$.

2. $ac \equiv bd \pmod n$.

3. *For each $m \in \mathbb{N}$, $a^m \equiv b^m \pmod n$.*

Proof. We will prove Parts (2) and (3) and leave the proof of Part (1) as an exercise. Let n be a natural number and let $a, b, c,$ and d be integers. Assume that $a \equiv b \pmod n$ and that $c \equiv d \pmod n$. This means that n divides $a - b$ and that n divides $c - d$. Hence, there exist integers k and q such that $a - b = nk$ and $c - d = nq$.

We can then write $a = b + nk$ and $c = d + nq$ and obtain

$$ac = (b + nk)(d + nq)$$
$$= bd + bnq + dnk + n^2kq$$
$$= bd + n(bq + dk + nkq)$$

By subtracting bd from both sides of the last equation, we see that

$$ac - bd = n(bq + dk + nkq).$$

Since $bq + dk + nkq$ is an integer, this proves that $n \mid (ac - bd)$, and hence we can conclude that $ac \equiv bd \pmod n$. This completes the proof of Part (2).

If we use the idea that a^m and b^m represent repeated multiplications, we see that Part (3) is actually a corollary of Part (2). For example, by using $c = a$ and $d = b$, we can use Part (2) to conclude that

$$a \cdot a \equiv b \cdot b \pmod n,$$

or that $a^2 \equiv b^2$ (mod n). We can then use this congruence and the congruence $a \equiv b$ (mod n) to conclude that

$$a^2 \cdot a \equiv b^2 \cdot b \quad (\text{mod } n),$$

or that $a^3 \equiv b^3$ (mod n). We can say that we can continue with this process to prove Part (3), but this is not considered to be a formal proof of this result. To construct a formal proof for this, we need to use a proof by mathematical induction. This will be studied in Chapter 5. See Exercise (9) in Section 5.1.   ∎

We will now use the results of Theorem 3.24 and Corollary 3.22 to provide another proof using cases of Proposition 3.20, which was proven in Activity 3.19. We will use Corollary 3.22 to form the cases. The idea is that any integer must be congruent modulo 3 to 0, 1, or 2. These will be the three cases used.

Proposition 3.25. *If n is an integer, then 3 divides $n^3 - n$.*

Proof. Let n be an integer. We will show that 3 divides $n^3 - n$ by examining three cases: $n \equiv 0$ (mod 3), $n \equiv 1$ (mod 3), or $n \equiv 2$ (mod 3).

Case 1: $n \equiv 0$ (mod 3). Using Part (3) of Theorem 3.24, we see that

$$n^3 \equiv 0^3 \quad (\text{mod } 3), \text{ or}$$
$$n^3 \equiv 0 \quad (\text{mod } 3).$$

Since $n^3 \equiv 0$ (mod 3) and $n \equiv 0$ (mod 3), we can apply Part (1) of Theorem 3.24 to obtain

$$\left(n^3 - n\right) \equiv (0 - 0) \quad (\text{mod } 3)$$
$$\left(n^3 - n\right) \equiv 0 \quad (\text{mod } 3).$$

The last congruence tells us that $3 \mid \left[\left(n^3 - n\right) - 0\right]$ or that $3 \mid \left(n^3 - n\right)$.

The other two cases are handled similarly. The computations using Theorem 3.24 are shown below.

Case 2: $n \equiv 1$ (mod 3). In this case, $n^3 \equiv 1^3$ (mod 3) or $n^3 \equiv 1$ (mod 3). So we obtain

$$\left(n^3 - n\right) \equiv (1 - 1) \quad (\text{mod } 3)$$
$$\left(n^3 - n\right) \equiv 0 \quad (\text{mod } 3).$$

<u>Case 3</u>: $n \equiv 2 \pmod 3$. In this case, $n^3 \equiv 2^3 \pmod 3$ or $n^3 \equiv 8 \pmod 3$. So we obtain

$$\left(n^3 - n\right) \equiv (8 - 2) \pmod 3$$
$$\left(n^3 - n\right) \equiv 6 \pmod 3$$

Since $6 \equiv 0 \pmod 3$, we can use the last equation to conclude that

$$\left(n^3 - n\right) \equiv 0 \pmod 3.$$

So in all three cases, we have proven that $\left(n^3 - n\right) \equiv 0 \pmod 3$, and this implies that $3 \mid \left(n^3 - n\right)$. Hence, we can conclude that if n is an integer, then 3 divides $n^3 - n$. ∎

Example 3.26. Theorem 3.24 provides tools to explore certain properties of natural numbers. For example, since we know that $3^4 = 81$, we can conclude that

$$3^4 \equiv 1 \pmod{10}.$$

This is not much of a surprise, but if we use this fact with Theorem 3.24, we can arrive at some not quite so obvious results. Using the second part of Theorem 3.24 with $a = c = 3^4$ and $b = d = 1$, we see that

$$3^4 \cdot 3^4 \equiv 1 \cdot 1 \pmod{10}$$
$$3^8 \equiv 1 \pmod{10}.$$

This tells us that the last digit in the decimal representation of 3^8 is 1. We can also use Part (3) of Theorem 3.24 and the fact that $3^4 \equiv 1 \pmod{10}$ to conclude that

$$\left(3^4\right)^{100} \equiv 1^{100} \pmod{10}$$
$$3^{400} \equiv 1 \pmod{10}.$$

This tells us that the last digit in the decimal representation of 3^{400} is 1.

Activity 3.27 (The Last Two Digits of a Large Integer).

1. Use the fact that $3^4 \equiv 81 \pmod{100}$ to prove that $3^{16} \equiv 21 \pmod{100}$. What does this tell you about the last two digits in the decimal representation of 3^{16}?

2. Use the two congruences in Part (1) and laws of exponents to determine r where $3^{20} \equiv r \pmod{100}$) and $r \in \mathbb{Z}$ with $0 \leq r < 100$. What does this tell you about the last two digits in the decimal representation of 3^{20}?

3. Determine the last two digits in the decimal representation of 3^{400}.

4. Determine the last two digits in the decimal representation of 4^{804}.

Hint: One way is to determine the "mod 100 values" for 4^2 , 4^4 , 4^8 , 4^{16} , 4^{32}, 4^{64}, and so on. Then use these values and laws of exponents to determine r, where $4^{804} \equiv r \pmod{100}$ and $r \in \mathbb{Z}$ with $0 \le r < 100$.

Exercises 3.4

1. In Preview Activities 2 and 3, we proved that if n is an integer, then $n^2 + n$ is an even integer. We define two integers to be **consecutive integers** if one of the integers is one more than the other integer. This means that we can represent consecutive integers as m and $m+1$, where m is some integer.

Explain why the result proven in Preview Activities 2 and 3 can be used to prove that the product of any two consecutive integers is divisible by 2.

2. Extending the idea in Exercise (1), we can represent three consecutive integers as m, $m + 1$, and $m + 2$, where m is an integer.

  (a) Explain why we can also represent three consecutive integers as $k - 1$, k, and $k + 1$ where k is an integer.

  (b) Explain why Proposition 3.20 proves that the product of any three consecutive integers is divisible by 3.

3. Prove that if u is an odd integer, then the equation $x^2 + x - u = 0$ has no solution that is an integer.

4. Prove that if n is an odd integer, then $n = 4k + 1$ for some integer k or $n = 4m - 1$ for some integer m.

5. **(a)** Prove the following proposition:

    Let a, b, and d be integers. If d divides a or d divides b, then d divides the product ab.

    Hint: Notice that the hypothesis is a disjunction. So use two cases.

  (b) Write the contrapositive of the proposition in Exercise (5a).

(c) Write the converse of the proposition in Exercise (5a). Is the converse true or false? Justify your conclusion.

6. (a) Let $n \in \mathbb{N}$ and let $a \in \mathbb{Z}$. Explain why $n \mid a$ if and only if $a \equiv 0 \pmod{n}$.

(b) Let $a \in \mathbb{Z}$. Explain why if $a \not\equiv 0 \pmod 3$, then $a \equiv 1 \pmod 3$ or $a \equiv 2 \pmod 3$.

(c) Is the following proposition true or false? Justify your conclusion.

Let $a \in \mathbb{Z}$. If $a \not\equiv 0 \pmod 3$, then $a^2 \equiv 1 \pmod 3$.

7. Prove the following proposition by proving its contrapositive. (<u>Hint</u>: Use case analysis. There are several cases.)

Let a and b be integers. If $ab \equiv 0 \pmod 3$, then $a \equiv 0 \pmod 3$ or $b \equiv 0 \pmod 3$.

8. Explain why the following proposition is equivalent to the proposition in Exercise (7).

Let a and b be integers. If $3 \mid ab$, then $3 \mid a$ or $3 \mid b$.

9. The proposition in Exercise (8) was proven in Exercise (7) using the concept of congruence. This result can be proven without using the concept of congruence. The idea is basically the same but we use the Division Algorithm directly instead of using congruence. Notice that if 3 does not divide an integer, then the remainder when that integer is divided by 3 must be 1 or 2. This allows us to use two cases for that integer.

Prove the proposition in Exercise (8) by using this idea. Hint: You will still have to prove the contrapositive and use case analysis.

10. Use the ideas presented in Activity 3.27 to complete the following:

(a) Determine the last two digits in the decimal representation of 3^{3356}.

(b) Determine the last two digits in the decimal representation of 7^{403}.

11. Is the following proposition true or false. Justify your conclusion with a proof or a counterexample.

Let n be a natural number. If 3 does not divide $\left(n^2 + 2\right)$, then n is not a prime number or $n = 3$.

12. **(a)** Is the following proposition true or false? Justify your conclusion with a counterexample or a proof.

For each integer n, if n is odd, then $8 \mid (n^2 - 1)$.

(b) Translate the statement in Part (a) into a corresponding statement dealing with congruence modulo 8.

13. Prove the following proposition:

Let $a, b \in \mathbb{Z}$. If 3 divides $(a^2 + b^2)$, then 3 divides a and 3 divides b.

14. Explore the statements in Exercises (14a) and (14b) by considering several examples where the hypothesis is true.

(a) If an integer a is divisible by both 4 and 6, then it divisible by 24.

(b) If an integer a is divisible by both 2 and 3, then it divisible by 6.

(c) What can you conclude from the examples in Exercise (14a)?

(d) What can you conclude from the examples in Exercise (14b)?

The proof of the following proposition [from Exercise (14b] uses cases. In this proof, however, we use cases and a proof by contradiction to prove that a certain integer cannot be odd. Hence, it must be even. Complete the proof of the proposition.

Proposition. Let $a \in \mathbb{Z}$. If 2 divides a and 3 divides a, then 6 divides a.

Proof: Let $a \in \mathbb{Z}$ and assume that 2 divides a and 3 divides a. We will prove that 6 divides a. Since 3 divides a, there exists an integer n such that

$$a = 3n.$$

The integer n is either even or it is odd. We will show that it must be even by obtaining a contradiction if it assumed to be odd. So, assume that n is odd. ...

3.5 Constructive Proofs

Preview Activity 1 (Systems of Linear Equations).

If x and y represent real numbers, solve each of the following systems of two linear equations in two unknowns:

1. $3x + 5y = 1$
$\quad\ 7x - y = 15$

2. $2x - 5y = 1$
$\quad\ -4x + 10y = 2$

3. $2x - 5y = 1$
$\quad\ -4x + 10y = -2$

Preview Activity 2 (Intersections of Circles).

Let $a, b \in \mathbb{R}$ and let r be a positive real number. Then the graph of an equation of the form

$$(x - a)^2 + (y - b)^2 = r^2$$

is a circle with center at the point (a, b) and radius r.

1. Using an appropriate pair of coordinate axes, carefully draw the graphs of the circles given by the following two equations.

$$x^2 + y^2 = 4$$
$$(x - 3)^2 + y^2 = 4$$

   Then solve the system of two equations in two unknowns to find all points of intersection of the two circles.

2. Using an appropriate pair of coordinate axes, carefully draw the graphs of the circles given by the following two equations.

$$x^2 + y^2 = 4$$
$$(x - 5)^2 + y^2 = 4$$

   Then solve the system of two equations in two unknowns to find all points of intersection of the two circles.

Preview Activity 1 was intended to be a review of methods of solving a system of two linear equations in two variables. The first system of equations has a unique solution $(x = 2, y = -1)$, the second system of equations has no solution, and the third system of equations has infinitely many solutions. These solutions can be expressed in parametric form as

$$x = t, y = \frac{2}{5}t - \frac{1}{5}.$$

We are going to focus mainly on systems of two linear equations in two variables similar to the first system of equations in Preview Activity 1.

For a more general situation, we are going to let a, b, c, d, r, and s be real numbers and assume that $ad - bc \neq 0$. The question we are going to investigate is this: Under these conditions, is it possible to find a solution for x and y for the following system of equations?

$$ax + by = r$$
$$cx + dy = s \qquad\qquad (1)$$

Activity 3.28 (A General System of Linear Equations).

1. For the system of equations given in Equation (1) (with $ad - bc \neq 0$), multiply the first equation by d and multiply the second equation by b.

2. For the two equations obtained in Part (1), subtract the second equation from the first equation and then solve the resulting equation for x. What assumption was needed in order to solve this equation for x?

This activity was intended to show that we can find a solution for x. We can do similar work to find a solution for y. This will be done in Theorem 3.29.

Note: In linear algebra class, we say that the system of linear equations *has a solution* (x, y). That is, the solution is an ordered pair of real numbers.

Theorem 3.29. *If a, b, c, d, r, and s are real numbers with $ad - bc \neq 0$, then for the system of equations*

$$ax + by = r$$
$$cx + dy = s,$$

there exists an ordered pair (x, y) that makes both equations simultaneously true.

Proof. Let $a, b, c, d, r,$ and s be real numbers and assume that $ad - bc \neq 0$. Now consider the following system of equations:

$$ax + by = r \tag{1}$$
$$cx + dy = s. \tag{2}$$

We will prove that there exists an ordered pair (x, y) such that both equations are simultaneously true by actually constructing the values for x and y. In Activity 3.28, we found a value for x that can be written as

$$x = \frac{dr - bs}{ad - bc}. \tag{3}$$

We will now find a value for y in a manner similar to the one used in Activity 3.28. By multiplying Equation (2) by a and Equation (1) by c, and then subtracting the two equations, we obtain

$$(ad - bc)\, y = as - cr. \tag{4}$$

From the hypothesis, $(ad - bc) \neq 0$. Hence, dividing both sides of Equation (4) by $(ad - bc)$, we see that

$$y = \frac{as - cr}{ad - bc}. \tag{5}$$

We now claim that the ordered pair $(x, y) = \left(\dfrac{dr - bs}{ad - bc}, \dfrac{as - cr}{ad - bc} \right)$ simultaneously satisfies both equations.

To verify this claim, we substitute the values of x and y given in Equations (3) and (5) into the left side of Equation (1). This gives

$$
\begin{aligned}
ax + by &= a \left(\frac{dr - bs}{ad - bc} \right) + b \left(\frac{as - cr}{ad - bc} \right) \\
&= \frac{a(dr - bs) + b(as - cr)}{ad - bc} \\
&= \frac{adr - abs + abs - bcr}{ad - bc} \\
&= \frac{adr - bcr}{ad - bc} \\
&= \frac{r(ad - bc)}{ad - bc} \\
&= r.
\end{aligned}
$$

This shows that the values for x and y given in Equations (3) and (5) satisfy Equation (1). A similar substitution shows that these values for x and y satisfy Equation (2). This completes the proof that there exists an ordered pair that makes both Equations (1) and (2) true simultaneously when $(ad - bc) \neq 0$.   ∎

Constructive Proofs

The proof given for Theorem 3.29 is often called a **constructive proof.** This is a technique that is often used to prove a so-called **existence theorem.** The simplest form of an existence theorem is

There exists an x such that $P(x)$.

The symbolic form is $(\exists x)(P(x))$. For a constructive proof of such a proposition, we actually name, describe, or explain how to construct some object in the universe that makes $P(x)$ true. This is what we did in Theorem 3.29.

The proposition in Theorem 3.29 has two variables (x and y), but it is basically stating that there exists a solution of the given system of equations. In this case, we named an object in the universe (a value for x and a value for y) that was an actual solution of the system of equations. Notice that although we discovered the solution essentially using the backward method, we did not claim that we had found a solution until we actually checked the solution.

Nonconstructive Proofs

Another type of proof that is often used to prove an existence theorem is the so-called **nonconstructive proof.** For this type of proof, we make an argument that an object in the universal set that makes $P(x)$ true must exist but we never construct or name the object that makes $P(x)$ true. The advantage of a constructive proof over a nonconstructive proof is that the constructive proof will yield a procedure or algorithm for obtaining the desired object.

The proof of the **Intermediate Value Theorem** from calculus is an example of an nonconstructive proof. The Intermediate Value Theorem can be stated as follows:

> If f is a continuous function on the closed interval $[a, b]$ and if q is any real number strictly between $f(a)$ and $f(b)$, then there exists a number c in the interval (a, b) such that $f(c) = q$.

The Intermediate Value Theorem can be used to prove that a solution to some equations must exist. This is shown in the next example.

Example 3.30. Let x represent a real number. We will use the Intermediate Value Theorem to prove that the equation $x^3 - x + 1 = 0$ has a real number solution.

To investigate solutions of the equation $x^3 - x + 1 = 0$, we will use the function

$$f(x) = x^3 - x + 1.$$

Notice that $f(-2) = -5$ and that $f(0) = 1$. Since $f(-2) < 0$ and $f(0) > 0$, the Intermediate Value Theorem tells us that there is a real number c between -2 and 0 such that $f(c) = 0$. This means that there exists a real number c between -2 and 0 such that

$$c^3 - c + 1 = 0,$$

and hence c is a real number solution of the equation $x^3 - x + 1 = 0$. This proves that the equation $x^3 - x + 1 = 0$ has at least one real number solution.

Notice that this proof does not tell us how to find the exact value of c. It does, however, suggest a method for approximating the value of c. This can be done by finding smaller and smaller intervals $[a, b]$ such that $f(a)$ and $f(b)$ have opposite signs.

Activity 3.31 (Intersections of Circles). In Preview Activity 2, we investigated the points of intersection of two different pairs of circles. For the first pair, there were two points of intersection, and for the second pair, there were no points of intersection. That is, the circles did not intersect. The two examples from this preview activity are related to the following proposition:

> **Proposition:** Let a be a nonzero real number and let r be a positive real number. If $|a| < 2r$, then there exist two points of intersection for the two circles whose equations are $x^2 + y^2 = r^2$ and $(x - a)^2 + y^2 = r^2$.

Prove this proposition by using a constructive proof and stating the points of intersection.

<u>Hint</u>: Try to proceed in this general case as you did in the specific cases in Preview Activity 2. Remember that $\sqrt{a^2} = |a|$.

Exercises 3.5

1. Prove the following proposition:

  If $p, q \in \mathbb{Q}$ with $p < q$, then there exists an $x \in \mathbb{Q}$ with $p < x < q$.

2. Is the following proposition true or false? Justify your conclusion.

  Let $m, b \in \mathbb{R}$ with $m \neq 0$. If $u \in \mathbb{R}$, then there exists an $x \in \mathbb{R}$ such that $mx + b = u$.

3. Prove the following proposition:

  If $p, q \in \mathbb{Q}$ with $q \neq 0$, then $\frac{p}{q}$ is a rational number.

4. If $n \in \mathbb{Z}$ and $m = n + 1$, then m and n are said to be **consecutive integers**. Is the following proposition true or false? Justify your conclusion.

  If m and n are two consecutive integers, then 4 divides $\left(m^2 + n^2 - 1\right)$.

5. Let $x \in \mathbb{R}$. Prove that there exists a real number δ with $\delta > 0$ such that $|(2x + 1) - 7| < 0.01$ whenever $|x - 3| < \delta$.

  Note: The symbol δ is the lowercase letter delta from the Greek alphabet. It is often used in mathematics to represent a small positive real number.

6. **(a)** Prove that there exists a real number x such that $x^3 - 4x^2 = 7$.

  (b) Prove that there is no integer x such that $x^3 - 4x^2 = 7$.

7. Classify each of your proofs in Exercises (2) through (6a) as a constructive proof or a nonconstructive proof.

8. Are the following propositions true or false? Justify your conclusion.

  (a) There exist integers x and y such that $4x + 6y = 2$.

  (b) There exist integers x and y such that $6x + 15y = 2$.

  (c) There exist integers x and y such that $6x + 15y = 9$.

9. Let y_1, y_2, y_3, y_4 be real numbers. The mean, $\bar{y}$, of these four numbers is defined to be the sum of the four numbers divided by 4. That is,

$$\bar{y} = \frac{y_1 + y_2 + y_3 + y_4}{4}.$$

For this preview activity, we will use the following sets:

$$A = \{-3, -2, -1, 0, 1, 2, 3\} \qquad B = \{x \in \mathbb{Z} | \, x^2 < 12\}$$
$$C = \{x \in \mathbb{Z} | \, 2x - 3 \leq 7\} \qquad D = \{y \in \mathbb{Z} | \, |y| \leq 5\}$$
$$S = \{x \in \mathbb{R} | \, 2x - 3 \leq 7\} \qquad T = \{y \in \mathbb{R} | \, |y| \leq 5\}$$

1. (a) Write the elements of the set B using the roster method.

   (b) Is the set A equal to the set B? Explain.

   (c) Is A a subset of B? Explain.

   (d) Is B a subset of A? Explain.

2. (a) Write the elements of the sets C and D using the roster method.

   (b) Is the set C equal to the set D? Explain.

   (c) Is C a subset of D? Explain.

   (d) Is D a subset of C? Explain.

3. (a) Is it possible to write the elements of the sets S and T using the roster method?

   (b) Is the set S equal to the set T? Explain.

   (c) Is S a subset of T? Explain.

   (d) Is T a subset of S? Explain.

4. Complete the following sentence in English without using the symbols for quantifiers:

   "The set X is not a subset of the set Y provided that ... "

Preview Activity 2 (Subsets of a Given Set).

When a set contains no elements, we say that the set is the **empty set**. The empty set is usually designated by the symbol $\emptyset$. For example, the set of all real numbers whose square is equal to -1 contains no elements, and hence is the empty set. Using set builder notation, this can be written as $\{x \in \mathbb{R} \mid x^2 = -1\} = \emptyset$.

1. Let A be a subset of some universal set U.

   (a) Is the statement "If $x \in \emptyset$, then $x \in A$" true or false? Explain.

   (b) Is the statement "$\emptyset \subseteq A$" true or false? Explain.

2. Determine all the subsets of the set $\{a\}$. How many subsets does the set $\{a\}$ have?

3. Determine all the subsets of the set $\{a, b\}$. How many subsets does the set $\{a, b\}$ have? <u>Hint</u>: Start with the sets listed in Part (2) and then create the subsets of $\{a, b\}$ that are not subsets of $\{a\}$.

4. Determine all the subsets of the set $\{a, b, c\}$. How many subsets does the set $\{a, b, c\}$ have? <u>Hint</u>: Start with the sets listed in Part (3) and then create the subsets of $\{a, b, c\}$ that are not subsets of $\{a, b\}$.

5. Based on your work in Parts (2), (3), and (4), how many subsets do you think a set with four elements will have? How many subsets do you think a set with five elements will have? Let $n \in \mathbb{N}$. How many subsets do you think a set with n elements will have?

Preview Activity 3 (Truth Sets).

In Section 2.1, we studied the truth set of a predicate (open sentence). The **truth set** of a predicate is the collection of objects in the universal set that can be substituted to make the predicate a true statement. For example,

- If the universal set is $\mathbb{R}$, then the truth set of "$x^2 - 3x - 10 = 0$" is $\{-2, 5\}$.

- If the universal set is $\mathbb{N}$, then the truth set of "$\sqrt{n} \in \mathbb{N}$" is $\{1, 4, 9, 16, \dots\}$.

For this activity, the universal set is $U = \{0, 1, 2, 3, \dots, 10\}$. We will use the following subsets of U:

$$A = \{0, 1, 2, 9\} \qquad\qquad B = \{2, 3, 4, 5, 6\}.$$

Use the roster method to list all of the elements of U that are in the truth set of each of the following predicates. Remember that the connective "or" is always used in the inclusive sense.

1. $x \in A$ and $x \in B$ 3. $x \notin A$

2. $x \in A$ or $x \in B$ 4. $x \notin B$

5. $x \in A$ and $x \notin B$

6. $x \notin A$ or $x \notin B$

7. $\neg\, (x \in A$ and $x \in B)$

8. $\neg\, (x \in A$ or $x \in B)$

Set Equality, Subsets, and Proper Subsets

In Preview Activity 1, we introduced some basic definitions used in set theory. We will now give more formal definitions of these concepts.

Definition. Let A and B be two sets contained in some universal set U.

The sets A and B are **equal**, written $A = B$, when they have precisely the same elements. Specifically,

$A = B$ provided that for all $x \in U, x \in A$ if and only if $x \in B$.

We write $A \neq B$ when the two sets A and B are not equal.

The set A is **contained** in a set B if each element of A is an element of B. In this case, we write $A \subseteq B$ and say that A is a **subset** of B. In particular,

$A \subseteq B$ provided that for all $x \in U$, if $x \in A$, then $x \in B$.

When A is not a subset of B, we write $A \nsubseteq B$.

The set A is a **proper subset** of B provided that $A \subseteq B$ and $A \neq B$.

One reason for the definition of proper subset is that each set is a subset of itself. That is,

If A is a set, then $A \subseteq A$.

However, sometimes we need to indicate that a set X is a subset of Y but $X \neq Y$. For example, if

$$X = \{1, 2\} \text{ and } Y = \{0, 1, 2, 3\},$$

then $X \subset Y$. We know that $X \subseteq Y$ since each element of X is an element of Y, but $X \neq Y$ since $0 \in Y$ and $0 \notin X$. (Also, $3 \in Y$ and $3 \notin X$.)

Notice that the notations $A \subset B$ and $A \subseteq B$ are used in a manner similar to inequality notation for numbers ($a < b$ and $a \leq b$).

It is often very important to be able to describe precisely what it means to say that one set is not a subset of the other. (We have already done this in Preview Activity 1.) In the preceding example, Y is not a subset of X since there exists an element of Y (namely, 0) that is not in X.

In general, the subset relation is described with the use of a universal quantifier since $A \subseteq B$ means that for each element x of U, if $x \in A$, then $x \in B$. So when we negate this, we use an existential quantifier as follows:

$$A \subseteq B \qquad \text{means} \qquad (\forall x)\,[(x \in A) \rightarrow (x \in B)].$$

$$A \not\subseteq B \qquad \text{means} \qquad \neg\,(\forall x)\,[(x \in A) \rightarrow (x \in B)]$$
$$(\exists x)\,\neg\,[(x \in A) \rightarrow (x \in B)]$$
$$(\exists x)\,[(x \in A) \wedge (x \notin B)].$$

So we see that $A \not\subseteq B$ means that there exists an x such that $x \in A$ and $x \notin B$.

We defined two sets to be equal when they have precisely the same elements. For example,

$$\left\{ x \in \mathbb{R} \mid x^2 = 4 \right\} = \{-2,\, 2\}.$$

If the two sets A and B are equal, then it must be true that every element of A is an element of B, that is, $A \subseteq B$, and it must be true that every element of B is an element of A, that is, $B \subseteq A$. Conversely, if $A \subseteq B$ and $B \subseteq A$, then A and B must have precisely the same elements. This gives us the following test for set equality:

Theorem 4.1. *Let A and B be subsets of some universal set U. Then $A = B$ if and only if $A \subseteq B$ and $B \subseteq A$.*

We will see later that this provides a useful method for proving that two sets are equal.

Activity 4.2 (Using Set Notation). Let the universal set be $U = \{1, 2, 3, 4, 5, 6\}$. Let

$$A = \{1, 2, 4\}, \quad B = \{1, 2, 3, 5\}, \quad C = \left\{ x \in U \mid x^2 \leq 2 \right\}.$$

In each of the following, fill in the blank with one or more of the symbols $\subset, \subseteq, =, \neq, \in,$ or $\notin$ so that the resulting statement is true. For each blank, include all symbols that result in a true statement. If none of these symbols makes a true statement, write nothing in the blank.

A	_____	B		$\emptyset$	_____	A
5	_____	B		$\{5\}$	_____	B
A	_____	C		$\{1,2\}$	_____	C
$\{1,2,2\}$	_____	A		$\{4,2,1\}$	_____	A
6	_____	A		B	_____	$\emptyset$

The Power Set

The symbol $\in$ is used to describe a relationship between an element of the universal set and a subset of the universal set, and the symbol $\subseteq$ is used to describe a relationship between two subsets of the universal set. For example, the number 5 is an integer, and so it is appropriate to write $5 \in \mathbb{Z}$. It is not appropriate, however, to write $5 \subseteq \mathbb{Z}$ since 5 is not a set. It is important to distinguish between 5 and $\{5\}$. The difference is that 5 is an integer and $\{5\}$ is a set consisting of one element. Consequently, it is appropriate to write $\{5\} \subseteq \mathbb{Z}$, but it is not appropriate to write $\{5\} \in \mathbb{Z}$. The distinction between these two symbols (5 and $\{5\}$) is important when we discuss what is called the power set of a given set.

Definition. If A is a subset of a universal set U, then the set whose members are all the subsets of A is called the **power set** of A. We denote the power set of A by $\mathcal{P}(A)$. Symbolically, we write

$$\mathcal{P}(A) = \{X \subseteq U \mid X \subseteq A\}.$$

That is, $X \in \mathcal{P}(A)$ if and only if $X \subseteq A$.

When dealing with the power set of A, we must always remember that $\emptyset \subseteq A$ and $A \subseteq A$. For reference, we state this as Theorem 4.3.

Theorem 4.3. *For any set A, $\emptyset \subseteq A$ and $A \subseteq A$. That is, $\emptyset \in \mathcal{P}(A)$ and $A \in \mathcal{P}(A)$.*

We actually explored the idea of a power set in Preview Activity 2 without calling it the power set. For example, if $A = \{a, b\}$, then the subsets of A are

$$\emptyset, \{a\}, \{b\}, \{a, b\}.$$

We can write this as

$$\mathcal{P}(A) = \{\emptyset, \{a\}, \{b\}, \{a, b\}\}.$$

In Preview Activity 2, we also saw that if $B = \{a, b, c\}$, then

$$\mathcal{P}(B) = \{\emptyset, \{a\}, \{b\}, \{a,\ b\}, \{c\}, \{a,c\}, \{b,c\}, \{a,b,c\}\}.$$

Notice, for example, that we could write

$$\{a,c\} \subseteq B \text{ or that } \{a,c\} \in \mathcal{P}(B).$$

Also, notice that if A has two elements, then $\mathcal{P}(A)$ has four elements, and if B has three elements, then $\mathcal{P}(B)$ has eight elements. Now, let n be a nonnegative integer. Although we cannot prove it at this time, it is true that

> If A contains n elements, then $\mathcal{P}(A)$ contains 2^n elements. That is, A has 2^n subsets.

The Cardinality of a Finite Set

In our discussion of the power set, we were concerned with the number of elements in a set. In fact, the number of elements in a finite set is a distinguishing characteristic of the set, so we give it the following name.

Definition. The number of elements in a finite set A is called the **cardinality** of A and is denoted by $|A|$.

For example, $|\emptyset| = 0$; $|\{a,b\}| = 2$; $|\mathcal{P}(\{a,b\})| = 4$.

A word about notation: We are using the notation $|A|$ to denote the cardinality of a set A. Do not confuse this with the absolute value of a real number. It is common practice in mathematics to use the same notation for two different concepts when there is little chance of confusion. We must be careful to use (and understand) the notation in the proper context. In set theory, $|A|$ is the cardinality of the set A, and in the algebra of real numbers, $|x|$ is the absolute value of the real number x.

Operations on Sets

In Section 2.2, we used logical operators (conjunction, disjunction, negation) to form new statements from existing statements. In a similar manner, there are several ways to create new sets from sets that have already been defined. The most common set operations are given in the following definitions.

Definition. Let A and B be subsets of some universal set U. The **intersection** of A and B, written $A \cap B$ and read "A intersect B," is the set of all elements that are in both A and B. That is,

$$A \cap B = \{x \in U \mid x \in A \text{ and } x \in B\}.$$

Definition. Let A and B be subsets of some universal set U. The **union** of A and B, written $A \cup B$ and read "A union B," is the set of all elements that are in A or in B. That is,

$$A \cup B = \{x \in U \mid x \in A \text{ or } x \in B\}.$$

Definition. Let A be a subset of some universal set U. The **complement** of the set A, written A^c and read "the complement of A," is the set of all elements of U that are not in A. That is,

$$A^c = \{x \in U \mid x \notin A\}.$$

Definition. Let A and B be subsets of some universal set U. The **set difference** of A and B, or **relative complement** of B with respect to A, written $A - B$ and read "A minus B" or "the complement of B with respect to A," is the set of all elements in A that are not in B. That is,

$$A - B = \{x \in U \mid x \in A \text{ and } x \notin B\}.$$

These definitions are related to some of the truth sets in Preview Activity 3.

Example 4.4. In Preview Activity 3, we used

$$U = \{0, 1, 2, 3, \ldots, 10\}, \quad A = \{0, 1, 2, 9\}, \quad B = \{2, 3, 4, 5, 6\}.$$

So in this case,

$$A \cap B = \{x \in U \,|\, x \in A \text{ and } x \in B\} = \{2\}$$

$$A \cup B = \{x \in U \,|\, x \in A \text{ or } x \in B\} = \{0, 1, 2, 3, 4, 5, 6, 9\}$$

$$A^c = \{x \in U \,|\, x \notin A\} = \{3, 4, 5, 6, 7, 8, 10\}$$

$$A - B = \{x \in A \,|\, x \notin B\} = \{0, 1, 9\}.$$

We can also use these definitions in combinations to form other sets. For example,

$$B^c = \{x \in U \,|\, x \notin B\} = \{0, 1, 7, 8, 9, 10\}$$

$$A^c \cup B^c = \{x \in U \,|\, x \in A^c \text{ or } x \in B^c\} = \{0, 1, 3, 4, 5, 6, 7, 8, 9, 10\}$$

$$(A \cap B)^c = \{x \in U \,|\, x \notin A \cap B\} = \{0, 1, 3, 4, 5, 6, 7, 8, 9, 10\}.$$

Example 4.5. Let $U = \mathbb{Z}$, $A = \{1, 2, 3, \ldots, 10\}$, and $B = \{x \in \mathbb{Z} \,|\, x \leq 4\}$. Then,

$$A \cap B = \{1, 2, 3, 4\} \qquad\qquad A \cup B = \{x \in \mathbb{Z} \,|\, x \leq 10\}$$
$$A - B = \{5, 6, 7, 8, 9, 10\} \qquad\qquad B - A = \{x \in \mathbb{Z} \,|\, x \leq 0\}$$
$$A^c = \{x \in \mathbb{Z} \,|\, x < 1 \text{ or } x > 10\} \qquad\qquad B^c = \{x \in \mathbb{Z} \,|\, x > 4\}$$

There is another way to write a symbolic representation of A^c. Notice that there are two distinct parts to A^c, one where $x < 1$ and the other where $x > 10$. In this case, it is possible to write a rule (or roster) for each part and then use the operation of union, rather than trying to find one rule (or roster) for the entire set. So we could write

$$A^c = \{x \in \mathbb{Z} \,|\, x < 1 \text{ or } x > 10\}, \text{ or}$$
$$A^c = \{x \in \mathbb{Z} \,|\, x < 1\} \cup \{x \in \mathbb{Z} \,|\, x > 10\}.$$

In some of these examples, we used set builder notation and in others we used the roster method. Either one is suitable to use in these examples.

For example, if it helps understand what is happening, we could write $A = \{1, 2, 3, \ldots, 10\}$ and $B = \{\ldots, -1, 0, 1, 2, 3, 4\}$. Then, we could write

$$A \cup B = \{\ldots, -1, 0, 1, 2, 3, 4, 5, 6, 7, 8, 9, 10\},$$
$$B - A = \{\ldots, -3, -2, -1, 0\},$$
$$A^c = \{\ldots, -3, -2, -1, 0\} \cup \{11, 12, 13, \ldots\},$$
$$B^c = \{5, 6, 7, \ldots\}.$$

Notice that in this case, it is better to write A^c as a union of two sets.

Venn Diagrams

So far, we have been working with verbal and symbolic definitions of set relationships and set operations. When doing so, it is useful to study specific examples such as the last one to help us understand the general definitions. It is also helpful to have a visual representation of sets. In **Venn diagrams**, sets are represented by circles (or some other closed geometric shape) drawn inside a rectangle. The rectangle represents the universal set U, and the elements of a set are represented by the points inside the circle that represents the set. For example, Figure 4.1 is a Venn diagram showing two sets.

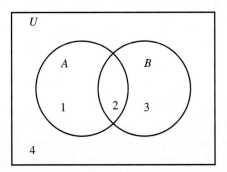

Figure 4.1: Venn Diagram for Two Sets

In this diagram, the elements of A are represented by the points inside the left circle, and the elements of B are represented by the points inside the right circle. The four distinct regions in the diagram are numbered for reference purposes only. (The numbers do not represent elements in a set.) The following table describes the four regions in the diagram.

Region	Elements of U	Set
1	In A and not in B	$A - B$
2	In A and in B	$A \cap B$
3	In B and not in A	$B - A$
4	Not in A and not in B	$A^c \cap B^c$

We can use these regions to represent other sets. For example, the set $A \cup B$ is represented by regions 1, 2, and 3 or the shaded region in Figure 4.2.

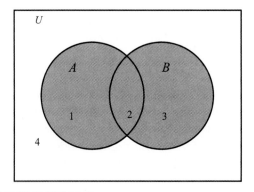

Figure 4.2: Venn Diagram for $A \cup B$

We can, of course, include more than two sets in a Venn diagram. Figure 4.3 shows a general Venn diagram for three sets. There are eight distinct regions, and each region has a unique reference number. For example, the set A is represented by the combination of regions 1, 2, 4, and 5, whereas the set C is represented by the combination of regions 4, 5, 6, and 7. This means that the set $A \cap C$ is represented by the combination of regions 4 and 5. This is shown as the shaded region in Figure 4.3.

Finally, Venn diagrams can also be used to illustrate special relationships between sets. For example, if $A \subseteq B$, then the circle representing A should be completely contained in the circle for B. So if $A \subseteq B$, and we know nothing about any relationship between the set C and the sets A and B, we could use the Venn diagram shown in Figure 4.4.

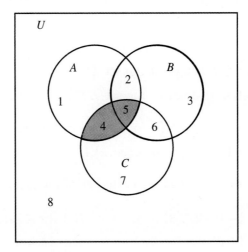

Figure 4.3: Venn Diagram for $A \cap C$

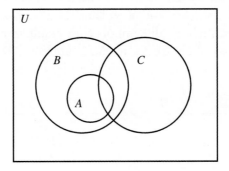

Figure 4.4: Venn Diagram showing $A \subseteq B$

Exercises 4.1

1. Assume the universal set is the set of real numbers. Let

$$A = \{-3, -2, 2, 3\}, \qquad B = \{x \in \mathbb{R} \mid x^2 = 4 \text{ or } x^2 = 9\},$$
$$C = \{x \in \mathbb{R} \mid x^2 + 2 = 0\}, \qquad D = \{x \in \mathbb{R} \mid x > 0\}.$$

Respond to each of the following questions. In each case, explain your answer.

(a) Is the set A equal to the set B?

(b) Is the set A a subset of the set B?

(c) Is the set C equal to the set D?

(d) Is the set C a subset of the set D?

(e) Is the set A a subset of the set D?

2. **(a)** Explain why the set $\{a, b\}$ is equal to the set $\{b, a\}$.

   (b) Explain why the set $\{a, b, b, a, c\}$ is equal to the set $\{b, c, a\}$.

3. Assume that the universal set is the set of integers. Let

$$A = \{-3, -2, 2, 3\}, \qquad B = \{x \in \mathbb{Z} \mid x^2 \leq 9\},$$
$$C = \{x \in \mathbb{Z} \mid x \geq -3\}, \qquad D = \{1, 2, 3, 4\}.$$

In each of the following, fill in the blank with one or more of the symbols $\subset$, $\subseteq$, $\not\subseteq$, $=$, $\neq$, $\in$, or $\notin$ so that the resulting statement is true. For each blank, include all symbols that result in a true statement. If none of these symbols makes a true statement, write nothing in the blank.

A _____ B		$\emptyset$ _____ A
5 _____ C		$\{5\}$ _____ C
A _____ C		$\{1, 2\}$ _____ B
$\{1, 2\}$ _____ A		$\{3, 2, 1\}$ _____ D
4 _____ B		D _____ $\emptyset$
$|A|$ _____ $|D|$		$|A|$ _____ $|B|$
A _____ $\mathcal{P}(A)$		A _____ $\mathcal{P}(B)$

4. Write all of the proper subset relations that are possible using the sets of numbers $\mathbb{N}$, $\mathbb{Z}$, $\mathbb{Q}$, $\mathbb{R}$, and $\mathbb{C}$.

5. For each statement, write a brief, clear explanation why the statement is true or why it is false.

   (a) The set $\{a, b\}$ is a subset of $\{a, c, d, e\}$.

   (b) The set $\{-2, 0, 2\}$ is equal to $\{x \in \mathbb{Z} \mid x \text{ is even and } x^2 < 5\}$.

   (c) The empty set $\emptyset$ is a subset of $\{1\}$.

   (d) If $A = \{a, b\}$, then the set $\{a\}$ is a subset of $\mathcal{P}(A)$.

6. Let $U = \{1, 2, 3, 4, 5, 6, 7, 8, 9, 10\}$, and let

$$A = \{3, 4, 5, 6, 7\}, \qquad\qquad B = \{1, 5, 7, 9\},$$
$$C = \{3, 6, 9\}, \qquad\qquad D = \{2, 4, 6, 8\}.$$

Use the roster method to list all of the elements of each of the following sets.

(a) $A \cap B$

(b) $A \cup B$

(c) $(A \cup B)^c$

(d) $A^c \cap B^c$

(e) $(A \cup B) \cap C$

(f) $A \cap C$

(g) $B \cap C$

(h) $(A \cap C) \cup (B \cap C)$

(i) $B \cap D$

(j) $(B \cap D)^c$

(k) $A - D$

(l) $B - D$

(m) $(A - D) \cup (B - D)$

(n) $(A \cup B) - D$

7. In previous mathematics courses, we have frequently used subsets of the real numbers called **intervals**. There are some common names and notations for intervals. These are given in the following table, where it is assumed that a and b are real numbers and $a < b$.

Interval Notation	Set Notation	Name
$(a, b) =$	$\{x \in \mathbb{R} \mid a < x < b\}$	Open interval from a to b
$[a, b] =$	$\{x \in \mathbb{R} \mid a \leq x \leq b\}$	Closed interval from a to b
$[a, b) =$	$\{x \in \mathbb{R} \mid a \leq x < b\}$	Half-open interval
$(a, b] =$	$\{x \in \mathbb{R} \mid a < x \leq b\}$	Half-open interval
$(a, +\infty) =$	$\{x \in \mathbb{R} \mid x > a\}$	Open ray
$(-\infty, b) =$	$\{x \in \mathbb{R} \mid x < b\}$	Open ray
$[a, +\infty) =$	$\{x \in \mathbb{R} \mid x \geq a\}$	Closed ray
$(-\infty, b] =$	$\{x \in \mathbb{R} \mid x \leq b\}$	Closed ray

(a) Is (a, b) a proper subset of $(a, b]$? Explain.

(b) Is $[a, b]$ a subset of $(a, +\infty)$? Explain.

(c) Use interval notation to describe the intersection of the interval $[-3, 7]$ with the interval $(5, 9]$.

(d) Write the set $\{x \in \mathbb{R} \mid |x| \leq 0.01\}$ using interval notation.

(e) Write the set $\{x \in \mathbb{R} \mid |x| > 2\}$ as the union of two intervals.

8. Using the definitions of intersection and union, we can say that

   - $x \in A \cap B$ if and only if $x \in A$ and $x \in B$.
   - $x \in A \cup B$ if and only if $x \in A$ or $x \in B$.
   - $x \in A - B$ if and only if $x \in A$ and $x \notin B$.

   Write a useful negation of each of these sentences. That is, complete each of the following sentences:

   (a) $x \notin A \cap B$ if and only if ...
   (b) $x \notin A \cup B$ if and only if ...
   (c) $x \notin A - B$ if and only if ...

9. Let P, Q, R, and S be subsets of a universal set U. Assume that $(P - Q) \subseteq (R \cap S)$.

   (a) Complete the following sentence:
       For all $x \in U$, if $x \in (P - Q)$, then ...
   (b) Write a useful negation of the statement in Part (9a).
   (c) Write the contrapositive of the statement in Part (9a).

10. Let A, B, and C be subsets of some universal set U. Consider the following statement:

$$\text{If } A \subseteq B, \text{ then } B^c \subseteq A^c.$$

   (a) Write the negation of this statement.
   (b) Write the contrapositive of this statement.
   (c) Identify three conditional statements in the given statement.
   (d) Write the given statement in symbolic form using quantifiers.
   (e) Write the negation of the statement in Part (10d) in symbolic form.
   (f) Write the contrapositive of the statement in Part (10d) in symbolic form.

11. Let A, B, and C be subsets of some universal set U. Draw a Venn diagram for each of the following situations.

(a) $A \subseteq C$

(b) $A \cap B = \emptyset$

(c) $A \not\subseteq B, B \not\subseteq A, C \subseteq A$, and $C \not\subseteq B$

(d) $A \subseteq B, C \subseteq B$, and $A \cap C = \emptyset$

12. Let A, B, and C be subsets of some universal set U. For each of the following, draw a general Venn diagram for the three sets and then shade the indicated region.

(a) $A \cap B$

(b) $A \cap C$

(c) $(A \cap B) \cup (A \cap C)$

(d) $B \cup C$

(e) $A \cap (B \cup C)$

(f) $(A \cap B) - C$

4.2 Proving Set Relationships

Preview Activity 1 (Working with Two Specific Sets).

Let S be the set of all integers that are multiples of 6, and let T be the set of all even integers.

1. List at least four different positive elements of S and at least four different negative elements of S. Are all of these integers even?

2. Use the roster method to specify the sets S and T. (See Section 2.1 for a review of the roster method.) Does there appear to be any relationship between these two sets? That is, does it appear that the sets are equal or that one set is a subset of the other set?

3. Use set builder notation to specify the sets S and T. (See Section 2.1 for a review of the set builder notation.)

4. Using appropriate definitions, describe what it means to say that an integer x is a multiple of 6 and what it means to say that an integer y is even.

5. How do we prove that the set S is a subset of the set T?

Preview Activity 2 (Proving a Subset Relationship).

Review the definition of subset in Section 4.1. In Preview Activity 1, we considered the set S of all integers that are multiples of 6 and the set T of all even integers.

Complete the following know-show table for the proposition that S is a subset of T.

This table is in the form of a proof method called the **choose an element method.** This method is frequently used when we encounter a universal quantifier in a statement in the backward process. (In this case, this is Step $Q1$.) The key is that we have to prove something about all elements in S. When we see this, we can add something to the forward process by choosing an arbitrary element from the set S. (This is done in Step $P1$.) This does not mean that we can choose a specific element of S. Rather, we must give the arbitrary element a name and use only the properties it has by being a member of the set S. In this case, it is a multiple of 6.

Step	Know	Reason
P	S is the set of all integers that are multiples of 6. T is the set of all even integers.	Hypothesis
$P1$	Let $x \in S$.	Choose an arbitrary element of S.
$P2$	$(\exists m \in \mathbb{Z})\,(x = 6m)$	Definition of "multiple"
$\vdots$	$\vdots$	$\vdots$
$Q2$	x is an element of T.	x is even.
$Q1$	$(\forall x)\,[(x \in S) \to (x \in T)]$	Step $P1$ and Step $Q2$
Q	$S \subseteq T$.	Definition of "subset"
Step	**Show**	**Reason**

Preview Activity 3 (Working with Venn Diagrams).

1. Draw a Venn diagram for two sets, A and B, with the assumption that $A \subseteq B$.

2. On this Venn diagram, lightly shade the area corresponding to A^c.

3. Determine the region on the Venn diagram that corresponds to B^c. What appears to be the relationship between A^c and B^c? Explain.

4. Draw a general Venn diagram for two sets, A and B.

5. On the Venn diagram, shade the region corresponding to $A - B$ and shade the region corresponding to $A \cap B^c$. What appears to be the relationship between these two sets? Explain.

In this section, we will learn how to prove certain relationships about sets. Two of the most basic types of relationships between sets are the equality relation and the subset relation. So if we are asked a question of the form "How are the sets A and B related?", we can answer the question if we can prove that the two sets are equal or that one set is a subset of the other set. There are other ways to answer this, but we will concentrate on these two for now.

The Choose an Element Method

The method of proof we will introduce in this section is called the **choose an element method.** This method is frequently used when we encounter a universal quantifier in a statement in the backward process. This statement often has the form

For each element with a given property, something happens.

Since most statements with a universal quantifier can be expressed in the form of a conditional statement, this statement could have the following equivalent form:

If an element has a given property, then something happens.

We will illustrate this with the proposition from Preview Activity 2. This proposition could be stated as follows:

Let S be the set of all integers that are multiples of 6, and let T be the set of all even integers. Then S is a subset of T.

In Preview Activity 2, we worked on a know-show table for this proposition. The key was that in the backward process, we encountered the following statement:

Each element of S is an element of T or, more precisely, if $x \in S$, then $x \in T$.

In this case, the "element" is an integer, the "given property" is that it is an element of S, and the "something that happens" is that the element is also an element of T.

One way to approach this is to create a list of all elements with the given property and verify that for each one, the "something happens." When the list is short, this may be a reasonable approach. However, as in this case, when the list is infinite (or even just plain long), this approach is not practical.

We overcome this difficulty by using the **choose an element method**, where we choose an arbitrary element with the given property. So in this case, we choose an integer x that is a multiple of 6. We cannot use a specific multiple of 6 (such as 12 or 24), but rather the only thing we can assume is that the integer satisfies the property that it is a multiple of 6. This is the key part of this method.

> *Whenever we choose an arbitrary element with a given property, we are not selecting a specific element. Rather, the only thing we can assume about the element is the given property.*

It is important to realize that once we have chosen the arbitrary element, we have added information to the forward process. So in the know-show table for this proposition, we added the statement, "Let $x \in S$" to the forward process.

Following is a completed proof of this proposition following the outline of the know-show table from Preview Activity 2.

Proposition 4.6. *Let S be the set of all integers that are multiples of 6, and let T be the set of all even integers. Then S is a subset of T.*

Proof. Let S be the set of all integers that are multiples of 6, and let T be the set of all even integers. We will show that S is a subset of T by showing that if an integer x is an element of S, then it is also an element of T.

Let $x \in S$. (<u>Note</u>: The use of the word "let" is often an indication that the we are choosing an arbitrary element.) This means that x is a multiple of 6. Therefore, there exists an integer m such that

$$x = 6m.$$

However, we know that $6 = 2 \cdot 3$, and hence this equation can be written in the form

$$x = 2\,(3m).$$

By closure properties of the integers, $3m$ is an integer. Hence, this last equation proves that x must be even.

Therefore, we have shown that if x is an element of S, then x is an element of T, and hence that $S \subseteq T$.   ∎

Activity 4.7 (Using the Choose an Element Method). The Venn diagram in Preview Activity 3 suggests that the following proposition is true.

Proposition 4.8. *Let A and B be subsets of the universal set U. If $A \subseteq B$, then $B^c \subseteq A^c$.*

1. Since the conclusion of the conditional statement is $B^c \subseteq A^c$, explain why we should try the choose method to prove this proposition.

2. Complete the following know-show table for this proposition and explain exactly where the choose method is used.

Step	Know	Reason
P	$A \subseteq B$	Hypothesis
$P1$	Let $x \in B^c$.	Choose an arbitrary element of B^c.
$P2$	If $x \in A$, then $x \in B$.	Definition of "subset"
⋮	⋮	⋮
$Q1$	If $x \in B^c$, then $x \in A^c$.	
Q	$B^c \subseteq A^c$	Definition of "subset"
Step	**Show**	**Reason**

Proving Set Equality

One way to prove set equality is to use Theorem 4.1 and prove two subset relationships. In particular, let A and B be subsets of some universal set. Theorem 4.1 states that $A = B$ if and only if $A \subseteq B$ and $B \subseteq A$.

In Preview Activity 3, we created a Venn diagram that indicated that $A - B = A \cap B^c$. Following is a proof of this result. Notice where the choose an element method is used in each case.

Proposition 4.9. *Let A and B be subsets of some universal set. Then $A - B = A \cap B^c$.*

Proof. Let A and B be subsets of some universal set. We will prove that $A - B = A \cap B^c$ by proving that $A - B \subseteq A \cap B^c$ and that $A \cap B^c \subseteq A - B$.
First, let $x \in A - B$. This means that

$$x \in A \text{ and } x \notin B.$$

But since $x \notin B$, we know that $x \in B^c$. Hence, we can conclude that

$$x \in A \text{ and } x \in B^c.$$

This means that $x \in A \cap B^c$, and hence we have proven that $A - B \subseteq A \cap B^c$.
Now choose $y \in A \cap B^c$. This means that

$$y \in A \text{ and } y \in B^c.$$

But because $y \in B^c$, we know that $y \notin B$. Hence, we can conclude that

$$y \in A \text{ and } y \notin B.$$

This means that $y \in A - B$, and hence we have proven that $A \cap B^c \subseteq A - B$.
Since we have proven that $A - B \subseteq A \cap B^c$ and $A \cap B^c \subseteq A - B$, we conclude that $A - B = A \cap B^c$. ∎

Using the Choose an Element Method in a Different Setting

We have used the choose an element method to prove Propositions 4.6 and 4.9. These types of proofs using the are sometimes referred to as **element chasing proofs.** This name is used since the basic method is to choose an arbitrary element from one set and "chase it" until you prove it must be in another set.

This method, however, is a general proof technique and can be used in settings other than set theory. It is often used whenever we encounter a universal quantifier in a statement in the backward process. Consider the following proposition.

Proposition 4.10. *Let a, b, and d be integers. If d divides a and d divides b, then for all integers x and y, d divides ax + by.*

Notice that the conclusion of the conditional statement in this proposition involves the universal quantifier. So in the backward process, we would have

Q: For all integers x and y, d divides $ax + by$.

The "elements" in this sentence are the integers x and y. In this case, these integers have no "given property." (They are just integers.) The "something that happens" is that d divides $ax+by$. Of course, this quantified sentence could be written as a conditional statement as follows:

Q: If x and y are integers, then d divides $ax + by$.

This means that in the forward process, we can use the hypothesis of the proposition and choose integers x and y. That is, in the forward process, we could have

P: a, b, and d are integers, d divides a and d divides b.

$P1$: Let $x \in \mathbb{Z}$ and let $y \in \mathbb{Z}$.

The proof of this proposition is part of Activity 4.11.

Activity 4.11 (Exploring and Proving Proposition 4.10).

1. Whenever we encounter a new proposition, it is a good idea to explore the proposition by looking at specific examples. For example, let $a = 20$, $b = 12$, and $d = 4$. In this case, $d \mid a$ and $d \mid b$. In each of the following cases, determine the value of $ax + by$ and determine if d divides $ax + by$.

   (a) $x = 1, y = 1$ **(c)** $x = 2, y = 2$ **(e)** $x = -2, y = 3$

   (b) $x = 1, y = -1$ **(d)** $x = 2, y = -3$ **(f)** $x = -2, y = -5$

2. Repeat Part (1) with $a = 21$, $b = -6$, and $d = 3$.

3. We started the forward-backward process for the proof of Proposition 4.10 following the discussion of this proposition. Complete the following proof of Proposition 4.10.

   Proposition 4.10. *Let a, b, and d be integers. If d divides a and d divides b, then for all integers x and y, d divides $ax + by$.*

   Proof. Let a, b, and d be integers, and assume that d divides a and d divides b. We will prove that for all integers x and y, d divides $ax+by$.

   So let $x \in \mathbb{Z}$ and let $y \in \mathbb{Z}$. Since d divides a, there exists an integer m such that .... ∎

Disjoint Sets

Earlier in this section, we discussed the set relations of equality and containment. There are other possible relationships between two sets; one is that the sets are disjoint. Basically, two sets are disjoint if and only if they have nothing in common. We express this formally in the following definition.

Definition. Let A and B be subsets of the universal set U. The sets A and B are said to be **disjoint** provided that $A \cap B = \emptyset$.

Example 4.12. The Venn diagram in Figure 4.5 shows two sets A and B with $A \subseteq B$. The shaded region is the region that represents B^c. From the

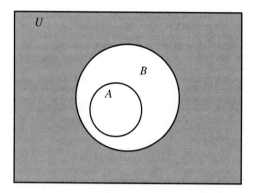

Figure 4.5: Venn Diagram with $A \subseteq B$

Venn diagram, it appears that $A \cap B^c = \emptyset$. This means that A and B^c are disjoint.

The preceding example suggests that the following proposition is true:

$$\text{If } A \subseteq B, \text{ then } A \cap B^c = \emptyset.$$

If we would like to prove this proposition, a reasonable "backward question" is, "How do we prove that a set is equal to the empty set?"

Since this question seems difficult to answer, this is an instance where proving the contrapositive or a proof by contradiction could be reasonable approaches. To illustrate these methods, let us assume the proposition we are trying to prove is of the following form:

$$\text{If } P, \text{ then } T = \emptyset.$$

If we choose to use prove the contrapositive or use a proof by contradiction, we need to assume that $T \neq \emptyset$. These methods can be outlined as follows:

- The contrapositive of "If P, then $T = \emptyset$" is, "If $T \neq \emptyset$, then $\neg P$." So in this case, we would assume $T \neq \emptyset$, and try to prove $\neg P$.

- Using a proof by contradiction, we would assume P and assume that $T \neq \emptyset$. From these two assumptions, we would attempt to derive a contradiction.

One advantage of these methods is that when we assume that $T \neq \emptyset$, then we know that there exists an element in the set T. We can then use that element in the rest of the proof.

We will prove Proposition 4.13 by proving its contrapositive.

Proposition 4.13. *Let A and B be subsets of some universal set. If $A \subseteq B$, then $A \cap B^c = \emptyset$.*

Proof. Let A and B be subsets of some universal set. We will prove that if $A \subseteq B$, then $A \cap B^c = \emptyset$, by proving its contrapositive. That is, we will prove

$$\text{If } A \cap B^c \neq \emptyset, \text{ then } A \not\subseteq B.$$

So assume that $A \cap B^c \neq \emptyset$. We will prove that $A \not\subseteq B$ by proving that there must exist an element x such that $x \in A$ and $x \notin B$.

Since $A \cap B^c \neq \emptyset$, there exists an element x that is in $A \cap B^c$. This means that

$$x \in A \text{ and } x \in B^c.$$

Now, the fact that $x \in B^c$ means that $x \notin B$. Hence, we can conclude that

$$x \in A \text{ and } x \notin B.$$

This means that $A \not\subseteq B$, and hence, we have proven the contrapositive of the proposition. This proves that if $A \subseteq B$, then $A \cap B^c = \emptyset$. ∎

Exercises 4.2

1. Let $A = \{x \in \mathbb{Z} \mid x \text{ is a multiple of } 9\}$ and let
   $B = \{x \in \mathbb{Z} \mid x \text{ is a multiple of } 3\}$.

   (a) Use the choose an element method to prove that $A \subseteq B$.

(b) Prove that $B \not\subseteq A$ by finding an integer that is in B but is not in A.

2. Let $A = \left\{ x \in \mathbb{R} \mid x^2 < 4 \right\}$ and let $B = \{ x \in \mathbb{R} \mid x < 2 \}$.

   (a) Is $A \subseteq B$? Justify your conclusion with a proof or a counterexample.

   (b) Is $B \subseteq A$? Justify your conclusion with a proof or a counterexample.

3. Let A, B, and C be subsets of a universal set U.

   (a) Draw a Venn diagram with $A \subseteq B$ and $B \subseteq C$. Does it appear that $A \subseteq C$?

   (b) Prove the following proposition:

   $$\text{If } A \subseteq B \text{ and } B \subseteq C, \text{ then } A \subseteq C.$$

   Note: This may seem like an obvious result. However, one of the reasons for this exercise is to provide practice at properly writing a proof that one set is a subset of another set. So we should start the proof by assuming that $A \subseteq B$ and $B \subseteq C$. Then we should choose an arbitrary element of A.

4. Let $A = \{ x \in \mathbb{Z} \mid x \equiv 7 \pmod 8 \}$ and $B = \{ x \in \mathbb{Z} \mid x \equiv 3 \pmod 4 \}$.

   (a) List at least five different elements of the set A.

   (b) List at least five different elements of the set B.

   (c) Is $A \subseteq B$? Justify your conclusion with a proof or a counterexample.

   (d) Is $B \subseteq A$? Justify your conclusion with a proof or a counterexample.

5. Let $C = \{ x \in \mathbb{Z} \mid x \equiv 7 \pmod 9 \}$ and $D = \{ x \in \mathbb{Z} \mid x \equiv 1 \pmod 3 \}$.

   (a) List at least five different elements of the set C.

   (b) List at least five different elements of the set D.

   (c) Is $C \subseteq D$? Justify your conclusion with a proof or a counterexample.

(d) Is $D \subseteq C$? Justify your conclusion with a proof or a counterexample.

6. Let A and B be subsets of some universal set U. Prove each of the following:

(a) $A \cap B \subseteq A$

(b) $A \subseteq A \cup B$

(c) $A \cap A = A$

(d) $A \cup A = A$

(e) $A \cap \emptyset = \emptyset$

(f) $A \cup \emptyset = A$

7. Let A and B be subsets of some universal set U. Is the following proposition true or false? Justify your conclusion with a proof or a counterexample.

$$\text{If } A \cap B^c = \emptyset, \text{ then } A \subseteq B.$$

8. Let A and B be subsets of some universal set U. Is the following proposition true or false? Justify your conclusion with a proof or a counterexample.

$$\text{The sets } A \cap B \text{ and } A - B \text{ are disjoint.}$$

9. Let A, B, and C be subsets of a universal set U. Prove each of the following:

(a) If $A \subseteq B$, then $A \cap C \subseteq B \cap C$.

(b) If $A \subseteq B$, then $A \cup C \subseteq B \cup C$.

10. Let A, B, and C be subsets of a universal set U. Are the following propositions true or false? Justify your conclusions.

(a) If $A \cap C \subseteq B \cap C$, then $A \subseteq B$.

(b) If $A \cup C \subseteq B \cup C$, then $A \subseteq B$.

11. Let a and b be integers. Is the following proposition true or false? Justify your conclusion.

$$\text{If } a \mid b \text{ , then for all } x \in \mathbb{Z} \text{ with } x \neq 0, \ ax \text{ divides } bx.$$

12. One of the most famous unsolved problems in mathematics is a conjecture made by Christian Goldbach in a letter to Leonhard Euler in 1742. The conjecture made in this letter is now known as **Goldbach's Conjecture**. The conjecture is

> *Every even integer greater than 2 can be expressed as the sum of two (not necessarily distinct) prime numbers.*

As of January 1, 2002, it is not known if this conjecture is true or false, although most mathematicians believe it to be true.

    (a) Describe one way to prove that Goldbach's Conjecture is false.

    (b) Prove the following:

> If there exists an odd integer greater than 5 that is not the sum of three prime numbers, then Goldbach's Conjecture is false.

4.3   Properties of Set Operations

Preview Activity 1 (Working with Venn Diagrams).
Let A and B be subsets of some universal set U.

1. Draw two general Venn diagrams for the sets A and B. On one, shade the region that represents $(A \cup B)^c$, and on the other, shade the region that represents $A^c \cap B^c$. Carefully explain how you determined these regions.

2. Based on the Venn diagrams in Part (1), what appears to be the relationship between the sets $(A \cup B)^c$ and $A^c \cap B^c$?

3. Draw two general Venn diagrams for the sets A and B. On one, shade the region that represents $(A \cap B)^c$, and on the other, shade the region that represents $A^c \cup B^c$. Carefully explain how you determined these regions.

4. Based on the Venn diagrams in Part (3), what appears to be the relationship between the sets $(A \cap B)^c$ and $A^c \cup B^c$?

Preview Activity 2 (Working with Definitions).

Some of the properties of set operations are closely related to some of the logical equivalencies we studied in Section 2.3. This is due to the fact that set intersection is defined using a conjunction (and), and set union is defined using a disjunction (or). For example, consider one of De Morgan's Laws for statements P and Q.

$$\neg (P \vee Q) \equiv \neg P \wedge \neg Q.$$

Now, if A and B are subsets of some universal set U, then an element x is in $A \cup B$ if and only if $x \in A$ or $x \in B$.

1. Carefully explain what it means to say that an element x is not in $A \cup B$.

2. What does it mean to say that an element x is in A^c? What does it mean to say that an element x is in B^c?

3. Carefully explain what it means to say that an element x is in $A^c \cap B^c$.

4. Compare your response in Part (1) to your response in Part (3). Are they equivalent? Explain.

5. How do you think the sets $(A \cup B)^c$ and $A^c \cap B^c$ are related? Is this consistent with the Venn diagrams from Preview Activity 1?

Preview Activity 3 (Proving that Statements are Equivalent).

1. Let X, Y, and Z be statements. Complete a truth table for
$[(X \to Y) \wedge (Y \to Z)] \to (X \to Z)$.

2. Assume that P, Q, and R are statements and that we have proven that the following conditional statements are true:

   - If P then Q. $(P \to Q)$
   - If Q then R. $(Q \to R)$
   - If R then P. $(R \to P)$

   Explain why each of the following statements is true.

   (a) P if and only if Q. $(P \leftrightarrow Q)$.

(b) Q if and only if R. $(Q \leftrightarrow R)$.

(c) R if and only if P. $(R \leftrightarrow P)$.

Remember that $X \leftrightarrow Y$ is logically equivalent to $(X \to Y) \wedge (Y \to X)$.

This section contains many results concerning the properties of the set operations. Many of these results may seem quite obvious, and we have already proven some of the results. Others will be proven in this section or in the exercises. The primary purpose of this section is to have in one place many of the properties of set operations that we may use in later proofs. These results are part of what is known as the **algebra of sets** or as **set theory**.

Theorem 4.14 (Algebra of Set Operations). *Let A, B, and C be subsets of some universal set U. Then all of the following equalities hold.*

1. *Properties of the Empty Set and the Universal Set.*
$$A \cap \emptyset = \emptyset \qquad\qquad\qquad A \cup \emptyset = A$$
$$A \cap U = A \qquad\qquad\qquad A \cup U = U$$

2. *Idempotent Laws*
$$A \cap A = A \qquad\qquad\qquad A \cup A = A$$

3. *Commutative Laws*
$$A \cap B = B \cap A \qquad\qquad\qquad A \cup B = B \cup A$$

4. *Associative Laws*
$$(A \cap B) \cap C = A \cap (B \cap C) \qquad (A \cup B) \cup C = A \cup (B \cup C)$$

5. *Distributive Laws*
$$A \cap (B \cup C) = (A \cap B) \cup (A \cap C)$$
$$A \cup (B \cap C) = (A \cup B) \cap (A \cup C)$$

Many of these results should be intuitively obvious. In order to be complete with the development of set theory, we should prove all of them. We choose to prove only some of them.

In Section 4.2, we learned that we can prove that two sets are equal by proving that each one is a subset of the other one. However, we also know that if S and T are sets, then

$$S = T \text{ if and only if } (\forall x)\,(x \in S \text{ if and only if } x \in T).$$

We can use this to prove that two sets are equal by choosing an element from one set and chasing the element to the other set through a sequence of "if and only if" statements. We now use this idea to prove one of the commutative laws.

Proof of One of the Commutative Laws in Theorem 4.14.

Proof. We will prove that $A \cap B = B \cap A$. Let $x \in A \cap B$. Then

$$x \in A \cap B \text{ if and only if } x \in A \text{ and } x \in B. \tag{1}$$

However, we know that if P and Q are statements, then $P \wedge Q$ is logically equivalent to $Q \wedge P$. Consequently, we can conclude that

$$x \in A \text{ and } x \in B \text{ if and only if } x \in B \text{ and } x \in A. \tag{2}$$

Now, we know that

$$x \in B \text{ and } x \in A \text{ if and only if } x \in B \cap A. \tag{3}$$

This means that we can use (1), (2), and (3) to conclude that

$$x \in A \cap B \text{ if and only if } x \in B \cap A,$$

and hence, we have proven that $A \cap B = B \cap A$. ∎

Activity 4.15 (Working with Venn Diagrams for Three Sets).
We can use Venn diagrams to explore the more complicated properties in Theorem 4.14, such as the associative and distributive laws. To that end, let A, B, and C be subsets of some universal set U.

1. Draw two general Venn diagrams for the sets A, B, and C. On one, shade the region that represents $A \cup (B \cap C)$, and on the other, shade the region that represents $(A \cup B) \cap (A \cup C)$. Carefully explain how you determined these regions.

2. Based on the Venn diagrams in Part (1), what appears to be the relationship between the sets $A \cup (B \cap C)$ and $(A \cup B) \cap (A \cup C)$?

Proof of One of the Distributive Laws in Theorem 4.14.

We will now prove the distributive law explored in Activity 4.15. Notice that we will prove two subset relations, and that for each subset relation, we will begin by choosing an arbitrary element from a set. Also notice how nicely a proof dealing with the union of two sets can be broken into cases.

Proof. Let A, B, and C be subsets of some universal set U. We will prove that $A \cup (B \cap C) = (A \cup B) \cap (A \cup C)$ by proving that each set is a subset of the other set.

We will first prove that $A \cup (B \cap C) \subseteq (A \cup B) \cap (A \cup C)$. We let $x \in A \cup (B \cap C)$. Then $x \in A$ or $x \in B \cap C$.

Case 1:   If $x \in A$, then $x \in A \cup B$ and $x \in A \cup C$. This means that $x \in (A \cup B) \cap (A \cup C)$.

Case 2:   If $x \in B \cap C$, then $x \in B$ and $x \in C$. But, $x \in B$ implies that $x \in A \cup B$, and $x \in C$ implies that $x \in A \cup C$. Thus, $x \in (A \cup B) \cap (A \cup C)$.

So, in both cases, we see that $x \in (A \cup B) \cap (A \cup C)$, and this proves that $A \cup (B \cap C) \subseteq (A \cup B) \cap (A \cup C)$.

We next prove that $(A \cup B) \cap (A \cup C) \subseteq A \cup (B \cap C)$. So let $y \in (A \cup B) \cap (A \cup C)$. Then, $y \in A \cup B$ and $y \in A \cup C$. We must prove that $y \in A \cup (B \cap C)$. We will consider the following two cases:

Case 1: $y \in A$. In this case, $y \in A \cup (B \cap C)$.

Case 2: $y \notin A$. It has been established that $y \in A \cup B$ and $y \in A \cup C$. Since $y \notin A$ and $y \in A \cup B$, y must be an element of B. Similarly, since $y \notin A$ and $y \in A \cup C$, y must be an element of C. Thus, $y \in B \cap C$ and hence, $y \in A \cup (B \cap C)$.

In both cases, we have proven that $y \in A \cup (B \cap C)$. This proves that $(A \cup B) \cap (A \cup C) \subseteq A \cup (B \cap C)$.

The two subset relations establish the equality of the two sets. Thus, $A \cup (B \cap C) = (A \cup B) \cap (A \cup C)$.   ∎

Activity 4.16 (Comparison to Properties of the Real Numbers).

The following are some of the basic properties of addition and multiplication of real numbers.

| Commutative Laws: | $a + b = b + a$, for all $a, b \in \mathbb{R}$. |
| | $a \cdot b = b \cdot a$, for all $a, b \in \mathbb{R}$. |

Commutative Laws: $a + b = b + a$, for all $a, b \in \mathbb{R}$.
$a \cdot b = b \cdot a$, for all $a, b \in \mathbb{R}$.

Associative Laws: $(a + b) + c = a + (b + c)$, for all $a, b, c \in \mathbb{R}$.
$(a \cdot b) \cdot c = a \cdot (b \cdot c)$, for all $a, b, c \in \mathbb{R}$.

Distributive Law: $a \cdot (b + c) = a \cdot b + a \cdot c$, for all $a, b, c \in \mathbb{R}$.

Additive Identity: For all $a \in \mathbb{R}$, $a + 0 = a = 0 + a$.

Multiplicative Identity: For all $a \in \mathbb{R}$, $a \cdot 1 = a = 1 \cdot a$.

Additive Inverses: For all $a \in \mathbb{R}$, $a + (-a) = 0 = (-a) + a$.

Multiplicative Inverses: For all $a \in \mathbb{R}$ with $a \neq 0$,
$a \cdot a^{-1} = 1 = a^{-1} \cdot a$.

Discuss the similarities and differences among the properties of addition and multiplication of real numbers and the properties of union and intersection of sets.

Theorem 4.17. *Let A, B, and C be subsets of some universal set U. Then:*

- $A \cap B \subseteq A$ *and* $A \subseteq A \cup B$.

- *If* $A \subseteq B$, *then* $A \cap C \subseteq B \cap C$ *and* $A \cup C \subseteq B \cup C$.

Proof. The first part of this theorem was included in Exercise (6) from Section 4.2. The second part of the theorem was Exercise (9) from Section 4.2. ∎

The three main set operations are union, intersection, and complementation. Theorems 4.14 and 4.17 deal with properties of unions and intersections. The next theorem states some basic properties of complements and the important relations dealing with complements of unions and complements of intersections. Two relationships in the next theorem are known as **De Morgan's Laws** for sets and are closely related to De Morgan's Laws for statements.

Theorem 4.18. *Let A and B be subsets of some universal set U. Then the following are true:*

Basic Properties	$(A^c)^c = A$
	$A - B = A \cap B^c$

Empty Set and Universal Set	$A - \emptyset = A$ *and* $A - U = \emptyset$
	$\emptyset^c = U$ *and* $U^c = \emptyset$

De Morgan's Laws	$(A \cap B)^c = A^c \cup B^c$
	$(A \cup B)^c = A^c \cap B^c$

Subsets and Complements	$A \subseteq B$ *if and only if* $B^c \subseteq A^c$.

Proof. We will only prove one of De Morgan's Laws. The proofs of the other parst are left as exercises. Let A and B be subsets of some universal set U. We will prove that $(A \cup B)^c = A^c \cap B^c$ by proving that an element is in $(A \cup B)^c$ if and only if it is in $A^c \cap B^c$. So, let x be in the universal set U. Then

$$x \in (A \cup B)^c \text{ if and only if } x \notin A \cup B, \tag{1}$$

and

$$x \notin A \cup B \text{ if and only if } x \notin A \text{ and } x \notin B. \tag{2}$$

Combining (1) and (2), we see that

$$x \in (A \cup B)^c \text{ if and only if } x \notin A \text{ and } x \notin B. \tag{3}$$

In addition, we know that

$$x \notin A \text{ and } x \notin B \text{ if and only if } x \in A^c \text{ and } x \in B^c, \tag{4}$$

and this is true if and only if $x \in A^c \cap B^c$. So we can use (3) and (4) to conclude that

$$x \in (A \cup B)^c \text{ if and only if } x \in A^c \cap B^c,$$

and hence that $(A \cup B)^c = A^c \cap B^c$. ∎

Activity 4.19 (Using the Algebra of Sets).

1. Draw two general Venn diagrams for the sets A, B, and C. On one, shade the region that represents $(A \cup B) - C$, and on the other, shade the region that represents $(A - C) \cup (B - C)$. Carefully explain how you determined these regions.

2. It is possible to prove the relationship suggested in Part (1) by proving that each set is a subset of the other set. However, Theorems 4.14, 4.17, and 4.18 establish some of the basic results of set theory. These results can be used to prove other results about set operations. For example, we can use one of the basic properties in Theorem 4.18 to write

$$(A \cup B) - C = (A \cup B) \cap C^c.$$

Then we can use one of the commutative properties to write

$$(A \cup B) \cap C^c = C^c \cap (A \cup B).$$

We can then conclude that

$$(A \cup B) - C = C^c \cap (A \cup B).$$

Now, use a distributive property to rewrite the right side of the last equation. Then use a commutative property and Theorem 4.14 to prove that

$$(A \cup B) - C = (A - C) \cup (B - C).$$

Proving that a List of Statements Are Equivalent

When we have a list of three statements P, Q, and R such that each statement in the list is equivalent to the other two statements in the list, we say that the three statements are equivalent. This means that each of the statements in the list implies each of the other statements in the list.

The purpose of Preview Activity 3 was to provide one way to prove that three (or more) statements are equivalent. The basic idea is to prove a sequence of conditional statements so that there is an unbroken chain of conditional statements from each statement to every other statement. This method of proof will be used in Theorem 4.20.

Theorem 4.20. *Let A and B be subsets of some universal set U. The following are equivalent:*

1. $A \subseteq B$ *2.* $A \cap B^c = \emptyset$ *3.* $A^c \cup B = U$

Proof. To prove that these are equivalent conditions, we will prove that (1) implies (2), that (2) implies (3), and that (3) implies (1).

Let A and B be subsets of some universal set U. We have proven that (1) implies (2) in Proposition 4.13.

To prove that (2) implies (3), we will assume that $A \cap B^c = \emptyset$, and use the fact that $\emptyset^c = U$. We then see that

$$(A \cap B^c)^c = \emptyset^c.$$

Then, using one of De Morgan's Laws, we obtain

$$A^c \cup (B^c)^c = U$$
$$A^c \cup B = U.$$

This completes the proof that (2) implies (3).

We now need to prove that (3) implies (1). We assume that $A^c \cup B = U$ and will prove that $A \subseteq B$ by proving that every element of A must be in B.

So let $x \in A$. Then we know that $x \notin A^c$. However, $x \in U$ and since $A^c \cup B = U$, we can conclude that $x \in A^c \cup B$. Since $x \notin A^c$, we conclude that $x \in B$. This proves that $A \subseteq B$ and hence that (3) implies (1).

Since we have now proven that (1) implies (2), that (2) implies (3), and that (3) implies (1), we have proven that the three conditions are equivalent.

$\blacksquare$

Exercises 4.3

1. Let A be a subset of some universal set U. Prove each of the following (from Theorem 4.18):

   (a) $(A^c)^c = A.$ (c) $\emptyset^c = U.$

   (b) $A - \emptyset = A.$ (d) $U^c = \emptyset.$

2. Proposition 4.8 states the following:

   Let A and B be subsets of some universal set U. If $A \subseteq B$, then $B^c \subseteq A^c$.

   This was proven in Activity 4.7. Now, prove the following proposition:

   Let A and B be subsets of some universal set U. Then $A \subseteq B$ if and only if $B^c \subseteq A^c$.

3. Let A, B, and C be subsets of some universal set U. In this section, we proved one of the distributive laws. Prove the other one. That is, prove that

$$A \cap (B \cup C) = (A \cap B) \cup (A \cap C).$$

4. Let A, B, and C be subsets of some universal set U. In this section, we proved one of De Morgan's Laws. Prove the other one. That is, prove that

$$(A \cap B)^c = A^c \cup B^c.$$

5. Let A, B, and C be subsets of some universal set U.

   (a) Draw two general Venn diagrams for the sets A, B, and C. On one, shade the region that represents $A - (B \cup C)$, and on the other, shade the region that represents $(A - B) \cap (A - C)$. Based on the Venn diagrams, make a conjecture about the relationship between the sets $A - (B \cup C)$ and $(A - B) \cap (A - C)$.

   (b) Use the choose an element method to prove the conjecture from Exercise (5a).

   (c) Use the algebra of dets to prove the conjecture from Exercise (5a).

6. Let A, B, and C be subsets of some universal set U.

   (a) Draw two general Venn diagrams for the sets A, B, and C. On one, shade the region that represents $A - (B \cap C)$, and on the other, shade the region that represents $(A - B) \cup (A - C)$. Based on the Venn diagrams, make a conjecture about the relationship between the sets $A - (B \cap C)$ and $(A - B) \cup (A - C)$.

   (b) Use the choose an element method to prove the conjecture from Exercise (6a).

   (c) Use the algebra of sets to prove the conjecture from Exercise (6a).

7. Let A, B, and C be subsets of some universal set U.

   (a) Draw two general Venn diagrams for the sets A, B, and C. On one, shade the region that represents $A - (B - C)$, and on the other, shade the region that represents $(A - B) - C$. Based on the Venn diagrams, make a conjecture about the relationship between the sets $A - (B - C)$ and $(A - B) - C$.

   (b) Prove the conjecture from Exercise (7a).

8. Prove or disprove (with a counterexample) the following proposition:

If A, B, and C are subsets of some universal set U, then

$$(A \cap B) - C = (A - C) \cap (B - C).$$

9. Let A and B be subsets of some universal set U.

   (a) Prove that A and $B - A$ are disjoint sets.
   (b) Prove that $A \cup B = A \cup (B - A)$.

10. Let A and B be subsets of some universal set U.

   (a) Prove that $A - B$ and $A \cap B$ are disjoint sets.
   (b) Prove that $A = (A - B) \cup (A \cap B)$.

4.4   Cartesian Products

Preview Activity 1 (An Equation with Two Variables).
   In Section 2.1, we introduced the concept of the **truth set of a predicate with one variable.** This was defined to be the set of all elements in the universal set that can be substituted for the variable to make the predicate a true statement.
   Assume that x and y represent real numbers. Then the equation

$$2x + 3y = 12$$

is a predicate with two variables. An element of the truth set of this predicate (also called a solution of the equation) is an ordered pair (a, b) of real numbers so that when a is substituted for x and b is substituted for y, the predicate becomes a true statement (a true equation in this case).
Important Note: The order of the of the two numbers in the ordered pair is very important. We are using the convention that the first number is to be substituted for x and the second number is to be substituted for y. With this convention, $(3, 2)$ is a solution of the equation $2x + 3y = 12$, but $(2, 3)$ is not a solution of this equation.

   1. List five different elements of the truth set (often called the solution set) of this predicate with two variables.

   2. Sketch the graph of the equation $2x + 3y = 12$ in the xy-coordinate plane. What does the graph of the equation $2x + 3y = 12$ show?

3. Write a description of the solution set S of the equation $2x + 3y = 12$ using set builder notation.

Preview Activity 2 (The Cartesian Product of Two Sets).

Definition. If A and B are sets, then the **Cartesian product**, $A \times B$, of A and B is the set of all ordered pairs (x, y) where $x \in A$ and $y \in B$. Using set builder notation, we can write

$$A \times B = \{(x, y) \mid x \in A \text{ and } y \in B\}.$$

Let $A = \{1, 2, 3\}$ and $B = \{a, b\}$.

1. List all the elements of $A \times B$ using the roster method.

2. Is the ordered pair $(3, a)$ in the Cartesian product $A \times B$? Explain.

3. Is the ordered pair $(3, a)$ in the Cartesian product $A \times A$? Explain.

4. Is the ordered pair $(3, 1)$ in the Cartesian product $A \times A$? Explain.

5. List all of the elements of the set $A \times A$ using the roster method.

6. For any sets C and D, very carefully explain what it means to say that the ordered pair (x, y) is not in the Cartesian product $C \times D$.

In this section, we will study a new set operation that gives a way of combining elements from two given sets to form ordered pairs. In the Preview Activities, we worked with ordered pairs without providing a definition of an ordered pair. We instead relied on your previous work with ordered pairs, primarily from graphing equations with two variables.

Definition. An **ordered pair** is a single pair of objects, denoted by (a, b) , with an implied order. This means that for two ordered pairs to be equal, they must contain exactly the same objects in the same order. That is,

$$(a, b) = (c, d) \text{ if and only if } a = c \text{ and } b = d.$$

The objects in the ordered pair are called the **coordinates** of the ordered pair. In the ordered pair (a, b), a is the **first coordinate** and b is the **second coordinate**.

This may seem like a lengthy definition for something that we are familiar with, but in some areas of mathematics, an even more formal and precise definition of ordered pair is needed. We will explore this definition in Activity 4.24.

In Preview Activity 2, we formed the set of all possible ordered pairs such that the first coordinate was from a given set A and the second coordinate was from a given set B. That is, we formed the Cartesian product of A and B.

Definition. The **Cartesian product** of two sets A and B, written $A \times B$, is the set of all ordered pairs (a, b) where $a \in A$ and $b \in B$. That is,
$$A \times B = \{(a, b) \mid a \in A \text{ and } b \in B\}.$$
We frequently read $A \times B$ as "A cross B."

In the case where the two sets are the same, we will write A^2 for $A \times A$. That is,
$$A^2 = A \times A = \{(a, b) \mid a \in A \text{ and } b \in A\}$$

An important thing to remember is that a Cartesian product is a set. As a set, it consists of a collection of elements. In this case, the elements of a Cartesian product are ordered pairs. We should think of an ordered pair as a single object that consists of two other objects in a specified order. For example,

- If $a \neq 1$, then the ordered pair $(1, a)$ is not equal to the ordered pair $(a, 1)$. That is, $(1, a) \neq (a, 1)$.

- If $A = \{1, 2, 3\}$ and $B = \{a, b\}$, then the ordered pair $(3, a)$ is an element of the set $A \times B$. That is, $(3, a) \in A \times B$.

- If $A = \{1, 2, 3\}$ and $B = \{a, b\}$, then the ordered pair $(5, a)$ is not an element of the set $A \times B$. That is, $(5, a) \notin A \times B$.

Activity 4.21 (Relationships between Cartesian Products).

Let $A = \{1, 2, 3\}$, $T = \{1, 2\}$, $B = \{a, b\}$, and $C = \{a, c\}$. We can then form new sets from all of the set operations we have studied. For example, $B \cap C = \{a\}$, and so

$$A \times (B \cap C) = \{(1, a), (2, a), (3, a)\}.$$

1. Use the roster method to list all of the elements (ordered pairs) in each of the following sets:

   (a) $A \times B$

   (b) $T \times B$

   (c) $A \times C$

   (d) $A \times (B \cap C)$

   (e) $(A \times B) \cap (A \times C)$

   (f) $A \times (B \cup C)$

   (g) $(A \times B) \cup (A \times C)$

   (h) $A \times (B - C)$

   (i) $(A \times B) - (A \times C)$

   (j) $B \times A$

2. List all the relationships between the sets in Part (1) that you observe.

The Cartesian Plane

In Preview Activity 1, we sketched the graph of the equation $2x + 3y = 12$ in the xy-plane. This xy-plane, with which you are familiar, is a representation of the set $\mathbb{R} \times \mathbb{R}$ or $\mathbb{R}^2$. This plane is called the **Cartesian plane.**

The basic idea is that each ordered pair of real numbers corresponds to a point in the plane, and each point in the plane corresponds to an ordered pair of real numbers. This geometric representation of $\mathbb{R}^2$ is an extension of the geometric representation of $\mathbb{R}$ as a straight line whose points correspond to real numbers.

Since the Cartesian product $\mathbb{R}^2$ corresponds to the Cartesian plane, the Cartesian product of two subsets of $\mathbb{R}$ corresponds to a subset of the Cartesian plane. For example, if A is the interval $[1, 3]$, and B is the interval $[2, 5]$, then

$$A \times B = \{(x, y) \in \mathbb{R}^2 \mid 1 \leq x \leq 3 \text{ and } 2 \leq y \leq 5\}.$$

A graph of the set $A \times B$ can then be drawn in the Cartesian plane as shown in Figure 4.6.

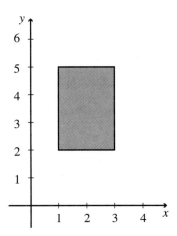

Figure 4.6: Cartesian Product $A \times B$

This illustrates the fact that the graph of a Cartesian product of two intervals of finite length in $\mathbb{R}$ corresponds to the interior of a rectangle and possibly some or all of its boundary. The solid line for the boundary in Figure 4.6 indicates that the boundary is included. In this case, the Cartesian product contained all of the boundary of the rectangle. When the graph does not contain a portion of the boundary, we usually draw that portion of the boundary with a dotted line.

A Caution about Notation

The standard notation for an open interval in $\mathbb{R}$ is the same as the notation for an ordered pair, which is an element of $\mathbb{R} \times \mathbb{R}$. We need to use the context in which the notation is used to determine which interpretation is intended. For example,

- If we write $\left(\sqrt{2}, 7\right) \in \mathbb{R} \times \mathbb{R}$, then we are using $\left(\sqrt{2}, 7\right)$ to represent an ordered pair of real numbers.

- If we write $(1, 2) \times \{4\}$, then we are interpreting $(1, 2)$ as an open interval. We could write

$$(1, 2) \times \{4\} = \{\, (x, 4) \mid 1 < x < 2 \,\}.$$

Activity 4.22 (Cartesian Products of Intervals). For this activity, we will use the following intervals that are subsets of $\mathbb{R}$.

$$A = [0,2] \quad T = (1,2) \quad B = [2,4) \quad C = (3,5]$$

1. Draw a graph of each of the following subsets of the Cartesian plane and write each subset using set builder notation.

   (a) $A \times B$

   (b) $T \times B$

   (c) $A \times C$

   (d) $A \times (B \cap C)$

   (e) $(A \times B) \cap (A \times C)$

   (f) $A \times (B \cup C)$

   (g) $(A \times B) \cup (A \times C)$

   (h) $A \times (B - C)$

   (i) $(A \times B) - (A \times C)$

   (j) $B \times A$

2. List all the relationships between the sets in Part (1) that you observe.

One purpose of the work in Activities 4.21 and 4.22 was to indicate the plausibility of many of the results contained in the next theorem.

Theorem 4.23. *Let A, B, and C be sets. Then*

1. $A \times (B \cap C) = (A \times B) \cap (A \times C)$.

2. $A \times (B \cup C) = (A \times B) \cup (A \times C)$.

3. $(A \cap B) \times C = (A \times C) \cap (B \times C)$.

4. $(A \cup B) \times C = (A \times C) \cup (B \times C)$.

5. $A \times (B - C) = (A \times B) - (A \times C)$.

6. $(A - B) \times C = (A \times C) - (B \times C)$.

7. *If $T \subseteq A$, then $T \times B \subseteq A \times B$.*

8. *If $Y \subseteq B$, then $A \times Y \subseteq A \times B$.*

We will not concern ourselves with proofs of all these results, but rather, we will prove Part (2) of Theorem 4.23 and leave some of the rest to the exercises. In constructing these proofs, we need to keep in mind that Cartesian products are sets, and so we follow many of the same principles to prove set relationships that we introduced in Sections 4.2 and 4.3. The other thing to remember is that the elements of a Cartesian product are ordered pairs.

So when we start a proof of a result such as Part (2) of Theorem 4.23, the primary goal is to prove that the two sets are equal. Here we will accomplish this by proving that each one is a subset of the other one. For example, if we want to prove that $A \times (B \cup C) \subseteq (A \times B) \cup (A \times C)$, we can start by choosing an arbitrary element of $A \times (B \cup C)$. The goal is then to show that this element must be in $(A \times B) \cup (A \times C)$. When we start by choosing an arbitrary element of $A \times (B \cup C)$, we could give that element a name. For example, we could start by letting

$$u \text{ be an element of } A \times (B \cup C).$$

We can then use the definition of ordered pair to conclude that there exists $x \in A$ and there exists $y \in B \cup C$ such that $u = (x, y)$. Following is a partial know-show table for the proof of $A \times (B \cup C) \subseteq (A \times B) \cup (A \times C)$.

Step	Know	Reason
P	Let $u \in A \times (B \cup C)$.	Choose an arbitrary element.
$P1$	$(\exists x \in A)$ $(\exists y \in B \cup C)\,[u = (x, y)]$.	Definition of $A \times (B \cup C)$
$\vdots$	$\vdots$	$\vdots$
$Q2$	$u \in A \times B$ or $u \in A \times C$.	
$Q1$	$u \in (A \times B) \cup (A \times C)$.	Definition of "union"
Q	$A \times (B \cup C) \subseteq (A \times B) \cup (A \times C)$.	Definition of "subset"
Step	**Show**	**Reason**

Look at Steps $P1$ and $Q2$. In Step $P1$, we know that $u = (x, y)$ and that $x \in A$ and $y \in B \cup C$. This is close to being Step $Q2$. For this step, in both cases, we need x to be in A. In one case, we need y to be in B, and in the other, we need y to be in C. But this is what we have in Step $P1$ since we know that $y \in B \cup C$. Since we are working with a union, we might want to use cases. Following is a proof of Part (2) of Theorem 4.23.

Theorem 4.23 (Part (2)). *Let A, B, and C be sets. Then*

$$A \times (B \cup C) = (A \times B) \cup (A \times C).$$

Proof. Let A, B, and C be sets. We will prove that $A \times (B \cup C)$ is equal to $(A \times B) \cup (A \times C)$ by proving that each set is a subset of the other set.

To prove that $A \times (B \cup C) \subseteq (A \times B) \cup (A \times C)$, we let $u \in A \times (B \cup C)$. Then, there exists $x \in A$ and there exists $y \in B \cup C$ such that $u = (x, y)$. Since $y \in B \cup C$, we know that $y \in B$ or $y \in C$.

In the case where $y \in B$, we have $u = (x, y)$, where $x \in A$ and $y \in B$. So in this case, $u \in A \times B$, and hence $u \in (A \times B) \cup (A \times C)$. Similarly, in the case where $y \in C$, we have $u = (x, y)$ where $x \in A$ and $y \in C$. So in this case, $u \in A \times C$, and hence $u \in (A \times B) \cup (A \times C)$.

In both cases, $u \in (A \times B) \cup (A \times C)$. Hence, we may conclude that if $u \in A \times (B \cup C)$, then $u \in (A \times B) \cup (A \times C)$, and this proves that

$$A \times (B \cup C) \subseteq (A \times B) \cup (A \times C). \tag{1}$$

We must now prove that $(A \times B) \cup (A \times C) \subseteq A \times (B \cup C)$. So let $v \in (A \times B) \cup (A \times C)$. Then $v \in (A \times B)$ or $v \in (A \times C)$.

In the case where $v \in (A \times B)$, we know that there exists $s \in A$ and there exists $t \in B$ such that $v = (s, t)$. But since $t \in B$, we know that $t \in B \cup C$, and hence $v \in A \times (B \cup C)$. Similarly, in the case where $v \in (A \times C)$, we know that there exists $s \in A$ and there exists $t \in C$ such that $v = (s, t)$. But because $t \in C$, we can conclude that $t \in B \cup C$, and hence $v \in A \times (B \cup C)$.

In both cases, $v \in A \times (B \cup C)$. Hence, we may conclude that if $v \in (A \times B) \cup (A \times C)$, then $v \in A \times (B \cup C)$, and this proves that

$$(A \times B) \cup (A \times C) \subseteq A \times (B \cup C). \tag{2}$$

The relationships in (1) and (2) prove that $A \times (B \cup C) = (A \times B) \cup (A \times C)$.

∎

Activity 4.24 (A Set Theoretic Definition of an Ordered Pair). In most of the mathematics that we will study, the notion of an ordered pair introduced at the beginning of this section will suffice. However, if we are interested in a formal development of the Cartesian product of two sets, we need a more precise definition of ordered pair. Following is one way to do this in terms of sets. This definition is credited to Kazimierz Kuratowski (1896 - 1980).

Let x be an element of the set A, and let y be an element of the set B. The **ordered pair** (x, y) is defined to be the set $\{\{x\}, \{x, y\}\}$. That is,

$$(x, y) = \{\{x\}, \{x, y\}\}.$$

1. Explain how this definition allows us to distinguish between the ordered pairs $(3, 5)$ and $(5, 3)$.

2. Let A and B be sets and let $a, c \in A$ and $b, d \in B$. Use this definition of an ordered pair and the concept of set equality to prove that $(a, b) = (c, d)$ if and only if $a = c$ and $b = d$.

An **ordered triple** can be thought of as a single triple of objects, denoted by (a, b, c), with an implied order. This means that in order for two ordered triples to be equal, they must contain exactly the same objects in the same order. That is, $(a, b, c) = (p, q, r)$ if and only if $a = p$, $b = q$ and $c = r$.

3. Let A, B, and C be sets, and let $x \in A, y \in B,$ and $z \in C$. Write a set theoretic definition of the ordered triple (x, y, z) similar to the set theoretic definition of ordered pair.

Exercises 4.4

1. Let $A = \{1, 2\}$, $B = \{a, b, c, d\}$, and $C = \{1, a, b\}$. Use the roster method to list all of the elements of each of the following sets:

   (a) $A \times B$ (e) $A \times (B \cap C)$
   (b) $B \times A$ (f) $(A \times B) \cap (A \times C)$
   (c) $A \times C$ (g) $A \times \emptyset$
   (d) A^2 (h) $B \times \{2\}$

2. Sketch a graph of each of the following Cartesian products in the Cartesian plane.

   (a) $[0, 2] \times [1, 3]$ (e) $\mathbb{R} \times (2, 4)$
   (b) $(0, 2) \times (1, 3]$ (f) $(2, 4) \times \mathbb{R}$
   (c) $[2, 3] \times \{1\}$ (g) $\mathbb{R} \times \{-1\}$
   (d) $\{1\} \times [2, 3]$ (h) $\{-1\} \times [1, +\infty)$

3. Prove Theorem 4.23, Part (1): $A \times (B \cap C) = (A \times B) \cap (A \times C)$.

4. Prove Theorem 4.23, Part (4): $(A \cup B) \times C = (A \times C) \cup (B \times C)$.

5. Prove Theorem 4.23, Part (5): $A \times (B - C) = (A \times B) - (A \times C)$.

6. Prove Theorem 4.23, Part (7): If $T \subseteq A$, then $T \times B \subseteq A \times B$.

7. Let $A = \{1\}, B = \{2\},$ and $C = \{3\}$.

(a) Explain why $A \times B \neq B \times A$.

(b) Explain why $(A \times B) \times C \neq A \times (B \times C)$.

8. Let A and B be nonempty sets. Prove that $A \times B = B \times A$ if and only if $A = B$.

9. Is the following proposition true or false? Justify your conclusion.

Let A, B, and C be sets with $A \neq \emptyset$. If $A \times B = A \times C$, then $B = C$.

Explain where the assumption that $A \neq \emptyset$ is needed.

Chapter 5

Mathematical Induction

5.1   The Principle of Mathematical Induction

Preview Activity 1 (A Conjecture about Divisibility).

In Section 2.1, we defined the **truth set** of a predicate $P(x)$ to be the collection of objects in the universal set that make the predicate a true statement when substituted for x.

Assume the universal set is $\mathbb{N}$, and let n be a natural number. Consider the following predicate:

$$4 \text{ divides } (5^n - 1).$$

1. Does this predicate become a true statement when $n = 1$? That is, is 1 in the truth set of this predicate?

2. Does this predicate become a true statement when $n = 2$? That is, is 2 in the truth set of this predicate?

3. Choose at least four more natural numbers and determine whether the predicate is true or false for each of your choices.

4. Based on this work, do you think the following proposition is true or false? Explain.

For each natural number n, 4 divides $(5^n - 1)$.

Preview Activity 2 (Exploring a Summation).

Let n be a natural number. Let $P(n)$ be the following open sentence:

$$1^2 + 2^2 + \cdots + n^2 = \frac{n(n+1)(2n+1)}{6}. \tag{1}$$

The left side of Equation (1) means to add the squares of the first n natural numbers. So, when $n = 1$, the left side of Equation (1) is 1^2. When $n = 2$, the left side of Equation (1) is $1^2 + 2^2$.

1. Does $P(n)$ become a true statement when $n = 1$? That is, is 1 in the truth set of $P(n)$?

2. Does $P(n)$ become a true statement when $n = 2$? That is, is 2 in the truth set of $P(n)$?

3. Does $P(n)$ become a true statement when $n = 3$? That is, is 3 in the truth set of $P(n)$?

4. Choose at least four more natural numbers and determine whether the open sentence is true or false for each of your choices. A table with the columns n, $1^2 + 2^2 + \cdots + n^2$, and $\dfrac{n(n+1)(2n+1)}{6}$ may help you organize your work.

5. Based on this work, do you think the following proposition is true or false? Explain.

For each natural number n, $1^2 + 2^2 + \cdots + n^2 = \dfrac{n(n+1)(2n+1)}{6}$.

Preview Activity 3 (A Property of the Natural Numbers).

Intuitively, the natural numbers begin with the number 1, and then there is 2, then 3, then 4, and so on. Does this process of "starting with 1" and "adding 1 repeatedly" result in all the natural numbers? We will explore this idea in this activity.

Consider the following property for a set T that is a subset of $\mathbb{Z}$, the set of all integers.

Property I: For every $k \in \mathbb{Z}$, if $k \in T$, then $k + 1 \in T$.

Answer each of the following questions. Do not worry about formal proofs.

1. Does the set $A = \{1, 2, 3, \ldots, 20\}$ satisfy Property I? Explain.

2. Does the set of natural numbers, $\mathbb{N}$, satisfy Property I? Explain.

3. Does the set $B = \{n \in \mathbb{N} \mid n \geq 5\}$ satisfy Property I? Explain.

4. Does the set $S = \{n \in \mathbb{Z} \mid n \geq -3\}$ satisfy Property I? Explain.

5. Does the set $R = \{n \in \mathbb{Z} \mid n \leq 100\}$ satisfy Property I? Explain.

6. Does the set of integers, $\mathbb{Z}$, satisfy Property I? Explain.

7. Now assume that $T \subseteq \mathbb{N}$ and assume that $1 \in T$ and that T satisfies Property I.

   (a) Is $2 \in T$? Explain.   (d) Is $100 \in T$? Explain.
   (b) Is $3 \in T$? Explain.   (e) Do you think that $T = \mathbb{N}$?
   (c) Is $4 \in T$? Explain.   Explain.

Preview Activity 1 and Preview Activity 2 each involved a predicate, $P(n)$, that appeared to be true for all values of n in the set of natural numbers, $\mathbb{N}$. That is, we were able to form conjectures of the form

$$(\forall n \in \mathbb{N})\,(P(n)).$$

In particular, the conjectures were

- For each natural number n, 4 divides $(5^n - 1)$.

- For each natural number n, $1^2 + 2^2 + \cdots + n^2 = \dfrac{n(n+1)(2n+1)}{6}$.

The key to proving statements of this form is suggested in Preview Activity 3. The idea is to prove that if one natural number makes the predicate true, then the next one also makes the predicate true. This is how we handle the phrase "and so on" when dealing with the natural numbers.

Inductive Sets

The examples in Preview Activity 3 dealt with a certain property of numbers. This property is given a name.

> **Definition.** A set T that is a subset of $\mathbb{Z}$ is an **inductive set** provided that if $k \in T$, then $k + 1 \in T$.

In Preview Activity 3, we saw several examples of inductive sets. In addition, the number systems $\mathbb{N}$ and $\mathbb{Z}$ are inductive. What we are trying to do is to somehow distinguish $\mathbb{N}$ from these other inductive sets. The way to do this was suggested in Part (7) of Preview Activity 3. Although we will not prove it, the following statement should seem true.

Statement 1: Let $T \subseteq \mathbb{N}$. If $1 \in T$ and T is inductive, then $T = \mathbb{N}$.

Notice that the integers, $\mathbb{Z}$, and the set $S = \{n \in \mathbb{Z} \mid n \geq -3\}$ both contain 1 and both are inductive, but they both contain numbers other than natural numbers. For example, the following statement is false:

Statement 2: Let $T \subseteq \mathbb{Z}$. If $1 \in T$ and T is inductive, then $T = \mathbb{Z}$.

For example, the set $T = \{x \in \mathbb{Z} \mid x \geq -1\} = \{-1, 0, 1, 2, 3, \ldots\}$ shows that this statement is false.

The Principle of Mathematical Induction

Although we could prove that Statement (2) is false, in this text, we will not prove that Statement (1) is true because we really do not have a formal definition of the natural numbers. However, we should be convinced that Statement (1) is true. We resolve this by making Statement (1) an axiom for the natural numbers so that this becomes one of the defining characteristics of the natural numbers.

> **The Principle of Mathematical Induction**
> If T is a subset of $\mathbb{N}$ such that
>
> 1. $1 \in T$, and
>
> 2. For every $k \in \mathbb{N}$, if $k \in T$, then $(k + 1) \in T$,
>
> then $T = \mathbb{N}$.

Using the Principle of Mathematical Induction

The primary use of the Principle of Mathematical Induction is to prove statements of the form

$$(\forall n \in \mathbb{N}) (P(n)),$$

where $P(n)$ is some predicate. Recall that a universally quantified statement like the one above is true if and only if the truth set T of the predicate $P(n)$ is the set $\mathbb{N}$. So our goal is to prove that $T = \mathbb{N}$, which is the conclusion of the Principle of Mathematical Induction. To verify the hypothesis of the Principle of Mathematical Induction, we must

1. Prove that $1 \in T$. That is, prove that $P(1)$ is true.

2. Prove that if $k \in T$, then $(k+1) \in T$. That is, prove that if $P(k)$ is true, then $P(k+1)$ is true.

The first step is called the **basis step** or the **initial step**, and the second step is called the **inductive step**. This means that a proof by mathematical induction will have the following form:

Procedure for a Proof by Mathematical Induction
To prove: $(\forall n \in \mathbb{N}) (P(n))$

   Basis step: Prove $P(1)$.

   Inductive step: Prove that for each $k \in \mathbb{N}$,
              if $P(k)$ is true, then $P(k+1)$ is true.

We can then conclude that $P(n)$ is true for all $n \in \mathbb{N}$.

Note that in the inductive step, we want to prove that the conditional statement $P(k) \rightarrow P(k+1)$ is true. So we will start the inductive step by assuming that $P(k)$ is true. This assumption is called the **inductive assumption** or the **inductive hypothesis.**

The key to constructing a proof by induction is to discover how $P(k+1)$ is related to $P(k)$ for an arbitrary natural number k. For example, in Preview Activity 2, the predicate, $P(n)$, was

$$1^2 + 2^2 + \cdots + n^2 = \frac{n(n+1)(2n+1)}{6}.$$

Sometimes it helps to look at some specific examples such as $P(2)$ and $P(3)$. The idea is not just to do the computations, but to see how the statements are related. This can sometimes be done by writing the details instead of immediately doing computations.

$$P(2) \quad \text{is} \quad 1^2 + 2^2 = \frac{2 \cdot 3 \cdot 5}{6}.$$

$$P(3) \quad \text{is} \quad 1^2 + 2^2 + 3^2 = \frac{3 \cdot 4 \cdot 7}{6}.$$

In this case, the key is the left side of each equation. The left side of $P(3)$ is obtained from the left side of $P(2)$ by adding one term, which is 3^2. This suggests that we might be able to obtain the equation for $P(3)$ by adding 3^2 to both sides of the equation in $P(2)$. Now for the general case, if $k \in \mathbb{N}$, we look at $P(k+1)$ and compare it to $P(k)$.

$$P(k) \text{ is } 1^2 + 2^2 + \cdots + k^2 = \frac{k(k+1)(2k+1)}{6}.$$

$$P(k+1) \text{ is } 1^2 + 2^2 + \cdots + (k+1)^2 = \frac{(k+1)\left[(k+1)+1\right]\left[2(k+1)+1\right]}{6}.$$

$P(k+1)$ was obtained by substituting $n = k+1$ into the equation for $P(n)$. The key is to look at the left side of the equation for $P(k+1)$ and realize what this notation means. It means that we are adding the squares of the first $(k+1)$ natural numbers. This means that we can write

$$1^2 + 2^2 + \cdots + (k+1)^2 = 1^2 + 2^2 + \cdots + k^2 + (k+1)^2.$$

This shows us that the left side of the equation for $P(k+1)$ is obtained from the left side of the equation for $P(k)$ by adding $(k+1)^2$. This is the motivation for proving the inductive step in the following proof.

Proposition 5.1. *For each natural number n,*

$$1^2 + 2^2 + \cdots + n^2 = \frac{n(n+1)(2n+1)}{6}.$$

Proof. We will use a proof by mathematical induction. For each natural number n, we let

$$P(n) \text{ be } 1^2 + 2^2 + \cdots + n^2 = \frac{n(n+1)(2n+1)}{6}.$$

We first prove that $P(1)$ is true. Notice that $\dfrac{1(1+1)(2 \cdot 1 + 1)}{6} = 1$. This shows that

$$1^2 = \frac{1(1+1)(2 \cdot 1 + 1)}{6}.$$

Hence, $P(1)$ is true.

For the inductive step, we prove that for all $k \in \mathbb{N}$, if $P(k)$ is true, then $P(k+1)$ is true. So let k be a natural number and assume that $P(k)$ is true. That is, assume that

$$1^2 + 2^2 + \cdots + k^2 = \frac{k(k+1)(2k+1)}{6}. \tag{1}$$

The goal now is to prove that $P(k+1)$ is true. That is, it must be proven that

$$1^2 + 2^2 + \cdots + (k+1)^2 = \frac{(k+1)[(k+1)+1][2(k+1)+1]}{6}. \tag{2}$$

To do this, we add $(k+1)^2$ to both sides of Equation (1) and algebraically rewrite the resulting equation. This gives

$$
\begin{aligned}
1^2 + 2^2 + \cdots + k^2 + (k+1)^2 &= \frac{k(k+1)(2k+1)}{6} + (k+1)^2 \\
&= \frac{k(k+1)(2k+1) + 6(k+1)^2}{6} \\
&= \frac{(k+1)[k(2k+1) + 6(k+1)]}{6} \\
&= \frac{(k+1)\left(2k^2 + 7k + 6\right)}{6} \\
&= \frac{(k+1)(k+2)(2k+3)}{6} \\
&= \frac{(k+1)((k+1)+1)(2(k+1)+1)}{6}.
\end{aligned}
$$

Comparing this result to Equation (2), we see that if $P(k)$ is true, then $P(k+1)$ is true. Hence, the inductive step has been established, and by the Principle of Mathematical Induction, we have proven that for each natural number n, $1^2 + 2^2 + \cdots + n^2 = \dfrac{n(n+1)(2n+1)}{6}$. ∎

Writing Guideline

The proof of Proposition 5.1 shows a standard way to write an induction proof. When writing a proof by mathematical induction, you should follow the guideline that we always keep the reader informed. This means that at the beginning of the proof, you should state that a proof by induction will be used. You should then clearly define the predicate $P(n)$ that will be used in the proof.

Summation Notation

The result in Proposition 5.1 could be written using summation notation as follows:

$$\text{For each natural number } n, \sum_{j=1}^{n} j^2 = \frac{n(n+1)(2n+1)}{6}.$$

In this case, we used j for the index for the summation, and the notation

$$\sum_{j=1}^{n} j^2$$

tells us to add all the values of j^2 for j from 1 to n, inclusive. That is,

$$\sum_{j=1}^{n} j^2 = 1^2 + 2^2 + \cdots + n^2.$$

So, in the proof of Proposition 5.1, we would let $P(n)$ be

$$\sum_{j=1}^{n} j^2 = \frac{n(n+1)(2n+1)}{6},$$

and we would use the fact that for each natural number k,

$$\sum_{j=1}^{k+1} j^2 = \left(\sum_{j=1}^{k} j^2 \right) + (k+1)^2.$$

Activity 5.2 (The Importance of the Basis Step).

Usually, most of the work done in constructing a proof by induction is in proving the inductive step. This was certainly the case in Proposition 5.1. However, the basis step is an essential part of the proof. Without it, the proof is incomplete. To see this, let $P(n)$ be

$$1 + 2 + \cdots + n = \frac{n^2 + n + 1}{2}.$$

1. Let $k \in \mathbb{N}$. Complete the following proof that if $P(k)$ is true, then $P(k+1)$ is true.

   Let $k \in \mathbb{N}$. Assume that $P(k)$ is true. That is, assume that

   $$1 + 2 + \cdots + k = \frac{k^2 + k + 1}{2}. \tag{1}$$

   The goal is to prove that $P(k+1)$ is true. That is, we need to prove that

   $$1 + 2 + \cdots + k + (k+1) = \frac{(k+1)^2 + (k+1) + 1}{2}. \tag{2}$$

   To do this, we add $(k+1)$ to both sides of Equation (1). This gives

   $$1 + 2 + \cdots + k + (k+1) = \frac{k^2 + k + 1}{2} + (k+1)$$

   $$= \cdots .$$

2. Is $P(1)$ true? Is $P(2)$ true? What about $P(3)$ and $P(4)$?

Some Comments about Mathematical Induction

1. As Activity 5.2 shows, the basis step is an essential part of a proof by mathematical induction. Activity 5.6 at the end of the exercises for this section will show that the inductive step is also an essential part of a proof by mathematical induction.

2. It is important to remember that the inductive step in an induction proof is a proof of a conditional statement. Although we did not explicitly use the forward-backward process in the inductive step for Proposition 5.1, it was implicitly used in the discussion prior to Proposition 5.1. The key question was, "How does knowing the sum of the first k squares help us find the sum of the first $(k+1)$ squares?"

3. The proof in Activity 5.2 is a legitimate proof of the proposition that if $P(k)$ is true, then $P(k+1)$ is true. This is a true conditional statement. The point is that even though this conditional statement is true, nothing has been proven about the individual statements $P(1)$, $P(2)$, $P(3)$, and so on.

4. When proving the inductive step in a proof by induction, the key question is,

"How does knowing $P(k)$ help us prove $P(k+1)$?"

In Proposition 5.1, we were able to see that the way to answer this question was to add a certain expression to both sides of the equation given in $P(k)$. Sometimes the relationship between $P(k)$ and $P(k+1)$ is not as easy to see. For example, in Preview Activity 1, the proposition was

For each natural number n, 4 divides $(5^n - 1)$.

This means that the predicate, $P(n)$, is "4 divides $(5^n - 1)$." So in the inductive step, we assume $k \in \mathbb{N}$ and that 4 divides $(5^k - 1)$. This means that there exists an integer m such that

$$5^k - 1 = 4m.$$

In the backward process, the goal is to prove that 4 divides $(5^{k+1} - 1)$. This can be accomplished if we can prove that there exists an integer s such that
$$5^{k+1} - 1 = 4s.$$

We now need to see if there is anything in the first equation that can be used in the second equation. The key is to find something in the equation $5^k - 1 = 4m$ that is related to something similar in the equation $5^{k+1} - 1 = 4s$. In this case, we notice that

$$5^{k+1} = 5 \cdot 5^k.$$

So if we can solve $5^k - 1 = 4m$ for 5^k, we could make a substitution for 5^k. This is done in the proof of the following proposition.

Proposition 5.3. *For every natural number n, 4 divides $(5^n - 1)$.*

Proof. [Proof by Mathematical Induction] For each natural number n, let $P(n)$ be "4 divides $(5^n - 1)$."

We first prove that $P(1)$ is true. Notice that when $n = 1$, $(5^n - 1) = 4$. Since 4 divides 4, $P(1)$ is true.

For the inductive step, we prove that for all $k \in \mathbb{N}$, if $P(k)$ is true, then $P(k+1)$ is true. So, let k be a natural number and assume that $P(k)$ is true. That is, assume that

$$4 \text{ divides } \left(5^k - 1\right).$$

This means that there exists an integer m such that

$$5^k - 1 = 4m.$$

Thus,

$$5^k = 4m + 1. \tag{1}$$

In order to prove that $P(k+1)$ is true, we must show that 4 divides $(5^{k+1} - 1)$. Since $5^{k+1} = 5 \cdot 5^k$, we can write

$$5^{k+1} - 1 = 5 \cdot 5^k - 1. \tag{2}$$

We now substitute the expression for 5^k from Equation (1) into Equation (2). This gives

$$\begin{aligned}
5^{k+1} - 1 &= 5 \cdot 5^k - 1 \\
&= 5(4m+1) - 1 \\
&= (20m+5) - 1 \\
&= 20m + 4 \\
&= 4(5m+1). \tag{3}
\end{aligned}$$

Since $(5m+1)$ is an integer, Equation (3) shows that 4 divides $(5^{k+1} - 1)$. Therefore, if $P(k)$ is true, then $P(k+1)$ is true and the inductive step has been established.

Thus, by the Principle of Mathematical Induction, for every natural number n, 4 divides $(5^n - 1)$. ∎

Note: We proved Proposition 5.3 using mathematical induction so that we could illustrate how to construct and write a proof by induction. However, there is another way to prove this result that uses the concept of congruence. See Exercise (14).

Activity 5.4 (The Derivatives of an Exponential Function).

Let a be a real number. We will explore the derivatives of the function $f(x) = e^{ax}$. By using the chain rule, we see that

$$\frac{d}{dx}(e^{ax}) = ae^{ax}.$$

Recall that the second derivative of a function is the derivative of the derivative function. Similarly, the third derivative is the derivative of the second derivative.

1. What is $\dfrac{d^2}{dx^2}(e^{ax})$, the second derivative of e^{ax}?

2. What is $\dfrac{d^3}{dx^3}(e^{ax})$, the third derivative of e^{ax}?

3. Let n be a natural number. Make a conjecture about the nth derivative of the function $f(x) = e^{ax}$. That is, what is $\dfrac{d^n}{dx^n}(e^{ax})$?

   Use mathematical induction to prove that your conjecture is correct.

Activity 5.5 (A Conjecture about Congruence Modulo 3).

In Section 3.1, we defined congruence modulo n for a natural number n. For $a, b \in \mathbb{Z}$,

$$a \equiv b \pmod{n} \text{ means that } n \text{ divides } (a - b).$$

In Section 3.4, we used the Division Algorithm to prove that each integer is congruent, modulo n, to precisely one of the integers $0, 1, 2, \ldots, n - 1$ (Corollary 3.22).

1. Find the value of r so that $4 \equiv r \pmod 3$ and $r \in \{0, 1, 2\}$.

2. Find the value of r so that $4^2 \equiv r \pmod 3$ and $r \in \{0, 1, 2\}$.

3. Find the value of r so that $4^3 \equiv r \pmod 3$ and $r \in \{0, 1, 2\}$.

4. For at least two other natural numbers n, find the value of r so that $4^n \equiv r \pmod 3$ and $r \in \{0, 1, 2\}$.

5. If $n \in \mathbb{N}$, make a conjecture concerning the value of r where $4^n \equiv r \pmod 3$ and $r \in \{0, 1, 2\}$.

6. Use mathematical induction to prove your conjecture.

Activity 5.6 (Regions of Circles).

Place n equally spaced points on a circle and connect each pair of points with the chord of the circle determined by that pair of points. See the following figure. Count the number of distinct regions within each circle. For example, with three points on the circle, there are four distinct regions. Organize your data in a table with two columns: "Number of Points on the Circle" and "Number of Distinct Regions in the Circle."

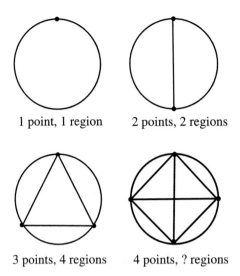

1 point, 1 region   2 points, 2 regions

3 points, 4 regions   4 points, ? regions

1. How many regions are there when there are four equally spaced points on the circle?

2. Based on the work so far, make a conjecture about how many distinct regions would you get with five equally spaced points.

3. Based on the work so far, make a conjecture about how many distinct regions would you get with six equally spaced points.

4. The following figure shows the figures associated with Parts (2) and (3). Count the number of regions in each case. Are your conjectures correct or incorrect?

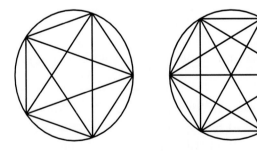

5. What is the purpose of this activity?

Exercises 5.1

1. Which of the following sets are inductive sets? Explain.

   (a) $\mathbb{Z}$

   (b) $\{x \in \mathbb{N} \mid x \geq 4\}$

   (c) $\{x \in \mathbb{Z} \mid x \leq 10\}$

   (d) $\{1, 2, 3, \ldots, 500\}$

2. (a) Can a finite, nonempty set be inductive? Explain.

   (b) Is the empty set inductive? Explain.

3. Use mathematical induction to prove each of the following:

   (a) For each natural number n, $1 + 2 + 3 + \cdots + n = \dfrac{n(n+1)}{2}$. Or, using summation notation, for each natural number n,

   $$\sum_{j=1}^{n} j = \frac{n(n+1)}{2}.$$

   (b) For each natural number n, $2 + 5 + 8 + \cdots + (3n - 1) = \dfrac{n(3n+1)}{2}$. Or, using summation notation, for each natural number n,

   $$\sum_{j=1}^{n} (3j - 1) = \frac{n(3n+1)}{2}.$$

   (c) For each natural number n,

   $$\sum_{j=1}^{n} j^3 = \left[\frac{n(n+1)}{2}\right]^2.$$

4. (a) Use Exercise (3a) to prove the following proposition:

   For each natural number n, $3 + 6 + 9 + \cdots + 3n = \dfrac{3n(n+1)}{2}$.

   (b) Based on Exercises (3a) and (3c), if $n \in \mathbb{N}$, is there any conclusion that can be made about the relationship between the sum $\left(1^3 + 2^3 + 3^3 + \cdots + n^3\right)$ and the sum $(1 + 2 + 3 + \cdots + n)$?

5. (a) Calculate $1 + 3 + 5 + \cdots + (2n - 1)$ for several natural numbers n.

(b) Based on your work in Exercise (5a), if $n \in \mathbb{N}$, make a conjecture about the value of the sum $1+3+5+\cdots+(2n-1) = \sum\limits_{j=1}^{n}(2j-1)$.

(c) Use mathematical induction to prove your conjecture in Exercise (5b).

6. Use mathematical induction to prove each of the following:

(a) For each natural number n, 3 divides $(4^n - 1)$.

(b) For each natural number n, 6 divides $(n^3 - n)$.

7. In Activity 5.5, we proved that for each natural number n, $4^n \equiv 1 \pmod 3$. Explain how this result is related to the proposition in Exercise (6a).

8. **(a)** Complete the following table:

n	$(5^n - 2^n)$	n	$(5^n - 2^n)$
1		5	
2		6	
3		7	
4		8	

(b) Based on your work in Exercise (8a), make a conjecture about the values of $5^n - 2^n$ for each natural number n.

(c) Use mathematical induction to prove your conjecture in Exercise (8b).

9. Prove Part (3) of Theorem 3.24 from Section 3.4. Let $n \in \mathbb{N}$ and let a and b be integers. If $a \equiv b \pmod n$ and $m \in \mathbb{N}$, then $a^m \equiv b^m \pmod n$.

10. **(a)** Why is it not possible to use mathematical induction to prove a proposition of the form

$$(\forall x \in \mathbb{Q})\,(P(x)),$$

where $P(x)$ is some predicate?

(b) Why is it not possible to use mathematical induction to prove a proposition of the form

For each real number x with $x \geq 1$, $P(x)$,

where $P(x)$ is some predicate?

11. Let i be the complex number whose square is -1, that is, $i^2 = -1$. Prove the following statement:

If x is a real number, then for each natural number n,

$$[\cos x + i(\sin x)]^n = \cos(nx) + i(\sin(nx)).$$

12. Let $y = \ln x$.

(a) Determine $\dfrac{dy}{dx}, \dfrac{d^2y}{dx^2}, \dfrac{d^3y}{dx^3},$ and $\dfrac{d^4y}{dx^4}$.

(b) Let n be a natural number. Formulate a conjecture for a formula for $\dfrac{d^ny}{dx^n}$. Then use mathematical induction to prove your conjecture.

13. Use mathematical induction to prove that the sum of the cubes of any three consecutive natural numbers is a multiple of 9.

14. Prove Proposition 5.3 using the concept of congruence modulo 4. <u>Hint</u>: Theorem 3.24 might help.

5.2 Other Forms of Mathematical Induction

Preview Activity 1 (Exploring a Proposition about Factorials).

> **Definition.** If n is a natural number, we define **n factorial**, denoted by $n!$, to be the product of the first n natural numbers. In addition, we define 0! to be equal to 1.

Using this definition, we see that

$$0! = 1 \qquad\qquad 3! = 1 \cdot 2 \cdot 3 = 6$$
$$1! = 1 \qquad\qquad 4! = 1 \cdot 2 \cdot 3 \cdot 4 = 24$$
$$2! = 1 \cdot 2 = 2 \qquad\qquad 5! = 1 \cdot 2 \cdot 3 \cdot 4 \cdot 5 = 120.$$

In general, we write $n! = 1 \cdot 2 \cdot 3 \cdot \cdots \cdot (n-1) \cdot n$ or $n! = n \cdot (n-1) \cdot \cdots \cdot 2 \cdot 1$. Notice that for $n \geq 1$, $n! = n \cdot (n-1)!$.

1. For each natural number n with $1 \leq n \leq 7$, compute the values of 2^n and $n!$.

2. Is the following proposition true or false?

   For each natural number n, $n! > 2^n$.

3. If the proposition in Part (2) is true, construct a proof. If it is false, rewrite it (by adding a condition on the natural number n) so that the new proposition appears to be true [based on the data in Part (1)].

Preview Activity 2 (Subsets of a Set with Four Elements).
    Review Preview Activity 2 from Section 4.1. In this activity, we saw that a set with one element has two subsets, a set with two elements has four subsets, and a set with three elements has eight subsets. The following list shows the eight subsets of the set $B = \{a, b, c\}$.

$\emptyset$ $\{a\}$ $\{b\}$ $\{c\}$
$\{a, b\}$ $\{a, c\}$ $\{b, c\}$ $\{a, b, c\}$

Now let $A = \{a, b, c, x\}$.

1. Are the eight subsets of B also subsets of A? Explain.

2. Create subsets of A by starting with the sets listed above and "adding" x to each subset. That is, if C is a subset of B (one of the eight listed above), then create a subset of A by using $C \cup \{x\}$.

3. How many subsets of A did you form in Part (2)? Are all of the subsets of A subsets of B or listed in Part (2)? Explain.

4. Use this work to explain why $\mathcal{P}(A) = \mathcal{P}(B) \cup \{C \cup \{x\} \mid C \in \mathcal{P}(B)\}$.

Preview Activity 3 (Prime Factors of a Natural Number).
    Recall that a natural number p is a **prime number** provided that it is greater than 1 and the only natural numbers that divide p are 1 and p. A natural number other than 1 that is not a prime number is a **composite number**. The number 1 is neither prime nor composite.

1. Give examples of four natural numbers that are prime and four natural numbers that are composite.

2. Write the number 20 as a product of prime numbers. Then write the number 40 as a product of prime numbers.

3. Repeat Part (2) using 50 and 150.

4. Do you think that any composite number can be written as a product of prime numbers?

5. Write a useful description of what it means to say that a natural number is a composite number (other than saying that it is not prime).

6. Based on your work in Parts (2) and (3), do you think it would be possible to use induction to prove that any composite number can be written as a product of prime numbers?

The Domino Theory

Mathematical induction is frequently used to prove statements of the form

$$(\forall n \in \mathbb{N}) \, (P(n)), \tag{1}$$

where $P(n)$ is a predicate. This means that we are proving that every statement in the following infinite list is true.

$$P(1), P(2), P(3), \ldots \tag{2}$$

One of the basic components of a proof by induction is to prove that if one statement in this infinite list of statements is true, then the next statement in the list must be true. This is the inductive step in an induction proof.

As we saw in Activity 5.2, this is not enough to prove the statement in (1). Imagine each statement in (2) as a domino in a chain of dominoes. When we prove the induction step, we are proving that if one domino is knocked over, it will knock over the next one in the chain. Even if the dominoes are set up so that when one falls, the next one will fall, no dominoes will fall unless we start by knocking one over. This is why we need the basis step in an induction proof. The basis step guarantees that we knock over the first domino. The inductive step, then, guarantees that all dominoes after the first one will also fall.

Now imagine what would happen if instead of knocking over the first domino, we knock over the sixth domino. If we have proven the inductive step, then we would know that every domino after the sixth domino would also fall. This is the idea of the Extended Principle of Mathematical Induction. It is not necessary for the basis step to be the proof that $P(1)$ is true. We can make the basis step be the proof of $P(M)$ where M is some natural number (greater than one).

The Extended Principle of Mathematical Induction is used to prove that a certain predicate is true for all natural numbers greater than or equal to some "starting" natural number M. Actually, this can be generalized somewhat by allowing M to be any integer. We are still only concerned with those integers that are greater than or equal to M. The set $\{n \in \mathbb{Z} \mid n \geq M\}$ is still an inductive set. This slight change allows use to use $M = 0$ as well as allowing M to be any natural number. In almost every case, we will use $M = 0$ or $M \in \mathbb{N}$.

The Extended Principle of Mathematical Induction
Let M be an integer. If T is a subset of $\mathbb{Z}$ such that

1. $M \in T$, and

2. For every $k \in \mathbb{Z}$ with $k \geq M$, if $k \in T$, then $(k+1) \in T$,

then T contains all integers greater than or equal to M. That is, $\{n \in \mathbb{Z} \mid n \geq M\} \subseteq T$.

Using the Extended Principle of Mathematical Induction

The primary use of the Principle of Mathematical Induction is to prove statements of the form

$$(\forall n \in \mathbb{Z}, \text{ with } n \geq M)(P(n)),$$

where M is an integer and $P(n)$ is some predicate. So our goal is to prove that the truth set T of the predicate $P(n)$ contains all integers greater than or equal to M. So, to verify the hypothesis of the Extended Principle of Mathematical Induction, we must

1. Prove that $M \in T$. That is, prove that $P(M)$ is true.

2. Prove that for every $k \in \mathbb{Z}$ with $k \geq M$, if $k \in T$, then $(k+1) \in T$. That is, prove that if $P(k)$ is true, then $P(k+1)$ is true.

As with "standard" induction, the first step is called the **basis step** or the **initial step**, and the second step is called the **inductive step**. This means that a proof using the Extended Principle of Mathematical Induction will have the following form:

> **Using the Extended Principle of Mathematical Induction**
> Let M be an integer. To prove $(\forall n \in \mathbb{Z}$ with $n \geq M)\,(P\,(n))$
>
>   Basis step: Prove $P\,(M)$.
>
>   Inductive step: Prove that for every $k \in \mathbb{Z}$ with $k \geq M$,
>   if $P\,(k)$ is true, then $P\,(k+1)$ is true.
>
> We can then conclude that $P\,(n)$ is true for all $n \in \mathbb{Z}$ with $n \geq M$.

This is basically the same procedure as the one for using the Principle of Mathematical Induction. The only difference is that the basis step uses an integer M other than 1. For this reason, when we write a proof that uses the Extended Principle of Mathematical Induction, we often simply say we are going to use a proof by mathematical induction. We will use the work from Preview Activity 1 to illustrate such a proof.

Proposition 5.7. *For each natural number n with $n \geq 4$, $n! > 2^n$.*

Proof. We will use a proof by mathematical induction. For this proof, we let

$$P\,(n) \text{ be ``} n! > 2^n.\text{''}$$

We first prove that $P\,(4)$ is true. Using $n = 4$, we see that $4! = 24$ and $2^4 = 16$. This means that $4! > 2^4$ and hence, $P\,(4)$ is true.

For the inductive step, we prove that for all $k \in \mathbb{N}$ with $k \geq 4$, if $P\,(k)$ is true, then $P\,(k+1)$ is true. Let k be a natural number greater than or equal to 4, and assume that $P\,(k)$ is true. That is, assume that

$$k! > 2^k. \tag{1}$$

The goal is to prove that $P\,(k+1)$ is true or that $(k+1)! > 2^{k+1}$. Multiplying both sides of Inequality (1) by $k + 1$ gives

$$(k+1) \cdot k! > (k+1) \cdot 2^k, \text{ or}$$
$$(k+1)! > (k+1) \cdot 2^k. \tag{2}$$

Now, $k \geq 4$. Thus, $k + 1 > 2$, and hence $(k+1) \cdot 2^k > 2 \cdot 2^k$. This means that

$$(k+1) \cdot 2^k > 2^{k+1}. \tag{3}$$

Inequalities (2) and (3) show that

$$(k+1)! > 2^{k+1},$$

and this means that $P(k+1)$ is true. Thus, the inductive step has been established, and so by the Extended Principle of Mathematical Induction, $n! > 2^n$ for each natural number n with $n \geq 4$.   ∎

In Preview Activity 2, we observed that a set with one element has two subsets, a set with two elements has four subsets, and a set with three elements has eight subsets. In this Preview Activity, we also studied a way to use the eight subsets of a set with three elements to create the 16 subsets of a set with four elements. This work suggests that the following proposition is true.

Proposition 5.8. *Let A and B be subsets of some universal set. If $A = B \cup \{x\}$, where $x \notin B$, then all the subsets of A are either subsets of B or of the form $C \cup \{x\}$, where C is a subset of B.*

Proof. Let A and B be subsets of some universal set, and assume that $A = B \cup \{x\}$ where $x \notin B$. Let Y be a subset of A. We need to show that Y is a subset of B or that $Y = C \cup \{x\}$ where C is some subset of B. There are two cases to consider: (1) x is not an element of Y, and (2) x is an element of Y.

Case 1: Assume that $x \notin Y$. Let $y \in Y$. Then $y \in A$ and $y \neq x$. Since

$$A = B \cup \{x\},$$

this means that y must be in B. Therefore, $Y \subseteq B$.

Case 2: Assume that $x \in Y$. In this case, let $C = Y - \{x\}$. Then every element of C is an element of B. Hence, we can conclude that $C \subseteq B$ and that $Y = C \cup \{x\}$.

Cases 1 and 2 show that if $Y \subseteq A$, then $Y \subseteq B$ or $Y = C \cup \{x\}$, where $C \subseteq B$.   ∎

The power set of A, $\mathcal{P}(A)$, is the set of all subsets of A. So,

$$Y \in \mathcal{P}(A) \text{ if and only if } Y \subseteq A.$$

Proposition 5.8 states that if $A = B \cup \{x\}$ and $x \notin B$, then $Y \subseteq A$ if and only if $Y \subseteq B$ or that $Y = C \cup \{x\}$, where C is some subset of B.

Using power sets and still assuming that $A = B \cup \{x\}$ and $x \notin B$, this means that $Y \in \mathcal{P}(A)$ if and only if $Y \in \mathcal{P}(B)$ or that $Y = C \cup \{x\}$, where $C \in \mathcal{P}(B)$. This gives us the following corollary of Proposition 5.8.

Corollary 5.9. *Let A and B be subsets of some universal set. If $x \notin B$ and $A = B \cup \{x\}$, then $\mathcal{P}(A) = \mathcal{P}(B) \cup \{C \cup \{x\} \, | \, C \in \mathcal{P}(B)\}$.*

Activity 5.10 (The Cardinality of a Power Set).

In this activity, we will use mathematical induction to prove the following proposition.

Proposition 5.11. *Let n be a nonnegative integer and let A be a subset of some universal set. If A is a finite set with n elements, then A has 2^n subsets. That is, if $|A| = n$, then $|\mathcal{P}(A)| = 2^n$.*

1. Verify that Proposition 5.11 is true when $n = 0$. That is, it is true when $A = \emptyset$. (This is the basis step for the induction proof.)

2. Verify that Proposition 5.11 is true when $n = 1$ and when $n = 2$.

3. Now assume that k is a nonnegative integer and assume that if a set has k elements, then that set has 2^k subsets. (This is the inductive assumption for the induction proof.)

   Let A be a subset of the universal set with $|A| = k + 1$, and let $x \in A$. Then, the set $B = A - \{x\}$ has k elements.

   Now use the inductive assumption to determine how many subsets B has. Then, use Proposition 5.8 to prove that A has twice as many subsets as B. This should help complete the inductive step for the induction proof.

The Second Principle of Mathematical Induction

Let $P(n)$ be

"n is a prime number or is a product of prime numbers."

(This is related to the work in Preview Activity 3.)

Suppose we would like to use induction to prove that $P(n)$ is true for all natural numbers greater than 1. We have seen that the idea of the inductive step in a proof by induction is to prove that if one statement in an infinite list of statements is true, then the next statement must also be true. The problem here is that when we factor a composite number (such as $40 = 2 \cdot 20$), we do not get to the previous case. In this situation, we know that we can factor 20 as a product of primes, but how do we handle this in a general case?

All of this work was intended to show the need for another principle of induction. In the inductive step of a proof by induction, we assume one statement is true and prove the next one is true. The idea of this new principle is to assume that *all* previous statements are true and use this assumption to prove the next statement is true. This is stated formally in terms of subsets of natural numbers in the Second Principle of Mathematical Induction.

The Second Principle of Mathematical Induction

Let M be an integer. If T is a subset of $\mathbb{Z}$ such that

  1. $M \in T$, and

  2. For every $k \in \mathbb{Z}$ with $k \geq M$, if $\{M, M+1, \ldots, k\} \subseteq T$, then $(k+1) \in T$,

then T contains all integers greater than or equal to M. That is, $\{n \in \mathbb{Z} \mid n \geq M\} \subseteq T$.

Rather than stating this principle in two versions, we simply stated the extended version. In many cases, we will use $M = 1$ or $M = 0$.

Using the Second Principle of Mathematical Induction

The primary use of the Principle of Mathematical Induction is to prove statements of the form

$$(\forall n \in \mathbb{Z}, \text{with } n \geq M)(P(n)),$$

where M is an integer and $P(n)$ is some predicate. So our goal is to prove that the truth set T of the predicate $P(n)$ contains all integers greater than or equal to M. So to verify the hypothesis of the Second Principle of Mathematical Induction, we must

1. Prove that $M \in T$. That is, prove that $P(M)$ is true.

2. Prove that for every $k \in \mathbb{N}$, if $k \geq M$ and $\{M, M+1, \ldots, k\} \subseteq T$, then $(k+1) \in T$. That is, prove that if $P(M), P(M+1), \ldots, P(k)$ are true, then $P(k+1)$ is true.

As before, the first step is called the **basis step** or the **Initial Step**, and the second step is called the **inductive step**. This means that a proof using the Second Principle of Mathematical Induction will have the following form:

Using the Second Principle of Mathematical Induction
Let M be an integer. To prove: $(\forall n \in \mathbb{Z}$ with $n \geq M)(P(n))$

Basis step: Prove $P(M)$.

Inductive step: Let $k \in \mathbb{Z}$ with $k \geq M$. Prove that if $P(M), P(M+1), \ldots, P(k)$ are true, then $P(k+1)$ is true.

We can then conclude that $P(n)$ is true for all $n \in \mathbb{Z}$ with $n \geq M$.

We will use this procedure to prove the proposition suggested in Preview Activity 3.

Theorem 5.12. *Each natural number greater than 1 is either a prime number or is a product of prime numbers.*

Proof. We will use the Second Principle of Mathematical Induction. We let $P(n)$ be

"n is either a prime number or is a product of prime numbers."

For the basis step, we show that $P(2)$ is true. Since 2 is a prime number, we know that $P(2)$ is true.

To prove the inductive step, we let k be a natural number with $k \geq 2$. We assume that $P(2), P(3), \ldots, P(k)$ are true. That is, we assume that each of the integers $2, 3, \ldots, k$ is a prime number or a product of prime numbers.

The goal is to prove that $P(k+1)$ is true or that $k+1$ is a prime number or a product of prime numbers.

Case 1: If $(k+1)$ is a prime number, then $P(k+1)$ is true.

<u>Case 2</u>: If $(k+1)$ is not a prime number, then $(k+1)$ can be factored into a product of natural numbers with each one being less than $(k+1)$. That is,

$$k + 1 = a \cdot b,$$

where $a, b \in \mathbb{N}$, $1 < a \le k$, and $1 < b \le k$. Using the inductive assumption, this means that $P(a)$ and $P(b)$ are both true. Consequently, a and b are prime numbers or are products of prime numbers. Since $k+1 = a \cdot b$, we conclude that $(k+1)$ is a product of prime numbers. That is, we conclude that $P(k+1)$ is true. This proves the inductive step.

Hence, by the Second Principle of Mathematical Induction, we conclude that $P(n)$ is true for all $n \in \mathbb{N}$ with $n \ge 2$, and hence that each natural number greater than 1 is either a prime number or is a product of prime numbers. ∎

Activity 5.13 (A Proposition about Natural Numbers).

In this activity, we will determine those natural numbers that can be written in the form $3x + 5y$, where x and y are nonnegative integers. We can word the question as follows:

> "For which natural numbers n do there exist nonnegative integers x and y such that $n = 3x + 5y$?"

To help answer this question, we will let

$$\mathbb{Z}^* = \{x \in \mathbb{Z} \mid x \ge 0\}$$

and let $P(n)$ be

> "There exist $x, y \in \mathbb{Z}^*$ such that $n = 3x + 5y$."

1. Explain why there do not exist nonnegative integers x and y such that $1 = 3x + 5y$. That is, explain why $P(1)$ is false.

2. Repeat Part (1) for $n = 2$, $n = 4$, $n = 6$, and $n = 7$.

3. Do there exist non-negative integers x and y such that $8 = 3x + 5y$? In other words, is $P(8)$ is true?

4. Repeat Part (3) for 9, 10, 11, 12, and 13.

5. Let $k \in \mathbb{N}$ with $k \ge 13$. Prove that if $P(8), P(9), \ldots, P(k)$ are true, then $P(k+1)$ is true.

   <u>Hint</u>: $k + 1 = 6 + (k - 5)$.

6. Is the following proposition true or false? Explain.

> If $n \in \mathbb{N}$ and $n \geq 8$, then there exist nonnegative integers x and y such that $n = 3x + 5y$.

Exercises 5.2

1. Use mathematical induction to prove each of the following:

   (a) For each natural number n with $n \geq 2$, $3^n > 1 + 2^n$.

   (b) For each natural number n with $n \geq 6$, $2^n > (n+1)^2$.

   (c) For each natural number n with $n \geq 3$, $\left(1 + \dfrac{1}{n}\right)^n < n$.

2. **(a)** Verify that $\left(1 - \dfrac{1}{4}\right) = \dfrac{3}{4}$ and that $\left(1 - \dfrac{1}{4}\right)\left(1 - \dfrac{1}{9}\right) = \dfrac{4}{6}$.

   (b) Verify that $\left(1 - \dfrac{1}{4}\right)\left(1 - \dfrac{1}{9}\right)\left(1 - \dfrac{1}{16}\right) = \dfrac{5}{8}$ and that
   $$\left(1 - \dfrac{1}{4}\right)\left(1 - \dfrac{1}{9}\right)\left(1 - \dfrac{1}{16}\right)\left(1 - \dfrac{1}{25}\right) = \dfrac{6}{10}.$$

   (c) For $n \in \mathbb{N}$ with $n \geq 2$, make a conjecture about a formula for the product $\left(1 - \dfrac{1}{4}\right)\left(1 - \dfrac{1}{9}\right)\left(1 - \dfrac{1}{16}\right) \cdots \left(1 - \dfrac{1}{n^2}\right)$.

   (d) Based on your work in Parts (2a) and (2b), state a proposition and then use the Extended Principle of Mathematical Induction to prove your proposition.

3. Is the following proposition true or false? Justify your conclusion.

> For each nonnegative integer n, $8^n \mid (4n)!$.

4. Can each natural number greater than or equal to 4 be written as the sum of at least two natural numbers, each of which is a 2 or a 3? Justify your conclusion.

> For example, $7 = 2 + 2 + 3$, and $17 = 2 + 2 + 2 + 2 + 3 + 3 + 3$.

5. Can each natural number greater than or equal to 6 be written as the sum of at least two natural numbers, each of which is a 2 or a 5? Justify your conclusion.

> For example, $6 = 2 + 2 + 2$, $9 = 2 + 2 + 5$, and $17 = 2 + 5 + 5 + 5$.

6. For which natural numbers n is it true that $n^2 < 2^n$? Justify your conclusion.

7. For which natural numbers n do there exist nonnegative integers x and y such that $n = 4x + 5y$? Justify your conclusion.

8. Use mathematical induction to prove the following proposition:

> Let x be a real number with $x > -1$. Then for each natural number n with $n \geq 2$, $(1 + x)^n > 1 + nx$.

Explain where the assumption that $x > -1$ was used in the proof.

9. Prove that for each odd natural number n with $n \geq 3$,

$$\left(1 + \frac{1}{2}\right)\left(1 - \frac{1}{3}\right)\left(1 + \frac{1}{4}\right) \cdots \left(1 + \frac{(-1)^n}{n}\right) = 1.$$

10. Prove that for each natural number n,

any set with n elements has $\dfrac{n(n-1)}{2}$ two-element subsets.

5.3 Induction and Recursion

Preview Activity 1 (Factorials and Recurrence Relations).
   In Preview Activity 1 in Section 5.2, we defined $n!$, read n **factorial**, for each natural number n as the product of the first n natural numbers. That is,

$$n! = 1 \cdot 2 \cdots \cdots n.$$

We also defined $0!$ to be equal to one. Define a sequence of numbers $a_0, a_1, a_2, \ldots$ as follows:

> $a_0 = 1$, and
>
> for each nonnegative integer n, $a_{n+1} = (n+1) \cdot a_n$.

Using $n = 0$, we see that this implies that $a_1 = 1 \cdot a_0 = 1 \cdot 1 = 1$.

1. Calculate a_2, a_3, a_4, a_5, and a_6.

2. Do you think that it is possible to calculate a_{20} and a_{100}? Explain.

3. Do you think it is possible to calculate a_n for any natural number n? Explain.

4. Compare the values of $a_0, a_1, a_2, a_3, a_4, a_5$, and a_6 with those of $0!, 1!, 2!, 3!, 4!, 5!$, and $6!$. What do you observe?

The boxed formulas in this Preview Activity provide another way to define $n!$ for each nonnegative integer n. This is an example of a **recursive definition.** In a recursive definition of a sequence, the value of a beginning term is specified (or beginning terms are specified), and succeeding terms are defined in terms of the previous values. In this case, a_0 is specified to be one, and for each nonnegative integer n, a_{n+1} is defined in terms of a_n. The relation that defines the succeeding terms is often called the **recurrence relation** for the recursive definition.

Preview Activity 2 (The Fibonacci Numbers).
The **Fibonacci numbers** are a sequence of natural numbers $f_1, f_2, f_3, \ldots, f_n, \ldots$ defined as follows:

- $f_1 = 1$ and $f_2 = 1$, and

- For each natural number n, $f_{n+2} = f_{n+1} + f_n$.

This means that

- $f_3 = f_2 + f_1 = 1 + 1 = 2$, and

- $f_4 = f_3 + f_2 = 2 + 1 = 3$.

1. Calculate f_5 through f_9.

2. Now calculate f_{10} through f_{20}.

3. Record any observations about the values of the Fibonacci numbers or any patterns that you observe in the sequence of Fibonacci numbers. (For example, which terms are even? Which terms are odd? Which terms are multiples of 3?) If necessary, compute more Fibonacci numbers.

Preview Activity 3 (Recursively Defined Sequences).

1. Define a sequence recursively as follows:

   $b_1 = 16$, and for each $n \in \mathbb{N}$, $b_{n+1} = \dfrac{1}{2}b_n$.

   Calculate b_2 through b_{10}. What seems to be happening to the values of b_n as n gets larger?

2. Define a sequence recursively as follows:

   $T_1 = 16$, and for each $n \in \mathbb{N}$, $T_{n+1} = 16 + \dfrac{1}{2}T_n$.

   Calculate T_2 through T_{10}. What seems to be happening to the values of T_n as n gets larger?

The sequences in Parts (1) and (2) can be generalized as follows: Let a and r be real numbers. Define two sequences recursively as follows:

   $a_1 = a$, and for each $n \in \mathbb{N}$, $a_{n+1} = r \cdot a_n$.
   $S_1 = a$, and for each $n \in \mathbb{N}$, $S_{n+1} = a + r \cdot S_n$.

3. Determine formulas (in terms of a and r) for a_2 through a_6. What do you think a_n is equal to (in terms of a, r, and n)?

4. Determine formulas (in terms of a and r) for S_2 through S_6. What do you think a_n is equal to (in terms of a, r, and n)?

Definition by Recursion

In a proof by mathematical induction, we "start with a first step" and then prove that we can always go from one step to the next step. We can use this same idea to define a sequence as well. For now, we can think of a **sequence** as an infinite list of objects that are indexed by the natural numbers (or some infinite subset of $\mathbb{N} \cup \{0\}$). We often write a sequence in the following form:

$$a_1, a_2, \ldots, a_n, \ldots.$$

The basic idea is to give a specific definition of the first term (or the first few terms) and then state, in general terms, how to determine a_{n+1} in terms of n and the first n terms $a_1, a_2, \ldots, a_n$. This process is known as **definition by recursion**. The specific definition of the first term is called the **initial condition**, and the general definition of a_{n+1} in terms of n and the first n

terms $a_1, a_2, \ldots, a_n$ is called the **recurrence relation**. (When more than one term is defined explicitly, we say that these are the initial conditions.)

In Preview Activity 1, we gave a specific definition of the first term (which was labeled a_0) and then the recurrence relation gave us the rule to compute a_{n+1} in terms of a_n. This was done as follows:

> Initial conditions: $a_0 = 1$.
> Recurrence relation: For all $n \in \mathbb{N} \cup \{0\}$, $a_{n+1} = (n+1) a_n$.

In Preview Activity 1, we saw that this definition gives $a_n = n!$ for all $n \in \mathbb{N} \cup \{0\}$. Actually, we need to prove this result using mathematical induction. This will be included in the exercises for this section.

The Fibonacci Numbers

A very famous example of a recurrence relation is contained in the well-known sequence investigated by Leonardo of Pisa (1170 - 1250), who is better known as Fibonacci. The **Fibonacci numbers** are defined as follows:

> Initial condition: $f_1 = 1, f_2 = 1$.
> Recurrence relation: For all $n \in \mathbb{N}$, $f_{n+2} = f_{n+1} + f_n$.

Fibonacci first introduced this sequence of numbers as a solution of the following problem:

> Suppose that a pair of adult rabbits (one male, one female) produces a pair of rabbits (one male, one female) each month. Also, suppose that newborn rabbits become adults in two months and produce another pair of rabbits. Starting with one adult pair of rabbits, how many pairs of rabbits will be produced each month for one year?

Since we start with one adult pair, there will be one pair produced the first month, and since there is still only one adult pair, one pair will also be produced in the second month (since the new pair produced in the first month is not yet mature). In the third month, two pairs will be produced, one by the original pair and one by the pair which was produced in the first month. In the fourth month, three pairs will be produced, and in the fifth month, five pairs will be produced.

The basic rule is that in a given month after the first two months, the number of adult pairs is the number of adult pairs one month ago plus the number of pairs born two months ago. This is summarized in Table 5.1, where the number of pairs produced is equal to the number of adult pairs.

Months	Adult Pairs	Newborn Pairs	Month-old Pairs
1	1	1	0
2	1	1	1
3	2	2	1
4	3	3	2
5	5	5	3
6	8	8	5
7	13	13	8
8	21	21	13
9	34	34	21
10	55	55	34
11	89	89	55
12	144	144	89

Table 5.1: Fibonacci Numbers

We see that the number of adult pairs follows the Fibonacci sequence of numbers that we developed in Preview Activity 2.

This may seem like an artificial (and even silly) way to generate a sequence. However, it turns out that this sequence occurs in nature frequently and has applications in computer science. There is even a scholarly journal, *The Fibonacci Quarterly*, devoted to the Fibonacci numbers.

The Fibonacci numbers are one of the most studied sequences in mathematics largely due to the many beautiful patterns they contain. Perhaps one observation you made in Preview Activity 2 is that every third Fibonacci number is even. This can be written as a proposition as follows:

For each natural number n, f_{3n} is an even natural number.

As with most propositions associated with definitions by recursion, we can prove this using mathematical induction. The first step is to define the appropriate predicate. For this, we can let

$P(n)$ be "f_{3n} is an even natural number."

Notice that $P(1)$ is true since $f_3 = 2$. We now need to prove the inductive step. To do this, we need to prove that for each $k \in \mathbb{N}$,

if $P(k)$ is true, then $P(k+1)$ is true.

That is, we need to prove that for each $k \in \mathbb{N}$,

if f_{3k} is even, then $f_{3(k+1)}$ is even.

So let's analyze this conditional statement using a know-show table.

Step	Know	Reason
P	f_{3k} is even.	Inductive hypothesis
$P1$	$(\exists m \in \mathbb{N})\,(f_{3k} = 2m)$	Definition of "even integer"
$\vdots$	$\vdots$	$\vdots$
$Q1$	$(\exists q \in \mathbb{N})\,(f_{3(k+1)} = 2q)$	
Q	$f_{3(k+1)}$ is even.	Definition of "even integer"
Step	Show	Reason

The key question now is, "Is there any relation between $f_{3(k+1)}$ and f_k?" We can use the recursion formula that defines the Fibonacci sequence to find such a relation.

The recurrence relation for the Fibonacci sequence states that a Fibonacci number (except for the first two) is equal to the sum of the two previous Fibonacci numbers. If we write $3\,(k+1) = 3k + 3$, then we get $f_{3(k+1)} = f_{3k+3}$. For f_{3k+3}, the two previous Fibonacci numbers are f_{3k+2} and f_{3k+1}. This means that

$$f_{3k+3} = f_{3k+2} + f_{3k+1}.$$

Combining all this, we can obtain the following:

$$
\begin{aligned}
f_{3(k+1)} &= f_{3k+3} \\
&= f_{3k+2} + f_{3k+1} \\
&= (f_{3k+1} + f_{3k}) + f_{3k+1}.
\end{aligned}
$$

The last equation states that $f_{3(k+1)} = 2f_{3k+1} + f_{3k}$. This equation can be used to complete the proof of the induction step.

Activity 5.14 (Every Third Fibonacci Number is Even).
Complete the proof of Proposition 5.15.

Proposition 5.15. *For each natural number n, the Fibonacci number f_{3n} is an even natural number.*

Hints: We have already defined the predicate $P(n)$ to be used in an induction proof and have proven the basis step. [The predicate, $P(n)$, is "f_{3n} is an even natural number."] Use the information in the preceding know-show table to help prove that if f_{3k} is even, then $f_{3(k+1)}$ is even.

Geometric Sequences

Let $a, r \in \mathbb{R}$. The following sequence was introduced in Preview Activity 3.

Initial condition:	$a_1 = a$.
Recurrence relation:	For each $n \in \mathbb{N}$, $a_{n+1} = r \cdot a_n$.

This is a recursive definition for a **geometric sequence** with **initial term** a and (common) **ratio** r. The basic idea is that the next term in the sequence is obtained by multiplying the previous term by the ratio r. The work in Preview Activity 3 suggests that the following proposition is true.

Proposition 5.16. *Let $a, r \in \mathbb{R}$. If a geometric sequence is defined by $a_1 = a$ and for each $n \in \mathbb{N}$, $a_{n+1} = r \cdot a_n$, then for each $n \in \mathbb{N}$, $a_n = a \cdot r^{n-1}$.*

The proof of this proposition will be included in the exercises.

Geometric Series

The other sequence that was introduced in Preview Activity 3 is related to geometric series.

Let $a, r \in \mathbb{R}$. The following sequence was introduced in Preview Activity 3.

Initial condition:	$S_1 = a$.
Recurrence relation:	For each $n \in \mathbb{N}$, $S_{n+1} = a + r \cdot S_n$.

For each $n \in \mathbb{N}$, the term S_n is a (finite) **geometric series** with **initial term** a and (common) **ratio** r. The work in Preview Activity 3 suggests that the following proposition is true.

Proposition 5.17. *Let $a, r \in \mathbb{R}$. If the sequence $S_1, S_2, \ldots, S_n, \ldots$ is defined by $S_1 = a$ and for each $n \in \mathbb{N}$, $S_{n+1} = a + r \cdot S_n$, then for each $n \in \mathbb{N}$, $S_n = a + a \cdot r + a \cdot r^2 + \cdots + a \cdot r^{n-1}$. That is, the geometric series S_n is the sum of the first n terms of the corresponding geometric sequence.*

The proof of Proposition 5.17 will be included in the exercises.

The recursive definition and Proposition 5.17 give two different ways to look at geometric series. Proposition 5.17 represents a geometric series as the sum of the first n terms of the corresponding geometric sequence. A third way to determine the terms of a geometric series is given in Proposition 5.18, where we will give a formula that does not use a summation for the terms of a geometric series.

Proposition 5.18. *Let $a, r \in \mathbb{R}$ and $r \neq 1$. If the sequence $S_1, S_2, \ldots, S_n, \ldots$ is defined by $S_1 = a$ and for each $n \in \mathbb{N}$, $S_{n+1} = a + r \cdot S_n$, then for each $n \in \mathbb{N}$, $S_n = a \left(\dfrac{1 - r^n}{1 - r} \right)$.*

Proof. [Proof by Mathematical Induction] Let $a, r \in \mathbb{R}$ and $r \neq 1$. Assume the sequence $S_1, S_2, \ldots, S_n, \ldots$ is defined by $S_1 = a$ and for each $n \in \mathbb{N}$, $S_{n+1} = a + r \cdot S_n$. Let $P(n)$ be

$$ S_n = a \left(\frac{1 - r^n}{1 - r} \right). $$

We first prove that $P(1)$ is true. For $n = 1$, $S_1 = a$ and

$$ a \left(\frac{1 - r^n}{1 - r} \right) = a \left(\frac{1 - r}{1 - r} \right) = a. $$

This proves that $P(1)$ is true.

For the inductive step, we prove that for each $k \in \mathbb{N}$, if $P(k)$ is true, then $P(k+1)$ is true. So let $k \in \mathbb{N}$ and assume that $P(k)$ is true. That is, assume that

$$ S_k = a \left(\frac{1 - r^k}{1 - r} \right). \tag{1} $$

The goal now is to prove that $P(k+1)$ is true or that

$$ S_{k+1} = a \left(\frac{1 - r^{k+1}}{1 - r} \right). $$

We start by using the recurrence relation for the sequence,

$$ S_{k+1} = a + r \cdot S_k. $$

We now substitute Equation (1) into this relation to find

$$S_{k+1} = a + r \cdot S_k$$
$$= a + r \cdot a \left(\frac{1 - r^k}{1 - r} \right). \tag{2}$$

We can now factor an a from the right side of Equation (2) and obtain

$$S_{k+1} = a \left(1 + r \left(\frac{1 - r^k}{1 - r} \right) \right). \tag{3}$$

Next, we rewrite the right side of Equation (3) as a single fraction by finding a common denominator and performing the following algebraic steps:

$$S_{k+1} = a \left(1 + r \left(\frac{1 - r^k}{1 - r} \right) \right)$$
$$= a \left(\frac{(1 - r) + r \left(1 - r^k \right)}{1 - r} \right)$$
$$= a \left(\frac{1 - r + r - r^{k+1}}{1 - r} \right)$$
$$= a \left(\frac{1 - r^{k+1}}{1 - r} \right).$$

This last equation shows that $P(k+1)$ is true. Hence, we have proven the inductive step, and so by mathematical induction, we have proven that for each $n \in \mathbb{N}$, $S_n = a \left(\frac{1 - r^n}{1 - r} \right)$. ∎

Activity 5.19 (Compound Interest).

Assume that R dollars is deposited in an account that has an interest rate of i for each compounding period. A compounding period is some specified time period such as a month or a year.

For each integer n with $n \geq 0$, let V_n be the amount of money in an account at the end of the nth compounding period. Then

$$V_1 = R + i \cdot R \qquad\qquad V_2 = V_1 + i \cdot V_1$$
$$= R(1 + i). \qquad\qquad = (1 + i) V_1$$
$$= (1 + i)^2 R.$$

1. Explain why $V_3 = V_2 + i \cdot V_2$. Then use the formula for V_2 to determine a formula for V_3 in terms of i and R.

2. Determine a recurrence relation for V_{n+1} in terms of i and V_n.

3. Write the recurrence relation in Part (2) so that it is in the form of a recurrence relation for a geometric sequence. What is the initial term of the geometric sequence and what is the common ratio?

4. Use Proposition 5.16 to determine a formula for V_n in terms of I, R, and n.

Activity 5.20 (The Future Value of an Ordinary Annuity).

For an **ordinary annuity**, R dollars is deposited in an account at the end of each compounding period. It is assumed that the interest rate, i, per compounding period for the account remains constant.

Let S_t represent the amount in the account at the end of the tth compounding period. S_t is frequently called the **future value** of the ordinary annuity.

So, $S_1 = R$. To determine the amount after two months, we first note that the amount after one month will gain interest and grow to $(1 + i) S_1$. In addition, a new deposit of R dollars will be made at the end of the second month. So

$$S_2 = R + (1 + i) S_1.$$

1. For each $n \in \mathbb{N}$, use a similar argument to determine a recurrence relation for S_{n+1} in terms of R, i, and S_n.

2. By recognizing this as a recursion formula for a geometric series, use Proposition 5.18 to determine a formula for S_n in terms of R, i, and n that does not use a summation. Then show that this formula can be written as

$$S_n = R \left(\frac{(1 + i)^n - 1}{i} \right).$$

3. What is the future value of an ordinary annuity in 20 years if $200 dollars is deposited in an account at the end of each month where the interest rate for the account is 6% per year compounded monthly? What is the amount of interest that has accumulated in this account during the 20 years?

Exercises 5.3

1. For the sequence $a_0, a_1, a_2, \ldots, a_n, \ldots$, assume that $a_0 = 1$ and that for each $n \in \mathbb{N} \cup \{0\}$, $a_{n+1} = (n+1)\, a_n$. Use mathematical induction to prove that for each $n \in \mathbb{N} \cup \{0\}$, $a_n = n!$.

2. Prove Proposition 5.16.

   Let $a, r \in \mathbb{R}$. If a geometric sequence is defined by $a_1 = a$ and for each $n \in \mathbb{N}$, $a_{n+1} = r \cdot a_n$, then for each $n \in \mathbb{N}$, $a_n = a \cdot r^{n-1}$.

3. Prove Proposition 5.17.

   Let $a, r \in \mathbb{R}$. If the sequence $S_1, S_2, \ldots, S_n, \ldots$ is defined by $S_1 = a$ and for each $n \in \mathbb{N}$, $S_{n+1} = a + r \cdot S_n$, then for each $n \in \mathbb{N}$, $S_n = a + a \cdot r + a \cdot r^2 + \cdots + a \cdot r^{n-1}$. That is, the geometric series S_n is the sum of the first n terms of the corresponding geometric sequence.

4. Assume that $f_1, f_2, \ldots, f_n, \ldots$ are the Fibonacci numbers. Prove each of the following:

   (a) For each $n \in \mathbb{N}$, f_{4n} is a multiple of 3.

   (b) For each $n \in \mathbb{N}$, f_{5n} is a multiple of 5.

   (c) For each $n \in \mathbb{N}$ with $n > 1$, $f_1 + f_2 + \cdots + f_{n-1} = f_{n+1} - 1$.

   (d) For each $n \in \mathbb{N}$, $f_1 + f_3 + \cdots + f_{2n-1} = f_{2n}$.

   (e) For each $n \in \mathbb{N}$, $f_2 + f_4 + \cdots + f_{2n} = f_{2n+1} - 1$.

   (f) For each $n \in \mathbb{N}$, $f_1^2 + f_2^2 + \cdots + f_n^2 = f_n f_{n+1}$.

   (g) For each $n \in \mathbb{N}$ such that $n \not\equiv 0 \pmod{3}$, f_n is an odd integer.

5. For the sequence $a_1, a_2, \ldots, a_n, \ldots$, assume that $a_1 = 1$ and that for each $n \in \mathbb{N}$, $a_{n+1} = \sqrt{5 + a_n}$.

   (a) Calculate, or approximate, a_2 through a_6.

   (b) Prove that for each $n \in \mathbb{N}$, $a_n < 3$.

6. For the sequence $a_1, a_2, \ldots, a_n, \ldots$, assume that $a_1 = 1$, $a_2 = 3$, and that for each $n \in \mathbb{N}$, $a_{n+2} = 3a_{n+1} - 2a_n$.

   (a) Calculate a_3 through a_6.

   (b) Make a conjecture for a formula for a_n for each $n \in \mathbb{N}$.

   (c) Prove that your conjecture in Exercise (6b) is correct.

7. For the sequence $a_1, a_2, \ldots, a_n, \ldots$, assume that $a_1 = 1$, $a_2 = 1$, and that for each $n \in \mathbb{N}$, $a_{n+2} = \dfrac{1}{2}\left(a_{n+1} + \dfrac{2}{a_n}\right)$.

(a) Calculate a_3 through a_6.

(b) Prove that for each $n \in \mathbb{N}$, $1 \le a_n \le 2$.

8. Let $a_1 = 1$, $a_2 = 1$, $a_3 = 1$, and for each natural number n, let

$$a_{n+3} = a_{n+2} + a_{n+1} + a_n.$$

(a) Compute a_4, a_5, a_6, and a_7.

(b) Prove that for each natural number n with $n > 1$, $a_n \le 2^{n-2}$.

9. (a) Compute $n!$ for the first ten natural numbers.

(b) Let $a_1 = 1$, and for each natural number k, let

$$a_{k+1} = a_k + k \cdot k!.$$

Compute a_n for the first ten natural numbers.

(c) Make a conjecture about a formula for a_n in terms of n that does not involve a summation or a recursion.

(d) Prove your conjecture in Part (c).

Chapter 6

Functions

6.1 Introduction to Functions

Preview Activity 1 (Functions from Previous Courses).

A **function** can be thought of as a procedure for associating with each element of some set, called the **domain of the function**, exactly one element of another set, called the **codomain of the function**. This can be thought of as an input-output-rule. The function takes the input, which is an element of the domain, and produces an output, which is an element of the codomain. In previous mathematics courses, when we wrote something like $f(x) = x^2 \sin x$, the notation meant the following:

- f is the name of the function.

- $f(x)$ is a real number. It is the output of the function when the input is the real number x. For example,

$$f\left(\frac{\pi}{2}\right) = \left(\frac{\pi}{2}\right)^2 \sin\left(\frac{\pi}{2}\right)$$
$$= \frac{\pi^2}{4} \cdot 1$$
$$= \frac{\pi^2}{4}.$$

- It is understood that the domain of the function is the set $\mathbb{R}$ of all real numbers. The domain can be considered to be the set of all possible inputs, and so this means that any real number can be used as an input. In general, we often think of the domain as the set of all possible real numbers x for which a real number output can be determined.

In this situation, we often consider the function f to be a rule that assigns to each real number input x a unique output $f(x)$. This is closely related to the equation $y = x^2 \sin x$. With this equation, we frequently think of x as the input and y as the output. In fact, we sometimes write $y = f(x)$.

Which of the following equations can be used to define a function with $x \in \mathbb{R}$ as the input and y as the output?

1. $y = x^2 - 2$ **5.** $x^2 + y^2 = 4$

2. $y^2 = x + 3$

6. $y = 2x - 1$
3. $y = \dfrac{1}{2}x^3 - 1$

4. $y = \dfrac{1}{2}x \sin x$ **7.** $y = \dfrac{x}{x - 1}$

Preview Activity 2 (The Birthday Function).

The functions in Preview Activity 1 and most of the functions we have studied in calculus and other mathematics courses had as domain and co-domain the set $\mathbb{R}$ of all real numbers, or some subset of $\mathbb{R}$. In most of these cases, the way in which the function associated elements of the domain with elements of the codomain was by a rule determined by an algebraic formula or some other mathematical expression. For example, when we say that f is the function such that

$$f(x) = \frac{x}{x - 1},$$

then the algebraic rule that determines the output of the function f when the input is x is $\dfrac{x}{x - 1}$. In this case, we would say that the domain of f is the set of all real numbers not equal to 1 since division by zero is not defined.

However, the concept of a function is much more general than this. The way in which a function associates elements of the domain with elements of the codomain can have many different forms. This input-output rule can be a formula, a graph, a table, a random process, a computer algorithm, or a verbal description. Following is such an example.

Let f be the function that assigns to each person his or her birthday (month and day). The domain of the function f is the set of all people and the codomain of f is the set of all days in a leap year (i.e., January 1 through December 31, including February 29).

1. Explain why f really is a function.

2. In 1995, Andrew Wiles became famous for publishing a proof of Fermat's Last Theorem. (See A. D. Aczel, *Fermat's Last Theorem: Unlocking the Secret of an Ancient Mathematical Problem*, Dell Publishing, New York, 1996). Andrew Wiles's birthday is April 11, 1953. Translate this fact into functional notation using the "birthday function" f. That is, fill in the spaces for the following question marks:

$$f(?) = ?.$$

3. Is the following statement true or false? Explain.

   For each day D of the year, there exists a person x such that $f(x) = D$.

4. Is the following statement true or false? Explain.

   For any people x and y, if x and y are different people, then $f(x) \neq f(y)$.

Preview Activity 3 (The Sum of the Divisors Function).

Let s be the function that associates with each natural number the sum of its distinct natural number factors. For example,

$$s(6) = 1 + 2 + 3 + 6$$
$$= 12.$$

1. What is the domain of the function s ?

2. Calculate $s(k)$ for each natural number k from 1 through 15.

3. Is $s\left(\sqrt{5}\right)$ defined? Explain. Is $s(\pi)$ defined? Is $s(-6)$ defined?

4. Does there exist a natural number n such that $s(n) = 5$? Justify your conclusion.

5. Is it possible to find two different natural numbers m and n such that $s(m) = s(n)$? Explain.

6. Are the following statements true or false?

   (a) For each $m \in \mathbb{N}$, there exists a natural number n such that $s(n) = m$.

   (b) For all $m, n \in \mathbb{N}$, if $m \neq n$, then $s(m) \neq s(n)$.

Review of Prior Work with Functions

One of the most important concepts in modern mathematics is that of a **function**. In previous mathematics courses, such as calculus, we have often thought of a function as some sort of input-output rule that assigns exactly one output to each input. In calculus and precalculus, the inputs and outputs were almost always real numbers.

In a more general sense, a function can be thought of as a way of associating with each element of some set, called the **domain of the function**, exactly one element of another set, called the **codomain of the function**. Using the input-output rule idea, the domain is the set of all inputs, and the codomain must contain all possible outputs. In calculus and precalculus, the domain and codomain of a function were almost always the set $\mathbb{R}$ of real numbers, or some subset of $\mathbb{R}$. For these functions, the way in which a function associates elements of the domain with exactly one element of the codomain was by means of some mathematical formula. For example, when we write something like $f(x) = x^2 + 3x$, we mean

- f is the name of the function, and

- $f(x)$ is a real number. It is the output of the function when the input is x.

So $f(2)$ is the real number output of the function when the input is 2. This means that

$$f(2) = 2^2 + 3 \cdot 2$$
$$= 10.$$

Even though the inputs and outputs are real numbers, in calculus, we frequently used algebraic representations of numbers as the inputs and outputs. For example, $f(a+h)$ represents the real number output of the function when the input is $a + h$. That is,

$$f(a+h) = (a+h)^2 + 3(a+h)$$
$$= (a^2 + 2ah + h^2) + (3a + 3h).$$

The key to remember is that a function must have exactly one output for each input as we saw in Preview Activity 1. When we write an equation such as

$$y = \frac{1}{2}x^3 - 1,$$

we can use this equation to define y as a function of x. This is because when we substitute a real number for x (the input), the equation produces exactly one real number for y (the output). We can give this function a name, such as g, and write

$$y = g\left(x\right) = \frac{1}{2}x^3 - 1.$$

However, as written, an equation such as

$$y^2 = x + 3$$

cannot be used to define y as a function of x since there are real numbers that can be substituted for x that will produce more than one possible value of y. For example, if $x = 1$, then $y^2 = 4$, and y could be -2 or 2.

Introduction to the General Concept of a Function

The concept of a function is much more general than the idea of a function used in calculus or precalculus. In particular, the domain and codomain do not have to be subsets of $\mathbb{R}$. In addition, the way in which a function associates elements of the domain with elements of the codomain can have many different forms. This input-output rule can be a formula, a graph, a table, a random process, a computer algorithm, or a verbal description. Two such examples were introduced in Preview Activity 2 and Preview Activity 3.

For the **birthday function** in Preview Activity 2, the domain would be the set of all people and the codomain would be the set of all days in a leap year. For the **sum of the divisors function** in Preview Activity 3, the domain is the set $\mathbb{N}$ of natural numbers, and the codomain could also be $\mathbb{N}$. In both of these cases, the input-output rule was a verbal description of how to assign an element of the codomain to an element of the domain.

We formally define the concept of a function as follows:

Definition. A **function** from a set A to a set B is a rule that associates with every element x of the set A exactly one element of the set B. A function from A to B is also called a **mapping** from A to B.

When we work with a function, we usually give it a name. The name is often a single letter, such as f or g. If f is a function from the set A to be the set B, we will write $f : A \to B$. This is simply shorthand notation for the fact that f is a function from the set A to the set B. In this case, we might also say that f maps A to B.

> **Definition.** Let $f : A \to B$. (This is read, "Let f be a function from A to B.") The set A is called the **domain** of the function f, and we write $A = \operatorname{dom}(f)$. The set B is called the **codomain** of the function f, and we write $B = \operatorname{codom}(f)$.
>
> If $x \in A$, then the element of B that is associated with x is denoted by $f(x)$ and is called the **image of x under f**. If $f(x) = y$, with $y \in B$, then x is called a **pre-image of y under f**. In this case, we would also say that x is the **independent variable** of the function f and y is the **dependent variable** of the function f.

Example 6.1. Consider the function $g : \mathbb{R} \to \mathbb{R}$ where $g(x)$ is defined by the formula

$$g(x) = x^2 - 2.$$

Note that this is indeed a function since given any input x in the domain, $\mathbb{R}$, there is exactly one output $g(x)$ in the codomain, $\mathbb{R}$. For example:

$$g(-2) = (-2)^2 - 2 = 2,$$
$$g(5) = 5^2 - 2 = 23,$$
$$g\left(\sqrt{2}\right) = \left(\sqrt{2}\right)^2 - 2 = 0,$$
$$g\left(-\sqrt{2}\right) = \left(-\sqrt{2}\right)^2 - 2 = 0.$$

So we say that the image of -2 is 2, the image of 5 is 23, and so on.

Notice in this case that the number 0 in the codomain has two pre-images, $-\sqrt{2}$ and $\sqrt{2}$. This does not violate the mathematical definition of a function since the definition only states that each input must produce one and only one output. That is, each element of the domain has exactly one image in the codomain. Nowhere does the definition stipulate that two different inputs cannot produce the same output.

Finding the pre-images of an element in the codomain can sometimes be difficult. In general, if y is in the codomain, to find its pre-images, we need to ask, "For which values of x in the domain will we have $y = g(x)$?" For example, for the function g, to find the pre-images of 5, we need to find all x for which $g(x) = 5$. In this case, since $g(x) = x^2 - 2$, we can do this by solving the equation

$$x^2 - 2 = 5.$$

The solutions of this equation are $-\sqrt{7}$ and $\sqrt{7}$. So for the function g, the pre-images of 5 are $-\sqrt{7}$ and $\sqrt{7}$.

Also notice that for this function, not every element in the codomain has a pre-image. For example, there is no input x such that $g(x) = -3$. This is true since for all real numbers x, $x^2 \geq 0$ and hence $x^2 - 2 \geq -2$. This means that for all x in $\mathbb{R}$,

$$g(x) \geq -2.$$

Finally, note that we introduced the function g with the sentence, "Consider the function $g : \mathbb{R} \to \mathbb{R}$, where $g(x)$ is defined by the formula $g(x) = x^2 - 2$." This is one correct way to do this, but we will frequently shorten this to, "Let $g : \mathbb{R} \to \mathbb{R}$ be defined by $g(x) = x^2 - 2$", or "Let $g : \mathbb{R} \to \mathbb{R}$, where $g(x) = x^2 - 2$."

Besides the domain and codomain, there is another important set associated with a function. The need for this was illustrated in Example 6.1 when it was noticed that there are elements in the codomain that have no pre-image or, equivalently, there are elements in the codomain that are not the image of any element in the domain. The set we are talking about is the subset of the codomain consisting of all images of the elements of the domain of the function, and it is called the range of the function.

Definition. Let $f : A \to B$. The set $\{f(x) \mid x \in A\}$ is called the **range of the function f** and is denoted by range (f). The range of f is sometimes called the **image of the function f** (or the **image of A under f**).

The range of $f : A \to B$ could equivalently be defined as follows:

$$\text{range}(f) = \{y \in B \mid y = f(x) \text{ for some } x \in A\}.$$

Activity 6.2 (Codomain and Range).

1. Let f be the function that assigns to each person his or her birthday (month and day).

   (a) What is the domain of this function?

   (b) What is a codomain for this function?

(c) In Preview Activity 2, we determined that the following statement is true:

For each day D of the year, there exists a person x such that $f(x) = D$.

What does this tell us about the range of the function f? Explain.

2. Let s be the function that associates with each natural number the sum of its distinct natural number factors.

(a) What is the domain of this function?

(b) What is a codomain for this function?

(c) In Preview Activity 3, we determined that the following statement is false:

For each $m \in \mathbb{N}$, there exists a natural number n such that $s(n) = m$.

Give an example of a natural number m that shows this statement is false, and explain what this tells us about the range of the function s.

The Graph of a Real Function

The function $g : \mathbb{R} \to \mathbb{R}$, where $g(x)$ is defined by $g(x) = x^2 - 2$, can be represented by a graph. We have graphed such functions in previous mathematics courses, and the familiar graph of this function is shown in Figure 6.1.

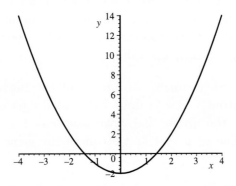

Figure 6.1: Graph of $y = g(x)$, where $g(x) = x^2 - 2$

Every point on this graph corresponds to an ordered pair (x, y) of real numbers where $y = g(x) = x^2 - 2$. Because we use the Cartesian plane when drawing this type of graph, we can only use this type of graph when both the domain and the codomain of the function are subsets of the real numbers $\mathbb{R}$. Such a function is sometimes called a **real function**. The graph of a real function is a visual way to communicate information about the function. This will be illustrated in the following activity, which is a review of material from previous mathematics courses.

Activity 6.3 (Using a Graph of a Real Function).
   The graph of a real function $f : \mathbb{R} \to \mathbb{R}$ is shown in Figure 6.2.

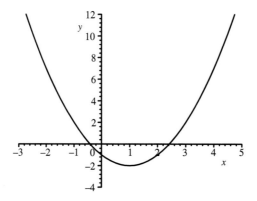

Figure 6.2: Graph of $y = f(x)$

1. We can use the graph to estimate the output for various inputs. This is done by estimating the y-coordinate for the point on the graph with a specified x-coordinate. On the graph, draw vertical lines at $x = -1$ and $x = 2$ and estimate the values of $f(-1)$ and $f(2)$.

2. Similarly, we can estimate the inputs of the function that produce a specified output. This is done by estimating the x-coordinates of the points on the graph that have a specified y-coordinate. Draw a horizontal line at $y = 4$ and estimate the value(s) of x such that $f(x) = 4$.

3. If certain conditions are satisfied, the graph also gives us an indication of the range of the function. The range is the set of all y-values that correspond to points on the graph. By stating that f is a function with

domain and codomain equal to $\mathbb{R}$, $f : \mathbb{R} \to \mathbb{R}$, we can assume that the graph shown is only a portion of the entire graph of the function. If we assume that the graph of the function continues the pattern shown on the left end and the right end, what is the range of this function? Note that we can only estimate the range since we will have to estimate the y-coordinate of the low point on the graph.

4. The graph shown could also be considered to be the entire graph for the function $k : [-3, 5] \to \mathbb{R}$. In this case, what is the range of the function k?

Example 6.4. Let $A = \{1, 2, 3\}$ and let $B = \{a, b\}$. Define the function $F : A \to B$ by
$$F(1) = a, F(2) = a, \text{ and } F(3) = b.$$

This is a function since each element of the domain is mapped to exactly one element in B. This function is defined by simply specifying the outputs for each input in the domain. A convenient way to illustrate or visualize this type of function is with an **arrow diagram** as shown in Figure 6.3.

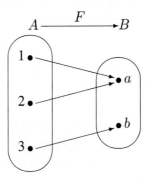

Figure 6.3: Arrow Diagram for a Function

This can be done when the domain and codomain of the function are finite (and small). We represent the elements of each set with points and then use arrows to show how the elements of the domain are associated with elements of the codomain. For example, the arrow from the point 2 in A to the point a in B represents the fact that $F(2) = a$.

Activity 6.5 (Creating Functions with Finite Domains).

Let $A = \{a, b, c, d\}$, $B = \{a, b, c\}$, and $C = \{s, t, u, v\}$. In each of the following exercises, draw an arrow diagram to represent your function when it is appropriate.

1. Create a function $f : A \to C$ whose range is the set C or explain why it is not possible to construct such a function.

2. Create a function $f : A \to C$ whose range is the set $\{u, v\}$ or explain why it is not possible to construct such a function.

3. Create a function $f : B \to C$ whose range is the set C or explain why it is not possible to construct such a function.

4. Create a function $f : A \to C$ whose range is the set $\{u\}$ or explain why it is not possible to construct such a function.

5. If possible, create a function $f : A \to C$ that satisfies the following condition:

$$\text{For all } x, y \in A, \text{ if } x \neq y, \text{ then } f(x) \neq f(y).$$

   If it is not possible to create such a function, explain why.

6. If possible, create a function $f : A \to \{s, t, u\}$ that satisfies the following condition:

$$\text{For all } x, y \in A, \text{ if } x \neq y, \text{ then } f(x) \neq f(y).$$

   If it is not possible to create such a function, explain why.

Example 6.6. In Section 4.4, we learned how to form the Cartesian product of two sets. Since a Cartesian product is a set, it could be used as the domain or codomain of a function. For example, we could use $\mathbb{Z} \times \mathbb{Z} = \mathbb{Z}^2$ as the domain of a function as follows:

$$\text{Let } f : \mathbb{Z} \times \mathbb{Z} \to \mathbb{Z} \text{ be defined by } f(m, n) = 2m + n.$$

A word about notation: Technically, an element of $\mathbb{Z} \times \mathbb{Z}$ is an ordered pair, and so we should write $f((m, n))$ for the output of the function f when the input is (m, n). However, the double parentheses seem unnecessary in this context and there should be no confusion if we write $f(m, n)$ for the output of the function f when the input is (m, n). So, for example, we simply write

$$f(3, 2) = 2 \cdot 3 + 2 = 8, \text{ and}$$
$$f(-4, 5) = 2 \cdot (-4) + 5 = -3.$$

Finding the pre-images of an element of the codomain, $\mathbb{Z}$, involves solving an equation with two variables. For example, to find the pre-images of $0 \in \mathbb{Z}$, we need to find all ordered pairs $(m, n) \in \mathbb{Z} \times \mathbb{Z}$ such that $f(m, n) = 0$. This means that

$$2m + n = 0.$$

Three such ordered pairs are $(0, 0)$, $(1, -2)$, and $(-1, 2)$. In fact, whenever we choose an integer value for m, we can find a corresponding integer n such that $2m + n = 0$. This means that 0 has infinitely many pre-images. One way to communicate this is to use set builder notation and say that the following set consists of all of the pre-images of 0:

$$\{(m, n) \in \mathbb{Z} \times \mathbb{Z} \mid 2m + n = 0\} = \{(m, n) \in \mathbb{Z} \times \mathbb{Z} \mid n = -2m\}.$$

The second formulation for this set was obtained by solving the equation $2m + n = 0$ for n.

Exercises 6.1

1. Let $f : \mathbb{R} \to \mathbb{R}$ be defined by $f(x) = x^2 - 2x$.

   (a) Evaluate $f(-3), f(-1), f(1),$ and $f(3)$.

   (b) Determine all of the pre-images of 0, and determine all of the pre-images of 4.

   (c) Determine all of the pre-images of -2.

   (d) Sketch a graph of the function f.

   (e) Determine the range of the function f.

2. Let $\mathbb{R}^* = \{x \in \mathbb{R} \mid x \geq 0\}$, and let $s : \mathbb{R} \to \mathbb{R}^*$ be defined by $s(x) = x^2$.

   (a) Evaluate $s(-3), s(-1), s(1),$ and $s(3)$.

   (b) Determine all of the pre-images of 0, and determine all of the pre-images of 4.

   (c) Determine all of the pre-images of 2.

(d) Sketch a graph of the function s.

(e) Determine the range of the function s.

3. Let $A = \{1, 2, 3, 4\}$ and let $B = \{a, b, c\}$. Which of the following arrow diagrams can be used to represent a function from A to B? Explain.

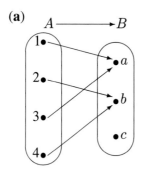 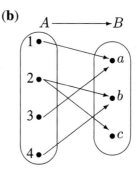

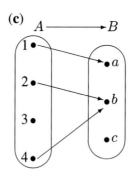

4. Let $f : \mathbb{Z} \to \mathbb{Z}$ be defined by $f(m) = 3 - m$.

(a) Evaluate $f(-7), f(-3), f(3),$ and $f(7)$.

(b) Determine all of the pre-images of 5, and determine all of the pre-images of 4.

(c) Determine the range of the function f.

(d) This function can be considered a real function since $\mathbb{Z} \subseteq \mathbb{R}$. So it is possible to sketch a graph of this function. The graph will be an infinite set of points that line on a line. However, it will not be a line since its domain is not $\mathbb{R}$ but is $\mathbb{Z}$.

5. Let $f : \mathbb{Z} \to \mathbb{Z}$ be defined by $f(m) = 2m + 1$.

(a) Evaluate $f(-7), f(-3), f(3)$, and $f(7)$.

(b) Determine all of the pre-images of 5, and determine all of the pre-images 4.

(c) Determine the range of the function f.

(d) Sketch a graph of the function f. See the comments in Exercise (4d).

6. **The Number of Divisors Function** Let d be the function that associates with each natural number the number of its natural number divisors. That is, $d : \mathbb{N} \rightarrow \mathbb{N}$ where $d(n)$ is the number of natural number divisors of n. For example, $d(6) = 4$ since 1, 2, 3, and 6 are the natural number divisors of 6.

(a) Calculate $d(k)$ for each natural number k from 1 through 12.

(b) Does there exist a natural number n such that $d(n) = 1$? If so, determine all pre-images of the natural number 1.

(c) Does there exist a natural number n such that $d(n) = 2$? If so, determine all pre-images of the natural number 2.

(d) Is the following statement true or false? Justify your conclusion.

$$\text{If } m, n \in \mathbb{N} \text{ and } m \neq n, \text{ then } d(m) \neq d(n).$$

(e) Calculate $d(2^k)$ for $k = 0$ and for each natural number k from 1 through 6.

(f) Based on your work in Exercise (6e), make a conjecture for a formula for $d(2^n)$ that depends only on n. Then use mathematical induction to prove that your conjecture is correct.

(g) Do you think the following statement is true or false?

For each $n \in \mathbb{N}$, there exists a natural number m such that $d(m) = n$.

7. Let $f : \mathbb{Z} \times \mathbb{Z} \rightarrow \mathbb{Z}$ be defined by $f(m, n) = m + 3n$.

(a) Calculate $f(-3, 4)$ and $f(-2, -7)$.

(b) Determine all the pre-images of 4 by using set builder notation to describe the set of all $(m, n) \in \mathbb{Z} \times \mathbb{Z}$ such that $f(m, n) = 4$.

8. Let $g : \mathbb{Z} \times \mathbb{Z} \rightarrow \mathbb{Z} \times \mathbb{Z}$ be defined by $g(m, n) = (2m, m - n)$.

(a) Calculate $g(3,5)$ and $g(-1,4)$.

(b) Determine all the pre-images of $(0,0)$. That is, find all $(m,n) \in \mathbb{Z} \times \mathbb{Z}$ such that $g(m,n) = (0,0)$.

(c) Determine all the pre-images of $(8,-3)$.

(d) Determine all the pre-images of $(1,1)$.

(e) Is the following proposition true or false? Justify your conclusion.

For each $(s,t) \in \mathbb{Z} \times \mathbb{Z}$, there exists an $(m,n) \in \mathbb{Z} \times \mathbb{Z}$ such that $f(m,n) = (s,t)$.

9. Recall that a **real function** is a function whose domain and codomain are subsets of the real numbers $\mathbb{R}$. Most of the functions used in calculus are real functions. Quite often, a real function is given by a formula or a graph with no specific reference to the domain or the codomain. In these cases, the usual convention is to assume that the domain of the real function f is the set of all real numbers x for which $f(x)$ is a real number, and that the codomain is $\mathbb{R}$. For example, if we define the (real) function f by

$$f(x) = \frac{x}{x-2},$$

we would be assuming that the domain is the set of all real numbers that are not equal to 2.

Determine the domain and range of each of the following real functions.

(a) The function k defined by $k(x) = \sqrt{x-3}$

(b) The function F defined by $F(x) = \ln(2x-1)$

(c) The function f defined by $f(x) = 3\sin(2x)$

(d) The function g defined by $g(x) = \dfrac{4}{x^2 - 4}$

6.2   More about Functions

Preview Activity 1 (A Function Defined by a Congruence).

Following are the statements of Theorem 3.21 and Corollary 3.22 from Section 3.4.

Theorem 3.21 *Let $n \in \mathbb{N}$ and let $a \in \mathbb{Z}$. If $a = nq + r$ and $0 \le r < n$ for some integers q and r, then $a \equiv r \pmod{n}$.*

This theorem says that an integer is congruent (mod n) to its remainder when it is divided by n. (Recall that we always mean the remainder guaranteed by the Division Algorithm, which is the least nonnegative remainder.) Since this remainder is unique and since the only possible remainders for division by n are $0, 1, 2, \ldots, n - 1$, we can state the following result.

Corollary 3.22 *If $n \in \mathbb{N}$, then each integer is congruent, modulo n, to precisely one of the integers $0, 1, 2, \ldots, n - 1$.*

1. Define the set $\mathbb{Z}_6$ to be $\mathbb{Z}_6 = \{0, 1, 2, 3, 4, 5\}$. For each $x \in \mathbb{Z}_6$, compute $x^2 + 3$ and then determine the value of r in $\mathbb{Z}_6$ so that

$$\left(x^2 + 3\right) \equiv r \pmod{6}.$$

   For example, $2^2 + 3 = 7$ and so $\left(2^2 + 3\right) \equiv 1 \pmod{6}$. Organize your results in a table with one column for the value of x and another column for the value of r, where $r \in \mathbb{Z}_6$ and $\left(x^2 + 3\right) \equiv r \pmod{6}$.

2. Explain how your work in Part (1) can be used to define a function from $\mathbb{Z}_6$ to $\mathbb{Z}_6$.

Preview Activity 2 (Derivatives).

In calculus, we learned how to find the derivatives of certain functions. For example, if

$$f(x) = x^2 (\sin x),$$

then we can use the product rule to obtain

$$f'(x) = 2x (\sin x) + x^2 (\cos x).$$

1. If possible, find the derivative of each of the following functions:

   (a) $f(x) = x^4 - 5x^3 + 3x - 7$ (d) $k(x) = e^{-x^2}$

   (b) $g(x) = \cos(5x)$ (e) $r(x) = |x|$

   (c) $h(x) = \dfrac{\sin x}{x}$

2. Is it possible to think of differentiation as a function? Explain. If so, what would be the domain of the function, what could be the codomain

of the function, and what is the rule for computing the element of the codomain (output) that is associated with a given element of the domain (input)?

Preview Activity 3 (The Number of Diagonals of a Polygon).

A **polygon** is a closed plane figure formed by the joining of three or more straight lines. For example, a **triangle** is a polygon that has three sides; a **quadrilateral** is a polygon that has four sides and includes squares, rectangles, and parallelograms; a **pentagon** is a polygon that has five sides; and an **octagon** is a polygon that has eight sides. A **regular polygon** is one that has equal length sides and congruent interior angles.

A **diagonal of a polygon** is a line segment that connects two non-adjacent vertices of the polygon. In this activity, we will assume that all polygons are convex polygons so that, except for the vertices, each diagonal lies inside the polygon.

1. How many diagonals does a triangle have? How many diagonals does a square have? How many diagonals does any quadrilateral have?

2. Let $D = \mathbb{N} - \{1, 2\}$. Define $d : D \to \mathbb{N} \cup \{0\}$ so that $d(n)$ is the number of diagonals of a convex polygon with n sides. Complete the following table, which shows the values of $d(n)$ for selected values of n.

n	$d(n)$	n	$d(n)$
3		6	
4		7	
5		8	

3. Let $f : \mathbb{R} \to \mathbb{R}$ be defined by

$$f(x) = \frac{x(x-3)}{2}.$$

Complete the following table, which shows the values of $f(x)$ for selected values of x.

x	$f(x)$	x	$f(x)$
0		5	
1		6	
2		7	
3		8	
4		9	

4. What (if any) are the differences between the functions described in Parts (2) and (3)? Explain.

In Section 6.1 and in the Preview Activities for this section, we have seen many examples of functions. Functions are everywhere in mathematics. We have also seen various ways to represent functions and to convey information about them. For example, we have seen that the rule for determining outputs of a function can be given by a formula, a graph, or a table of values. We have also seen that sometimes it is more convenient to give a verbal description of the rule for a function. In cases where the domain and codomain are small, finite sets, we used an arrow diagram to convey information about how inputs and outputs are associated without explicitly stating a rule.

Examples of Functions Using Verbal Descriptions

1. **The birthday function** from Preview Activity 2 from Section 6.1. In this case, f is the function that assigns to each person his or her birthday (month and day). The domain of the function f is the set of all people and the codomain of f is the set of all days in a leap year (i.e., January 1 through December 31, including February 29).

2. The function **balance** that assigns to each customer with a savings account at ManyBucksBank the balance in that customer's savings account at the end of the business day on March 3, 2002. The domain for this function would be the set of all savings account customers for the bank, and we could use the set of all rational numbers as the codomain.

   Note: For a given function, the codomain does not have to be the same as the range. In this case, it might be quite a hassle to determine the range. The range would consist of all of the actual savings account balances of the customers of ManyBucksBank at the end of the business day on March 3, 2002. We often have some freedom in selecting the codomain. In this case, since the savings balances are monetary values given in dollars and cents, it seemed reasonable to use the set of rational numbers as the codomain.

3. The **sum of the divisors function** introduced in Preview Activity 3 from Section 6.1 and studied further in Activity 6.2 is an example of a function that can be given by a formula, but the verbal description may be easier to understand and use.

Recall that $s : \mathbb{N} \to \mathbb{N}$, where for each $n \in \mathbb{N}$, $s(n)$ is the sum of the distinct natural number divisors of n. This is a function that is studied and used in number theory, where a formula is determined for $s(n)$. This formula can be written as

$$s(n) = \sum_{d \mid n} d.$$

We do not have to worry about this notation in this text. Basically, the formula says that $s(n)$ is the sum of all natural numbers d, where d divides n. Which formulation of $s(n)$ is easier to understand?

___

Functions Involving Congruences

The function in Preview Activity 1 is sort of a mix between a formula and a verbal description. Following is a similar example.

Let $\mathbb{Z}_4 = \{0, 1, 2, 3\}$. Define $g : \mathbb{Z}_4 \to \mathbb{Z}_4$ by $g(x) = r$, where $r \in \mathbb{Z}_4$ and $x^3 \equiv r \pmod 4$. Then

$$
\begin{array}{llll}
g(0) = 0 & \text{since} & 0^3 \equiv 0 & \pmod 4 \\
g(1) = 1 & \text{since} & 1^3 \equiv 1 & \pmod 4 \\
g(2) = 0 & \text{since} & 2^3 \equiv 0 & \pmod 4 \\
g(3) = 3 & \text{since} & 3^3 \equiv 3 & \pmod 4.
\end{array}
$$

This information about the outputs of this function could also be communicated by means of an arrow diagram or by means of a table of values.

The verbal description and the notation for the outputs of this function are a bit cumbersome, and we frequently use a more concise notation. Instead of writing, "$g : \mathbb{Z}_4 \to \mathbb{Z}_4$ by $g(x) = r$, where $r \in \mathbb{Z}_4$ and $x^3 \equiv r \pmod 4$," we will write

$$g : \mathbb{Z}_4 \to \mathbb{Z}_4 \text{ by } g(x) = x^3 \pmod 4.$$

Similarly, we could define the function from Preview Activity 1 as follows:

$$\text{Let } f : \mathbb{Z}_6 \to \mathbb{Z}_6 \text{ by } f(x) = (x^2 + 3) \pmod 6.$$

Equality of Functions

The idea of equality of functions has been in the background of our discussion of functions, and it is now time to discuss it explicitly. The preliminary work for this discussion was Preview Activity 3, in which $D = \mathbb{N} - \{1, 2\}$, and there were two functions:

- $d : D \to \mathbb{N} \cup \{0\}$, where $d(n) =$ the number of diagonals of a convex polygon with n sides

- $f : \mathbb{R} \to \mathbb{R}$, where $f(x) = \dfrac{x(x-3)}{2}$.

In Preview Activity 3, we saw that these two functions produced the same outputs for certain values of the input (independent variable). For example, we can verify that

$$d(3) = f(3) = 0,$$
$$d(4) = f(4) = 2,$$
$$d(5) = f(5) = 5,$$
$$d(6) = f(6) = 9.$$

Although the functions produce the same outputs for some inputs, these are two different functions. For example, the outputs of the function f are determined by a formula, and the outputs of the function d are determined by a verbal description. This is not enough, however, to say that these are two different functions. Based on the evidence from Preview Activity 3, we might make the following conjecture:

$$\text{For } n \geq 3, \, d(n) = \frac{n(n-3)}{2}.$$

While we do not know if this is true at this point, even if it is, we know these two functions are not the same. For example,

- $f(0) = 0$, but 0 is not in the domain of d;

- $f(\pi) = \dfrac{\pi(\pi - 3)}{2}$, yet π is not in the domain of d.

We thus see the importance of considering the domains and codomains of two functions we believe to be equal. This motivates the following definition.

Definition. Two functions f and g are **equal** provided that

- The domain of f equals the domain of g. That is, $\text{dom}(f) = \text{dom}(g)$.

- The codomain of f equals the codomain of g. That is, $\text{codom}(f) = \text{codom}(g)$.

- For each x in the domain of f (which equals the domain of g), $f(x) = g(x)$.

Sequences as Functions

So far, we have considered a sequence to be an infinite list of objects that are indexed (subscripted) by the natural numbers (or some infinite subset of $\mathbb{N} \cup \{0\}$). Using this idea, we often write a sequence in the following form:

$$a_1, a_2, \ldots, a_n, \ldots$$

In order to shorten our notation, we will often use the notation $\langle a_n \rangle$ to represent this sequence. Sometimes a formula can be used to represent the terms of a sequence, and we might include this formula as the nth term in the list for a sequence such as in the following example:

$$1, \frac{1}{2}, \frac{1}{3}, \ldots, \frac{1}{n}, \ldots$$

In this case, the nth term of the sequence is $\dfrac{1}{n}$. If we know a formula for the nth term, we often use this formula to represent the sequence. For example, we might say

> Define the sequence $\langle a_n \rangle$ by $a_n = \dfrac{1}{n}$ for each $n \in \mathbb{N}$.

This shows that this sequence is a function with domain $\mathbb{N}$. Given an element of the domain, we can consider a_n to be the output. In this case, we have used subscript notation to indicate the output rather than the usual function notation. We could just as easily write

$$a(n) = \frac{1}{n} \text{ instead of } a_n = \frac{1}{n}.$$

We make the following formal definition.

> **Definition.** An (infinite) **sequence** is a function whose domain is $\mathbb{N}$ or some infinite subset of $\mathbb{N} \cup \{0\}$.

Mathematical Processes as Functions

Certain mathematical processes can be thought of as functions. In Preview Activity 2, we reviewed how to find the derivatives of certain functions, and we considered whether or not we could think of this differentiation process as a function. In this case, if we use a differentiable function as the input and consider the derivative of that function to be the output, then we have the makings of a function. Computer algebra systems such as *Maple* and *Mathematica* have this derivative function as one of their predefined operators. Even some calculators now have a derivative function.

Following is an example of *Maple* code used to find the derivative function of the function given by $f(x) = x^2 (\sin x)$. The lines that start with the *Maple* prompt, $[>$, are the lines typed in by the user. The centered lines following these show the resulting *Maple* output. The first line defines the function f, and the second line uses the derivative function D to produce the derivative of the function f.

$[>$ f := x → x^2* sin(x);

$$f := x \to x^2 \sin(x)$$

$[>$ f1 := D(f);

$$f1 := x \to 2x \sin(x) + x^2 \cos(x)$$

We must be careful when determining the domain for the derivative function since there are functions that are not differentiable. To make things are reasonably easy, we just let S be the set of all real functions that are differentiable and call this the domain of the derivative function D. We will use the set T of all real functions as the codomain. So our function D is

$$D : S \to T \text{ by } D(f) = f'.$$

Activity 6.7 (Integration as a Function).

In calculus, we learned that if f is real function that is continuous on the closed interval $[a, b]$, then the definite integral $\int_a^b f(x)\, dx$ is a real number.

In fact, one form of the **Fundamental Theorem of Calculus** states that

$$\int_a^b f(x)\, dx = F(b) - F(a),$$

where F is any antiderivative of f, that is, where $F' = f$.

1. Let $[a, b]$ be a closed interval of real numbers and let $C[a, b]$ be the set of all real functions that are continuous on $[a, b]$. That is,

$$C[a, b] = \{f : [a, b] \to \mathbb{R} \mid f \text{ is continuous on } [a, b]\}.$$

   (a) Explain how the definite integral $\int_a^b f(x)\, dx$ can be used to define a function I from $C[a, b]$ to $\mathbb{R}$.

   (b) Let $[a, b] = [0, 2]$. Calculate $I(f)$ where $f(x) = x^2 + 1$.

   (c) Let $[a, b] = [0, 2]$. Calculate $I(g)$ where $g(x) = \sin(\pi x)$.

In calculus, we also learned how to determine the indefinite integral $\int f(x)\, dx$ of a continuous function f.

2. If $f(x) = x^2 + 1$, determine $\int f(x)\, dx$.

3. Let f be a continuous function on the closed interval $[0, 1]$ and let T be the set of all real functions. Let $A(f) = \int f(x)\, dx$. Does this define a function A from $C[0, 1]$ to T? Explain.

4. Another form of the Fundamental Theorem of Calculus states that if f is continuous on the interval $[a, b]$ and if

$$g(x) = \int_a^x f(t)\, dt$$

   for each x in $[a, b]$, then $g'(x) = f(x)$. That is, g is an antiderivative of f. Explain how this theorem can be used to define a function from $C[a, b]$ to T where the output of the function is an antiderivative of the input. (Recall that T is the set of all real functions.)

Some Standard Functions

The following are examples of functions that appear frequently in many areas of mathematics.

1. **The Identity Map**

   Let A be a nonempty set. The **identity map on the set A**, denoted by I_A , is the function $I_A : A \to A$ defined by $I_A(x) = x$ for every x in A. That is, for the identity map, the output is always equal to the input.

2. **Constant Functions**

   Any function whose range consists of a single element is called a **constant function**. This term is used because for such a function, the output is always the same. For example, let A and B be non-empty sets, and let $b \in B$. The function $f : A \to B$ defined by $f(x) = b$ for all $x \in A$ is a constant function.

3. **The Characteristic Function of a Set**

   Let A be a subset of some universal set U. The **characteristic function of the set A**, denoted by χ_A, is the function $\chi_A : U \to \{0, 1\}$ defined by

   $$\chi_A(x) = \begin{cases} 1 \text{ if } x \in A \\ 0 \text{ if } x \notin A \end{cases}$$

   The symbol χ is the lower case letter chi in the Greek alphabet. The characteristic function of A maps the elements of A to the number 1 and maps the elements of the complement of A to the element 0.

   If A is a subset of $\mathbb{R}$, then we can sketch a graph of its characteristic function. For example, if $A = [-1, 2]$, then its graph would be the graph shown in Figure 6.4.

4. **Projection Functions**

   Let A and B be two nonempty sets. There are two **projection functions** with domain $A \times B$, the Cartesian product of A and B. One projection function will map an ordered pair to its first coordinate, and the other projection function will map the ordered pair to its second coordinate. So we define

   $p_1 : A \times B \to A$ by $p_1(a, b) = a$ for every $(a, b) \in A \times B$; and

   $p_2 : A \times B \to B$ by $p_2(a, b) = b$ for every $(a, b) \in A \times B$.

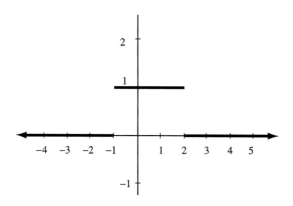

Figure 6.4: Graph of Characteristic Function of $A = [-1, 2]$

A Function as a Set of Ordered Pairs

When we graph a real function, we plot ordered pairs in the Cartesian plane where the first coordinate is the input of the function and the second coordinate is the output of the function. For example, if $g : \mathbb{R} \to \mathbb{R}$, then every point on the graph of g is an ordered pair (x, y) of real numbers where $y = g(x)$. This shows how we can generate ordered pairs from a function. It happens that we can do this with any function.

For example, let

$$A = \{1, 2, 3\} \text{ and } B = \{a, b\}.$$

Define the function $F : A \to B$ by

$$F(1) = a,$$
$$F(2) = b, \text{ and}$$
$$F(3) = b.$$

We can convert each of these to an ordered pair in $A \times B$ by using the input as the first coordinate and the output as the second coordinate. For example, $F(1) = a$ is converted to $(1, a)$, $F(2) = b$ is converted to $(2, b)$, and $F(3) = b$ is converted to $(3, b)$. So, we can think of this function as a set of ordered pairs, which is a subset of $A \times B$, and write

$$F = \{(1, a), (2, b), (3, b)\}.$$

In fact, any function can be represented as a set of ordered pairs. For example, if we have a real function, such as $g : \mathbb{R} \to \mathbb{R}$ by $g(x) = x^2 - 2$, then we can think of g as the following infinite subset of $\mathbb{R} \times \mathbb{R}$:

$$g = \left\{ (x,y) \in \mathbb{R} \times \mathbb{R} \mid y = x^2 - 2 \right\}.$$

If the context is clear, we usually simplify this to

$$g = \left\{ (x,y) \mid y = x^2 - 2 \right\} \quad \text{or} \quad g = \left\{ (x, x^2 - 2) \mid x \in \mathbb{R} \right\}.$$

Activity 6.8 (Defining a Function as a Set of Ordered Pairs).

So far, we have started with a function and generated a set of ordered pairs that we can use to represent the function. It is possible to reverse this process and first define a function as a special type of set of ordered pairs. For example, if we started with $A = \{1, 2, 3\}$, $B = \{a, b\}$, and defined

$$F = \{(1, a), (2, b), (3, b)\} \subseteq A \times B,$$

then we could think of F as a function from A to B with

$$F(1) = a,$$
$$F(2) = b, \text{ and}$$
$$F(3) = b.$$

1. Could we use the following subset of $A \times B$ to define a function from A to B? Explain.

$$f = \{(1, a), (2, a), (3, a), (1, b)\}$$

2. Could we use the following subset of $A \times B$ to define a function from A to B? Explain.

$$g = \{(1, a), (2, b), (3, a)\}$$

3. Could we use the following subset of $A \times B$ to define a function from A to B? Explain.

$$h = \{(1, a), (2, b)\}$$

In Section 6.1, we defined a function f from a set A to a set B to be a rule that associates with every element x of the set A exactly one element of the set B.

4. Carefully reformulate the definition of a function f from a set A to a set B to be a set of ordered pairs that is a subset of $A \times B$ that has some clearly defined property or properties. The examples in Parts (1) through (3) may be of help in specifying these properties.

   This definition should be equivalent to the definition of a function given in Section 6.1. This means that a function as defined in Section 6.1 should produce a subset of $A \times B$ with the specified properties, and a subset of $A \times B$ with the specified properties could be used to define a function with the first coordinate as the input and the second coordinate as the output.

<u>Note</u>: Many mathematicians believe that this ordered pair representation of a function is the most rigorous definition of a function. It allows us to use set theory to work with and compare functions. For example, equality of functions becomes a question of equality of sets. Therefore, many textbooks will use the ordered pair representation of a function as the definition of a function.

Exercises 6.2

1. Let $\mathbb{Z}_5 = \{0, 1, 2, 3, 4\}$. Define $f : \mathbb{Z}_5 \to \mathbb{Z}_5$ by $f(x) = x^4 \pmod 5$.

   (a) Calculate $f(0)$, $f(1)$, $f(2)$, $f(3)$, and $f(4)$.
   (b) Represent the function f with an arrow diagram.
   (c) Represent the function f as a set of ordered pairs.
   (d) Is the function f a constant function? Explain.

2. Let $\mathbb{Z}_5 = \{0, 1, 2, 3, 4\}$. Define $g : \mathbb{Z}_5 \to \mathbb{Z}_5$ by $g(x) = x^5 \pmod 5$.

   (a) Calculate $g(0)$, $g(1)$, $g(2)$, $g(3)$, and $g(4)$.
   (b) Represent the function g with an arrow diagram.
   (c) Represent the function g as a set of ordered pairs.
   (d) The function g is equal to what standard function on $\mathbb{Z}_5$? Explain.

3. Represent each of the following sequences as functions. In each case, state the domain, codomain, and rule for determining the outputs of the function. Also, determine if any of the sequences are equal.

(a) $1, \dfrac{1}{4}, \dfrac{1}{9}, \dfrac{1}{16}, \dots$

(b) $\dfrac{1}{3}, \dfrac{1}{9}, \dfrac{1}{27}, \dfrac{1}{81}, \dots$

(c) $1, -1, 1, -1, 1, -1, \dots$

(d) $\cos(0), \cos(\pi), \cos(2\pi), \cos(3\pi), \cos(4\pi), \dots$

4. Let $A = \{1, 2\}$ and let $B = \{x, y, z\}$.

   (a) Calculate all the outputs for all possible inputs for the projection function $p_1 : A \times B \to A$.

   (b) What is the range of this projection function?

   (c) Is the following statement true or false? Explain.

   For all $(m, n), (u, v) \in A \times B$, if $(m, n) \neq (u, v)$, then $p_1(m, n) \neq p_1(u, v)$.

5. In Exercise (6) from Section 6.1, we introduced the **number of divisors function** d. For this function, $d : \mathbb{N} \to \mathbb{N}$, where $d(n)$ is the number of natural number divisors of n.

   A function that is related to this function is the so-called **set of divisors function**. This can be defined as a function S that associates with each natural number the set of its distinct natural number factors. For example, $S(6) = \{1, 2, 3, 6\}$.

   (a) Discuss the function S by carefully stating its domain, codomain, and its rule for determining outputs.

   (b) Determine $S(n)$ for at least five different values of n.

   (c) Determine $S(n)$ for at least three different prime number values of n.

   (d) Does there exist a natural number n such that $|S(n)| = 1$? Explain. [Recall that $|S(n)|$ represents the cardinality of the set $S(n)$.]

   (e) Does there exist a natural number n such that $|S(n)| = 2$? Explain.

   (f) Write the output for the function d in terms of the output for the function S. That is, write $d(n)$ in terms of $S(n)$.

   (g) Is the following statement true or false? Justify your conclusion.

For all natural numbers m and n, if $m \neq n$, then $S(m) \neq S(n)$.

(**h**) Is the following statement true or false? Justify your conclusion.

For all sets T that are subsets of N, there exists a natural number n such that $S(n) = T$.

6. Let $D = \mathbb{N} - \{1, 2\}$ and define $d : D \to \mathbb{N} \cup \{0\}$ by $d(n) =$ the number of diagonals of a convex polygon with n sides. In Preview Activity 3, we showed that for values of n from 3 through 8,

$$d(n) = \frac{n(n-3)}{2}.$$

Use mathematical induction to prove that for all $n \in D$,

$$d(n) = \frac{n(n-3)}{2}.$$

Hint: To get an idea of how to handle the inductive step, use a pentagon. First, form all the diagonals that can be made from four of the vertices. Then consider how to make new diagonals when the fifth vertex is used. This may generate an idea of how to proceed from a polygon with k sides to a polygon with $k + 1$ sides.

7. A **2 by 2 matrix over** $\mathbb{R}$ is a rectangular array of four real numbers arranged in two rows and two columns. We usually write this array inside brackets (or parentheses) as follows:

$$A = \begin{bmatrix} a & b \\ c & d \end{bmatrix},$$

where a, b, c, and d are real numbers. The **determinant** of the 2 by 2 matrix A, denoted by $\det(A)$, is defined as

$$\det(A) = ad - bc.$$

(**a**) Calculate the determinant of each of the following matrices:

$$\begin{bmatrix} 3 & 5 \\ 4 & 1 \end{bmatrix}, \begin{bmatrix} 1 & 0 \\ 0 & 7 \end{bmatrix}, \text{ and } \begin{bmatrix} 3 & -2 \\ 5 & 0 \end{bmatrix}.$$

(b) Let $\mathcal{M}_{2,2}$ represent the set of all 2 by 2 matrices over $\mathbb{R}$. The mathematical process of finding the determinant of a 2 by 2 matrix over $\mathbb{R}$ can be thought of as a function. Carefully explain how to do so including a clear statement of the domain and codomain of this function.

8. Using the notation from Exercise (7), let

$$A = \begin{bmatrix} a & b \\ c & d \end{bmatrix}$$

be a 2 by 2 matrix over $\mathbb{R}$. The **transpose of the matrix** A, denoted by A^T, is the 2 by 2 matrix over $\mathbb{R}$ defined by

$$A^T = \begin{bmatrix} a & c \\ b & d \end{bmatrix}.$$

(a) Calculate the transpose of each of the following matrices:

$$\begin{bmatrix} 3 & 5 \\ 4 & 1 \end{bmatrix}, \begin{bmatrix} 1 & 0 \\ 0 & 7 \end{bmatrix}, \text{ and } \begin{bmatrix} 3 & -2 \\ 5 & 0 \end{bmatrix}.$$

(b) Let $\mathcal{M}_{2,2}$ represent the set of all 2 by 2 matrices over $\mathbb{R}$. The mathematical process of finding the transpose of a 2 by 2 matrix over $\mathbb{R}$ can be thought of as a function. Carefully explain how to do so, including a clear statement of the domain and codomain of this function.

6.3 Types of Functions

Preview Activity 1 (Functions with Finite Domains).
Let $A = \{1, 2, 3\}$, $B = \{a, b, c, d\}$, and $C = \{s, t\}$. Define

$f : A \to B$ by	$g : A \to B$ by	$h : A \to C$ by
$f(1) = a$	$g(1) = a$	$h(1) = s$
$f(2) = b$	$g(2) = b$	$h(2) = t$
$f(3) = c$	$g(3) = a$	$h(3) = s$

1. Determine the range of each of these functions.

2. Which of these functions (if any) satisfy the following property for a function F?

   For all $x, y \in \text{dom}\,(F)$, if $x \neq y$, then $F(x) \neq F(y)$.

3. Which of these functions (if any) satisfy the following property for a function F?

   For all $x, y \in \text{dom}\,(F)$, if $F(x) = F(y)$, then $x = y$.

4. Which of these functions (if any) have their range equal to their codomain?

5. Which of the these functions (if any) satisfy the following property for a function F?

   For all $y \in \text{codom}\,(F)$, there exists an $x \in \text{dom}\,(F)$ such that $F(x) = y$.

Preview Activity 2 (Creating Functions with Finite Domains).

If you have not already done so, complete the following parts of Activity 6.5: Let $A = \{a, b, c, d\}$, $B = \{a, b, c\}$, and $C = \{s, t, u, v\}$. In each of the following exercises, draw an arrow diagram to represent your function when it is appropriate.

1. Create a function $f : A \rightarrow C$ whose range is the set C or explain why it is not possible to construct such a function.

2. Create a function $f : B \rightarrow C$ whose range is the set C or explain why it is not possible to construct such a function.

3. If possible, create a function $f : A \rightarrow C$ that satisfies the following condition:

   For all $x, y \in A$, if $x \neq y$, then $f(x) \neq f(y)$.

   If it is not possible to create such a function, explain why.

4. If possible, create a function $f : A \rightarrow \{s, t, u\}$ that satisfies the following condition:

   For all $x, y \in A$, if $x \neq y$, then $f(x) \neq f(y)$.

If it is not possible to create such a function, explain why.

Preview Activity 3 (A Function of Two Variables).
Define $f : \mathbb{Z} \times \mathbb{Z} \to \mathbb{Z}$ by $f(m, n) = m - 3n$.

<u>Note</u>: Since the domain of this function is $\mathbb{Z} \times \mathbb{Z}$ and each element of $\mathbb{Z} \times \mathbb{Z}$ is an ordered pair of integers, we frequently call this type of function a **function of two variables**.

1. Calculate $f(3, 2)$, $f(3, -2)$, $f(0, 2)$, $f(2, 0)$, and $f(-6, 0)$.

2. If $m \in \mathbb{Z}$, what is $f(m, 0)$? If $n \in \mathbb{Z}$, what is $f(0, n)$?

3. What is the range of the function f? Explain.

4. Is the following statement true or false? Explain.

   For every $y \in \mathbb{Z}$, there exists $(m, n) \in \mathbb{Z} \times \mathbb{Z}$ such that $f(m, n) = y$? Explain.

5. Is the following statement true or false? Explain.

   For all $(m, n), (p, q) \in \mathbb{Z} \times \mathbb{Z}$, if $(m, n) \neq (p, q)$, then $f(m, n) \neq f(p, q)$.

6. Is the following statement true or false? Explain.

   For all $(m, n), (p, q) \in \mathbb{Z} \times \mathbb{Z}$, if $f(m, n) = f(p, q)$, then $(m, n) = (p, q)$.

Functions are frequently used in mathematics to define and describe certain relationships between sets and other mathematical objects. In addition, functions can be used to impose certain mathematical structures on sets. Two special types of functions that are used to describe these relationships are called injections and surjections. An injection is often called a one-to-one function, and a surjection is often called an onto function. Before defining these types of functions, we will revisit what the definition of a function tells us.

Consequences of the Definition of a Function

Let A and B be sets. Given a function $f : A \to B$, we know the following:

- For every $x \in A$, $f(x) \in B$. That is, every element of A is an input for the function f. This could also be stated as follows: For each $x \in A$, there exists a $y \in B$ such that $y = f(x)$.

- For a given $x \in A$, there is exactly one $y \in B$ such that $y = f(x)$.

Notice that the definition of a function does not require that different inputs produce different outputs. That is, it is possible to have $x_1, x_2 \in A$ with $x_1 \neq x_2$ and $f(x_1) = f(x_2)$. The arrow diagram in Figure 6.5 illustrates such a function.

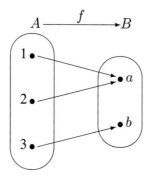

Figure 6.5: Arrow Diagram for a Function

Also, the definition of a function does not require that the range of the function must equal the codomain. The range is always a subset of the codomain, but these two sets are not required to be equal. That is, if $f : A \to B$, then it is possible to have a $y \in B$ such that $f(x) \neq y$ for all $x \in A$. The arrow diagram in Figure 6.6 illustrates such a function.

Injections

We have seen that there exist functions $f : A \to B$ for which there exist $x_1, x_2 \in A$ with $x_1 \neq x_2$ and $f(x_1) = f(x_2)$. This means that there exist different inputs that produce the same output. What does it mean to say that this does not happen? Using the input-output version of a function, it means that if the inputs are different then the outputs are different. More formally, it means that if $x_1 \neq x_2$, then $f(x_1) \neq f(x_2)$. Functions that satisfy this condition are given a special name.

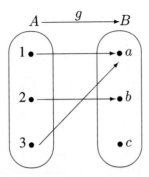

Figure 6.6: Arrow Diagram for a Function

Definition. Let $f : A \to B$ be a function from the set A to the set B. The function f is called an **injection** provided that

for all $x_1, x_2 \in A$, if $x_1 \neq x_2$, then $f(x_1) \neq f(x_2)$.

When f is an injection, we also say that f is a **one-to-one function**, or that f is an **injective function**.

The condition that specifies that a function f is an injection is given in the form of a conditional statement. As we shall see in the examples, it is usually easier to use the contrapositive of this conditional statement. We now summarize the conditions for f being an injection or not being an injection.

Let $f : A \to B$.

"The function f is an injection" **means that**

- For all $x_1, x_2 \in A$, if $x_1 \neq x_2$, then $f(x_1) \neq f(x_2)$; or

- For all $x_1, x_2 \in A$, if $f(x_1) = f(x_2)$, then $x_1 = x_2$.

"The function f is not an injection" **means that**

- There exist $x_1, x_2 \in A$ such that $x_1 \neq x_2$ and $f(x_1) = f(x_2)$.

We will see how to use these ideas in the activities and examples that follow the discussion of surjections.

Surjections

We have seen that there exist functions $f : A \to B$ for which there exists a $y \in B$ such that for all $x \in A$, $f(x) \neq y$. This means that there exists an element in the codomain that is not an output or, equivalently, that it is not in the range. Functions for which this does not happen are given a special name. What does it mean to say that this does not happen? One way to express this is to say that every element in the codomain is an output or that the codomain is equal to the range. Using quantifiers, this means that for every $y \in B$, there exists an $x \in A$ such that $f(x) = y$.

Definition. Let $f : A \to B$ be a function from the set A to the set B. The function f is called a **surjection** provided that the codomain of f equals the range of f. This means that

     for every $y \in B$, there exists an $x \in A$ such that $f(x) = y$.

When f is a surjection, we also say that f is an **onto function** or that f maps **A onto B**. We also say that f is a **surjective function**.

We now summarize the conditions for f being a surjection or not being a surjection.

Let $f : A \to B$.

"The function f is a surjection" means that

- range (f) = codom $(f) = B$; or

- For every $y \in B$, there exists an $x \in A$ such that $f(x) = y$.

"The function f is not a surjection" means that

- range $(f) \neq$ codom (f); or

- There exists a $y \in B$ such that for all $x \in A$, $f(x) \neq y$.

Definition. A **bijection** is a function that is both an injection and a surjection. If the function f is a bijection, we also say that f is **one-to-one and onto** and that f is a **bijective function**.

Activity 6.9 (Functions from Preview Activity 1).
Let $A = \{1, 2, 3\}$, $B = \{a, b, c, d\}$, and $C = \{s, t\}$. Define

$f : A \to B$ by	$g : A \to B$ by	$h : A \to C$ by
$f(1) = a$	$g(1) = a$	$h(1) = s$
$f(2) = b$	$g(2) = b$	$h(2) = t$
$f(3) = c$	$g(3) = a$	$h(3) = s$

1. Which of these functions are injections, and which of these functions are surjections?

2. Are your answers in Part (1) consistent with your answers in Preview Activity 1? Explain.

The Importance of the Domain and Codomain

The functions in the next two examples will illustrate why the domain and the codomain of a function are important when we need to determine if the function is a surjection.

Example 6.10. Let $f : \mathbb{R} \to \mathbb{R}$ be defined by $f(x) = x^2 + 1$. Notice that

$$f(2) = 5 \text{ and } f(-2) = 5.$$

This is enough to prove that the function f is not an injection since this shows that there exist two different inputs that produce the same output.

Since $f(x) = x^2 + 1$, we know that $f(x) \geq 1$ for all $x \in \mathbb{R}$. This means that the function f is not a surjection. For example, -2 is in the codomain of f and $f(x) \neq -2$ for all x in the domain of f.

Example 6.11. Let $T = \{y \in \mathbb{R} \mid y \geq 1\}$, and define $F : \mathbb{R} \to T$ by $F(x) = x^2 + 1$. As in Example 6.10, the function F is not an injection since $F(2) = F(-2)$.

Is the function F surjective? That is, does F map $\mathbb{R}$ onto T? As in Example 6.10, we do know that $F(x) \geq 1$ for all $x \in \mathbb{R}$.

To see if it is surjective, we must determine if it is true that for every $y \in T$, there exists an $x \in \mathbb{R}$ such that $F(x) = y$. So we choose $y \in T$. The goal is to determine if there exists an $x \in \mathbb{R}$ such that

$$F(x) = y, \text{ or}$$
$$x^2 + 1 = y.$$

One way to proceed is to work backward and solve the last equation (if possible) for x. Doing so, we get

$$x^2 = y - 1$$
$$x = \sqrt{y - 1} \text{ or } x = -\sqrt{y - 1}$$

Now, since $y \in T$, we know that $y \geq 1$ and hence that $y - 1 \geq 0$. This means that $\sqrt{y - 1} \in \mathbb{R}$. Hence, if we use $x = \sqrt{y - 1}$, then $x \in \mathbb{R}$, and

$$F(x) = F\left(\sqrt{y - 1}\right)$$
$$= \left(\sqrt{y - 1}\right)^2 + 1$$
$$= (y - 1) + 1$$
$$= y$$

This proves that F is surjective since we have shown that for all $y \in T$, there exists an $x \in \mathbb{R}$ such that $F(x) = y$. Notice that for each $y \in T$, this was a constructive proof of the existence of an $x \in \mathbb{R}$ such that $F(x) = y$.

An Important Lesson

In Examples 6.10 and 6.11, the same mathematical formula was used to determine the outputs for the functions. However, one function was not a surjection and the other one was a surjection. This illustrates the important fact that whether a function is surjective does not only depend on the formula that defines the output of the function but also on the domain and codomain of the function. The next example will show that whether or not a function is an injection also depends on the domain of the function.

Example 6.12.

Let $\mathbb{Z}^* = \{x \in \mathbb{Z} \mid x \geq 0\} = \mathbb{N} \cup \{0\}$. Define $g : \mathbb{Z}^* \to \mathbb{N}$ by $g(x) = x^2 + 1$. (Notice that this is the same formula used in Examples 6.10 and 6.11.)

Following is a table of values for some inputs for the function g.

x	$g(x)$	x	$g(x)$
0	1	3	10
1	2	4	17
2	5	5	26

Notice that the codomain is $\mathbb{N}$, and the table of values suggests that some natural numbers are not outputs of this function. So, it appears that the function g is not surjective.

To prove that g is not surjective, pick an element of $\mathbb{N}$ that does not appear to be in the range. We will use 3, and we will use a proof by contradiction to prove that there is no $x \in \mathbb{Z}^*$ such that $g(x) = 3$. So we assume that there exists an $x \in \mathbb{Z}^*$ with $g(x) = 3$. Then

$$x^2 + 1 = 3$$
$$x^2 = 2$$
$$x = \pm\sqrt{2}.$$

But this is not possible since $\sqrt{2} \notin \mathbb{Z}^*$. Therefore, there is no $x \in \mathbb{Z}^*$ with $g(x) = 3$. This means that for every $x \in \mathbb{Z}^*$, $g(x) \neq 3$. Therefore, 3 is not in the range of g, and hence g is not surjective.

The table of values suggests that different inputs produce different outputs, and hence that g is injective. To prove that g is injective, assume that $s, t \in \mathbb{Z}^*$ with $g(s) = g(t)$. Then

$$s^2 + 1 = t^2 + 1$$
$$s^2 = t^2.$$

Since $s, t \in \mathbb{Z}^*$, we know that $s \geq 0$ and $t \geq 0$. So the last equation implies that $s = t$. Hence, g is injective.

Another Important Lesson

The functions in the three preceding examples all used the same formula to determine the outputs. The functions in Examples 6.10 and 6.11 are not injections but the function in Example 6.12 is an injection. This illustrates the important fact that whether a function is injective not only depends on the formula that defines the output of the function but also on the domain of the function.

Example 6.13.

Let $f : \mathbb{Z} \to \mathbb{Z}$ be defined by $f(x) = 5x + 3$ for all $x \in \mathbb{Z}$. To prove that f is an injection, we must show that for every $a, b \in \mathbb{Z}$, if $f(a) = f(b)$, then $a = b$. So we let $a, b \in \mathbb{Z}$ and assume that $f(a) = f(b)$. This gives

$$5a + 3 = 5b + 3$$
$$5a = 5b$$
$$a = b.$$

Hence, we have proven that if $f(a) = f(b)$, then $a = b$, and hence that f is an injection.

Now notice that for each $x \in \mathbb{Z}$,

$$5x + 3 \equiv 3 \pmod 5.$$

This means that for each $x \in \mathbb{Z}$, $f(x) \equiv 3 \pmod 5$. Consequently, if $y \in \mathbb{Z}$, and $y \not\equiv 3 \pmod 5$, then for all $x \in \mathbb{Z}$, $f(x) \neq y$. Since there are integers that are not congruent to 3 modulo 5 (such as 0, 1, 2, and 4), this proves that f is not a surjection.

We can contrast this with a function with the same formula. Let $F : \mathbb{R} \to \mathbb{R}$ be defined by $F(x) = 5x + 3$ for all $x \in \mathbb{R}$. The function F is both an injection and a surjection. [See Exercise (4).]

Activity 6.14 (Injections, Surjections, and Bijections).

For each of the following functions, construct a table of values using at least five different elements of the domain. Then determine if the function is an injection, surjection, or bijection.

1. $f : \mathbb{R} \to \mathbb{R}$ by $f(x) = e^{-x}$

2. $f : \mathbb{R} \to (0, +\infty)$ by $f(x) = e^{-x}$

3. $f : \mathbb{R} \to \mathbb{R}$ by $f(x) = e^{-x^2}$

4. $f : \mathbb{R} \to [0, +\infty)$ by $f(x) = x^2$

5. $f : [0, +\infty) \to [0, +\infty)$ by $f(x) = x^2$

6. $g : \mathbb{R} \to \mathbb{R}$ by $g(x) = \sqrt[3]{2x - 3}$

7. $s : \mathbb{Z} \to \mathbb{Z}$ by $s(x) = 2x + 1$

8. $f : \left[\dfrac{-\pi}{2}, \dfrac{\pi}{2}\right] \to [-1, 1]$ by $f(x) = \sin x$

9. $g : \mathbb{Z} \to \{0, 1\}$ by
$$g(x) = \begin{cases} 0 \text{ if } x \text{ is even} \\ 1 \text{ if } x \text{ is odd} \end{cases}$$

For example, the first function is an injection but is not a surjection. Since this is a real function, a graph can be used to suggest these facts. To prove that it is an injection, we assume that $s, t \in \mathbb{R}$ with $f(s) = f(t)$. Then

$$e^{-s} = e^{-t}.$$

Taking the natural logarithm of both sides of the equation yields

$$\ln\left(e^{-s}\right) = \ln\left(e^{-t}\right)$$
$$-s = -t$$
$$s = t.$$

This proves that f is an injection. Also, since $e^{-x} > 0$ for all $x \in \mathbb{R}$, we can conclude that $f(x) > 0$ for all $x \in \mathbb{R}$. This means that the range of f is not equal to $\mathbb{R}$ and hence f is not surjective.

A Function of Two Variables

Define $f : \mathbb{R} \times \mathbb{R} \to \mathbb{R} \times \mathbb{R}$ by $f(a, b) = (2a + b, a - b)$ for all $(a, b) \in \mathbb{R} \times \mathbb{R}$.

Notice that both the domain and the codomain of this function is the set $\mathbb{R} \times \mathbb{R}$. Thus, the inputs and the outputs of this function are ordered pairs of real numbers. For example,

$$f(1, 1) = (3, 0),$$
$$f(-1, 2) = (0, -3).$$

To explore whether or not f is an injection, we assume that $(a, b) \in \mathbb{R} \times \mathbb{R}$, $(c, d) \in \mathbb{R} \times \mathbb{R}$, and $f(a, b) = f(c, d)$. This means that

$$(2a + b, a - b) = (2c + d, c - d).$$

Since this equation is an equality of ordered pairs, we see that

$$2a + b = 2c + d, \text{ and}$$
$$a - b = c - d.$$

By adding the two equations in this system, we obtain $3a = 3c$ and hence, $a = c$. Substituting $a = c$ into either equation in the system give us $b = d$. Since $a = c$ and $b = d$, we conclude that

$$(a, b) = (c, d).$$

Hence, we have shown that if $f(a, b) = f(c, d)$, then $(a, b) = (c, d)$. Therefore, f is an injection.

Now, to determine if f is a surjection, we let $(r, s) \in \mathbb{R} \times \mathbb{R}$, where (r, s) is considered to be an arbitrary element of the codomain of the function f. Can we find an ordered pair $(a, b) \in \mathbb{R} \times \mathbb{R}$ such that $f(a, b) = (r, s)$? Working backward, we see that in order to do this, we need

$$(2a + b, a - b) = (r, s).$$

That is, we need

$$2a + b = r, \text{ and}$$
$$a - b = s.$$

Solving this system for a and b yields

$$a = \frac{r + s}{3} \text{ and } b = \frac{r - 2s}{3}.$$

Since $r, s \in \mathbb{R}$, we can conclude that $a \in \mathbb{R}$ and $b \in \mathbb{R}$ and hence that $(a, b) \in \mathbb{R} \times \mathbb{R}$. We now need to verify that for these values of a and b, we get $f(a, b) = (r, s)$. So

$$f(a, b) = f\left(\frac{r + s}{3}, \frac{r - 2s}{3}\right)$$

$$= \left(2\left(\frac{r + s}{3}\right) + \frac{r - 2s}{3}, \frac{r + s}{3} - \frac{r - 2s}{3}\right)$$

$$= \left(\frac{2r + 2s + r - 2s}{3}, \frac{r + s - r + 2s}{3}\right)$$

$$= (r, s).$$

This proves that for all $(r, s) \in \mathbb{R} \times \mathbb{R}$, there exists $(a, b) \in \mathbb{R} \times \mathbb{R}$ such that $f(a, b) = (r, s)$. Hence, the function f is a surjection. Since f is both an injection and a surjection, it is a bijection.

Exercises 6.3

1. Let $f : A \to B$.

   (a) Carefully explain what it means to say that the function f is an injection.

   (b) Carefully explain what it means to say that the function f is not an injection.

2. Let $f : A \to B$.

(a) The function f is a surjection provided that range $(f) = $ codom (f). In terms of the elements of the sets A and B, carefully explain what it means to say that the function f is a surjection.

(b) The function f is not a surjection provided that range $(f) \neq$ codom (f). In terms of the elements of the sets A and B, carefully explain what it means to say that the function f is not a surjection.

3. (a) Draw an arrow diagram that represents a function that is an injection but is not a surjection.

(b) Draw an arrow diagram that represents a function that is an injection and is a surjection.

(c) Draw an arrow diagram that represents a function that is not an injection and is not a surjection.

(d) Draw an arrow diagram that represents a function that is not an injection but is a surjection.

(e) Draw an arrow diagram that represents a function that is not a bijection.

4. Let $F : \mathbb{R} \to \mathbb{R}$ be defined by $F(x) = 5x + 3$ for all $x \in \mathbb{R}$. Prove that the function F is a bijection.

5. (a) Let $f : \mathbb{Z} \to \mathbb{Z}$ be defined by $f(x) = 3x + 1$. Is the function f an injection? Is the function f a surjection? Justify your conclusions.

(b) Let $F : \mathbb{Q} \to \mathbb{Q}$ be defined by $F(x) = 3x + 1$. Is the function F an injection? Is the function F a surjection? Justify your conclusions.

6. (a) Let $g : \mathbb{R} \to \mathbb{R}$ be defined by $g(x) = x^3$. Is the function g an injection? Is the function g a surjection? Justify your conclusions.

(b) Let $f : \mathbb{Q} \to \mathbb{Q}$ be defined by $f(x) = x^3$. Is the function f an injection? Is the function f a surjection? Justify your conclusions.

7. Let $s : \mathbb{N} \to \mathbb{N}$, where for each $n \in \mathbb{N}$, $s(n)$ is the sum of the distinct natural number divisors of n. This is the **sum of the divisors function** that was introduced in Preview Activity 3 from Section 6.1. Is s an injection? Is s a surjection? Justify your conclusions.

8. Let $d : \mathbb{N} \to \mathbb{N}$, where $d(n)$ is the number of natural number divisors of n. This is the **number of divisors function** introduced in Exercise (6) from Section 6.1. Is the function d an injection? Is the function d a surjection? Justify your conclusions.

9. In Preview Activity 2 from Section 6.1 , we introduced the **birthday function**. Its definition is repeated here.

    Let f be the function that assigns to each person his or her birthday (month and day). The domain of the function f is the set of all people and the codomain of f is the set of all days in a leap year (i.e., January 1 through December 31, including February 29).

    Is the birthday function an injection? Is it a surjection? Justify your conclusions.

10. (a) Let $f : \mathbb{Z} \times \mathbb{Z} \to \mathbb{Z}$ be defined by $f(m, n) = 2m + n$. Is the function f an injection? Is the function f a surjection? Justify your conclusions.

    (b) Let $g : \mathbb{Z} \times \mathbb{Z} \to \mathbb{Z}$ be defined by $f(m, n) = 6m + 3n$. Is the function g an injection? Is the function g a surjection? Justify your conclusions.

11. (a) Let $f : \mathbb{R} \times \mathbb{R} \to \mathbb{R} \times \mathbb{R}$ be defined by $f(x, y) = (2x, x + y)$. Is the function f an injection? Is the function f a surjection? Justify your conclusions.

    (b) Let $g : \mathbb{Z} \times \mathbb{Z} \to \mathbb{Z} \times \mathbb{Z}$ be defined by $g(x, y) = (2x, x + y)$. Is the function g an injection? Is the function g a surjection? Justify your conclusions.

12. Let A be a nonempty set. The **identity map on the set A**, denoted by I_A, is the function $I_A : A \to A$ defined by $I_A(x) = x$ for every x in A. Is I_A an injection? Is I_A a surjection? Justify your conclusions.

13. Let A and B be two nonempty sets. Define

$$p_1 : A \times B \to A \text{ by } p_1(a, b) = a$$

for every $(a, b) \in A \times B$. This is the **first projection function** introduced in Section 6.2.

    (a) Is the function p_1 a surjection? Justify your conclusion.

(b) If $B = \{b\}$, is the function p_1 an injection? Justify your conclusion.

(c) Under what condition(s) is the function p_1 not an injection? Make a conjecture and prove it.

14. Let $\mathcal{M}_{2,2}$ represent the set of all 2 by 2 matrices over $\mathbb{R}$. Define $\det : \mathcal{M}_{2,2} \to \mathbb{R}$ by

$$\det \begin{bmatrix} a & b \\ c & d \end{bmatrix} = ad - bc.$$

This is the **determinant function** introduced in Exercise (7) from Section 6.2.

Is the determinant function an injection? Is the determinant function a surjection? Justify your conclusions.

15. Define tran : $\mathcal{M}_{2,2} \to \mathcal{M}_{2,2}$ by

$$\text{tran} \begin{bmatrix} a & b \\ c & d \end{bmatrix} = A^T = \begin{bmatrix} a & c \\ b & d \end{bmatrix}.$$

This is the **transpose function** introduced in Exercise (8) from Section 6.2.

Is the transpose function an injection? Is the transpose function a surjection? Justify your conclusions.

16. Let C be the set of all real functions that are continuous on the closed interval $[0, 1]$. Define the function $A : C \to \mathbb{R}$ as follows: For each $f \in C$,

$$A(f) = \int_0^1 f(x)\, dx.$$

Is the function A an injection? Is it a surjection? Justify your conclusions.

6.4   Composition of Functions

Preview Activity 1 (Constructing a New Function).

Let $A = \{a, b, c, d\}$, $B = \{p, q, r\}$, and $C = \{s, t, u, v\}$. The following arrow diagram shows two functions: $f : A \to B$ and $g : B \to C$.

Complete the following table. For example, if $x = a$, then $f(a) = p$, and $g(f(a)) = g(p) = t$.

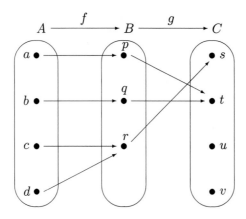

$$
\begin{array}{c|c|c}
x & f(x) & g(f(x)) \\
\hline
a & & \\
\hline
b & & \\
\hline
c & & \\
\hline
d & & \\
\end{array}
$$

Explain how this table defines a function from A to C.

Preview Activity 2 (A New Function from Graphs).

Figure 6.7 shows the graphs of two real functions: $f : \mathbb{R} \to \mathbb{R}$ and $g : \mathbb{R} \to \mathbb{R}$. It also shows the graph of the line $y = x$.

1. We will first use $x = 2$.

    (a) Draw the vertical line $x = 2$ so that it intersects the graph of f. Label this point P. This point of intersection is $(2, f(2))$. Although we could use the graph of f to approximate $f(2)$, we will not do so here.

    (b) Now, draw a horizontal line through the point P until it intersects the line $y = x$. Call this point of intersection Q. What are the coordinates of the point Q? Each coordinate should be expressed in terms of the function f.

    (c) Next, draw a vertical line through the point Q until it intersects the graph of g. Call this point of intersection R. What are the coordinates of the point R in terms of the functions f and g?

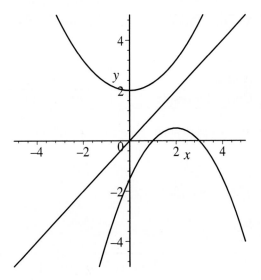

Figure 6.7: Graph of $y = g(x)$ and $y = f(x)$.

(d) Finally, draw a horizontal line through the point R until it intersects the vertical line $x = 2$ from Part (1a). Call this point of intersection S. What are the coordinates of the point S in terms of the functions f and g. Approximate the y-coordinate of the point S?

2. Repeat Part (1) starting with $x = 3$.

3. Repeat Part (1) starting with $x = 0$.

4. Explain how this process can be used to define a new function from $\mathbb{R}$ to $\mathbb{R}$.

Preview Activity 3 (Verbal Descriptions of Functions).

The outputs of most real functions we have studied in previous mathematics courses have been determined by mathematical expressions. In many cases, it is possible to use these expressions to give step-by-step verbal descriptions of how to compute the outputs. For example, if

$$f : \mathbb{R} \to \mathbb{R} \text{ is defined by } f(x) = (3x + 2)^3,$$

we could describe how to compute the outputs as follows:

Step	Verbal Description	Symbolic Result
1	Choose an input.	x
2	Multiply by 3.	$3x$
3	Add 2.	$3x + 2$
4	Cube the result.	$(3x + 2)^3$

Complete step-by-step verbal descriptions for each of the following functions.

1. $f : \mathbb{R} \to \mathbb{R}$ by $f(x) = \sqrt{3x^2 + 2}$.

2. $g : \mathbb{R} \to \mathbb{R}$ by $g(x) = \sin(3x^2 + 2)$.

3. $h : \mathbb{R} \to \mathbb{R}$ by $h(x) = e^{3x^2 + 2}$.

There are several ways to combine two existing functions to create a new function. In this section, we will focus on one way, the composition of two functions. We will also consider some results about the compositions of injections and surjections.

Composition of Functions

The basic idea of function composition is that when possible, the output of a function f is used as the input of a function g. This can be referred to as "f followed by g" and is called the composition of f and g.

The idea of the composition of two real functions is familiar from previous mathematics courses. For example, if $f(x) = 3x^2 + 2$ and $g(x) = \sin x$, then we can compute $g(f(x))$ as follows:

$$g(f(x)) = g(3x^2 + 2)$$
$$= \sin(3x^2 + 2).$$

In this case, $f(x)$, the output of the function f, was used as the input for the function g. This idea of using the output from one function as the input for another function can be illustrated nicely with arrow diagrams. This was done in Preview Activity 1, and another example is given here.

Let $A = \{a, b, c, d\}$, $B = \{p, q, r\}$, and $C = \{s, t, u, v\}$. The arrow diagram in Figure 6.8 shows two functions: $f : A \to B$ and $g : B \to C$.

If we follow the arrows from the set A to the set C, we will use the outputs of f as inputs of g, and get the arrow diagram from A to C shown in Figure 6.9. This diagram represents the composition of f and g and is denoted by $g \circ f$.

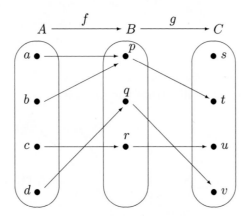

Figure 6.8: Arrow Diagram for Two Functions

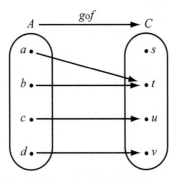

Figure 6.9: Arrow Diagram for $g \circ f : A \to C$

Definition. Let A, B, and C be nonempty sets, and let $f : A \to B$ and $g : B \to C$ be functions. The **composition of f and g** is the function $g \circ f : A \to C$ defined by

$$(g \circ f)(x) = g(f(x))$$

for all $x \in A$. We often refer to the function $g \circ f$ as a **composite function**

It is helpful to think of the composite function $g \circ f$ as "**f followed by g.**" We then refer to f as the **inner function** and g as the **outer function**.

In Preview Activity 2, we actually used the graphs of the real functions f and g to construct outputs for the composite function $g \circ f$. Following is an outline of the procedure for finding $g\left(f\left(x\right)\right)$ for a given value of x. Figure 6.10 will be used to illustrate the process.

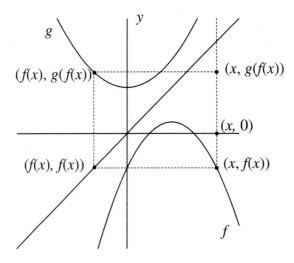

Figure 6.10: Composition of Functions

- We start with an input (labeled x on the x-axis). The vertical line through $(x, 0)$ intersects the graph of f at the point $(x, f\left(x\right))$.

- Then the horizontal line through this point intersects the line $y = x$ at the point $(f\left(x\right), f\left(x\right))$.

- The vertical line through this point will intersect the graph of g at the point $(f\left(x\right), g\left(f\left(x\right)\right))$. So the y-coordinate of this point is the output of the composite function $g \circ f$. Notice that the use of the line $y = x$ allowed us to use the output of the function f as the input for the function g.

- In order to obtain a point on the graph of $y = g\left(f\left(x\right)\right)$, we now draw a horizontal line from the point $(f\left(x\right), g\left(f\left(x\right)\right))$ until it intersects the vertical line through the point $(x, 0)$.

- The coordinates of this point will be $(x, g\left(f\left(x\right)\right))$, and hence this will be a point on the graph of $g \circ f$.

Activity 6.15 (Decomposing Functions).

We use the **chain rule** in calculus to find the derivative of a composite function. The first step in the process is to recognize a given function as a composite function. This can be done in many ways, but the work in Preview Activity 3 can be used to decompose a function in a way that works well with the chain rule. The use of the terms "inner function" and "outer function" can also be helpful. The idea is that the last step in the process represents the outer function, and the steps prior to that represent the inner function. So for the function,

$$f : \mathbb{R} \to \mathbb{R} \text{ by } f(x) = (3x + 2)^3,$$

the last step in the verbal description table was to cube the result. This means that the outer function g is the cubing function, and the prior steps are the inner function. We will denote the inner function by h. So we let $h : \mathbb{R} \to \mathbb{R}$ by $h(x) = 3x + 2$ and $g : \mathbb{R} \to \mathbb{R}$ by $g(x) = x^3$. Then

$$
\begin{aligned}
(g \circ h)(x) &= g(h(x)) \\
&= g(3x + 2) \\
&= (3x + 2)^3 \\
&= f(x).
\end{aligned}
$$

We see that $g \circ h = f$, and hence we have "decomposed" the function f.

1. Write each of the following functions as the composition of two functions.

   (a) $f : \mathbb{R} \to \mathbb{R}$ defined by $f(x) = \sqrt{3x^2 + 2}$.

   (b) $g : \mathbb{R} \to \mathbb{R}$ defined by $g(x) = \sin(3x^2 + 2)$.

   (c) $h : \mathbb{R} \to \mathbb{R}$ defined by $h(x) = e^{3x^2 + 2}$.

   (d) $k : \mathbb{R} \to \mathbb{R}$ defined by $k(x) = \ln(3x^2 + 2)$.

2. Let $h : \mathbb{R} \to \mathbb{R}$ be defined by $h(x) = 3x + 2$ and $g : \mathbb{R} \to \mathbb{R}$ be defined by $g(x) = x^3$. Determine formulas for the composite functions $g \circ h$ and $h \circ g$. What does this tell you about the operation of composition of functions?

Activity 6.16 (Compositions of Injections and Surjections).

Although other representations of functions can be used, it might be helpful to use arrow diagrams to represent the functions in this activity. When possible, use an example where the finite domains and codomains have different numbers of elements.

1. What type of function results from the composition of two injections?

2. What type of function results from the composition of two surjections?

3. What type of function results from the composition of two bijections?

Activity 6.17 (Exploring Composite Functions).

Let A, B, and C be nonempty sets and let $f : A \to B$ and $g : B \to C$. For this activity, it may be useful to draw your arrow diagrams in a triangular arrangement as follows:

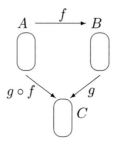

1. Is it possible to construct an example where $g \circ f$ is an injection, f is an injection, but g is not an injection? Either construct such an example or explain why it is not possible.

2. Is it possible to construct an example where $g \circ f$ is an injection, g is an injection, but f is not an injection? Either construct such an example or explain why it is not possible.

3. Is it possible to construct an example where $g \circ f$ is a surjection, f is a surjection, but g is not a surjection? Either construct such an example or explain why it is not possible.

4. Is it possible to construct an example where $g \circ f$ is surjection, g is a surjection, but f is not a surjection? Either construct such an example or explain why it is not possible.

Theorems about Composite Functions

In the previous two activities, we explored some properties of composite functions related to injections, surjections, and bijections. The following two theorems contain results that these explorations were intended to illustrate. Many of the proofs will be included in the exercises.

Theorem 6.18. *Let A, B, and C be nonempty sets and let $f : A \to B$ and $g : B \to C$.*

1. *If f and g are both injections, then $g \circ f$ is an injection.*

2. *If f and g are both surjections, then $g \circ f$ is a surjection.*

3. *If f and g are both bijections, then $g \circ f$ is a bijection.*

The results of Theorem 6.18 are related to the explorations in Activity 6.16. Part (3) of Theorem 6.18 is a direct consequence of the first two parts. We will discuss constructing a proof of Part (2). Using the forward-backward process, we first look at the conclusion of the conditional statement in Part (2). The goal is to prove that $g \circ f$ is a surjection. Since $g \circ f : A \to C$, this is equivalent to proving that

> For all $c \in C$, there exists an $a \in A$ such that $(g \circ f)(a) = c$.

Since this statement in the backward process uses a universal quantifier, we will start by selecting an arbitrary element c in the set C. The goal now is to find an $a \in A$ such that $(g \circ f)(a) = c$.

Now we can look at the hypotheses. In particular, we are assuming that both $f : A \to B$ and $g : B \to C$ are surjections.

Since we have chosen $c \in C$, and $g : B \to C$ is a surjection, we know that

> there exists a $b \in B$ such that $g(b) = c$.

Now, $b \in B$ and $f : A \to B$ is a surjection. Hence

> there exists an $a \in A$ such that $f(a) = b$.

If we now compute $(g \circ f)(a)$, we will see that

$$(g \circ f)(a) = g(f(a)) = g(b) = c.$$

We can now write the proof as follows:

Proof of Theorem 6.18, Part (2).

Proof. Let A, B, and C be nonempty sets and assume that $f : A \to B$ and $g : B \to C$ are both surjections. We will prove that $g \circ f : A \to C$ is a surjection.

Let c be an arbitrary element of C. We will prove there exists an $a \in A$ such that $(g \circ f)(a) = c$. Since $g : B \to C$ is a surjection, we conclude that

there exists a $b \in B$ such that $g(b) = c$.

Now, $b \in B$ and $f : A \to B$ is a surjection. Hence

there exists an $a \in A$ such that $f(a) = b$.

We now see that

$$(g \circ f)(a) = g(f(a))$$
$$= g(b)$$
$$= c.$$

We have now shown that for every $c \in C$, there exists an $a \in A$ such that $(g \circ f)(a) = c$, and this proves that $g \circ f$ is a surjection. ∎

Theorem 6.19. *Let A, B, and C be nonempty sets and let $f : A \to B$ and $g : B \to C$.*

1. *If $g \circ f : A \to C$ is an injection, then $f : A \to B$ is an injection.*

2. *If $g \circ f : A \to C$ is a surjection, then $g : B \to C$ is a surjection.*

The results of Theorem 6.19 are related to the explorations in Activity 6.17. The proof of both parts of Theorem 6.19 will be included in the exercises.

Exercises 6.4

1. In our definition of the composition of two functions, f and g, we required that the domain of g be equal to the codomain of f. However, it is sometimes possible to form the composite function $g \circ f$ even though $\mathrm{dom}\,(g) \neq \mathrm{codom}\,(f)$. For example, let

$$f : \mathbb{R} \to \mathbb{R} \qquad \text{be defined by} \quad f(x) = x^2 + 1, \text{ and let}$$
$$g : \mathbb{R} - \{0\} \to \mathbb{R} \quad \text{be defined by} \quad g(x) = \frac{1}{x}.$$

(a) Is it possible to determine $(g \circ f)(x)$ for all $x \in \mathbb{R}$? Explain.

(b) In general, let $f : A \to T$ and $g : B \to C$. Find a condition on the domain of g (other than $B = T$) that results in a meaningful definition of the composite function $g \circ f : A \to C$.

2. Following are formulas for certain real functions. Write each of these real functions as the composition of two functions. That is, decompose each of the functions.

(a) $F(x) = \cos(e^x)$

(c) $H(x) = \dfrac{1}{\sin x}$

(b) $G(x) = e^{\cos(x)}$

(d) $K(x) = \cos\left(e^{-x^2}\right)$

3. (a) Let $f : \mathbb{R} \to \mathbb{R}$ be defined by $f(x) = x^2$, let $g : \mathbb{R} \to \mathbb{R}$ be defined by $g(x) = \sin x$, and let $h : \mathbb{R} \to \mathbb{R}$ be defined by $h(x) = \sqrt[3]{x}$.

Determine formulas for $[(h \circ g) \circ f](x)$ and $[h \circ (g \circ f)](x)$.

Does this prove that $(h \circ g) \circ f = h \circ (g \circ f)$ for these particular functions? Explain.

(b) Let $f : A \to B$, $g : B \to C$, and $h : C \to D$. Prove that $(h \circ g) \circ f = h \circ (g \circ f)$. That is, prove that function composition is an associative operation.

4. In Section 6.2, we defined the **identity map**, I_S, (identity function) on a set S as follows: $I_S : S \to S$ by $I_S(x) = x$ for all $x \in S$.

Let $f : A \to B$.

(a) For each $x \in A$, determine $(f \circ I_A)(x)$ and use this to prove that $f \circ I_A = f$.

(b) Prove that $I_B \circ f = f$.

5. Prove Part (1) of Theorem 6.18

Let A, B, and C be nonempty sets and let $f : A \to B$ and $g : B \to C$. If f and g are both injections, then $g \circ f$ is an injection.

6. Prove Part (1) of Theorem 6.19

Let A, B, and C be nonempty sets and let $f : A \to B$ and $g : B \to C$. If $g \circ f : A \to C$ is an injection, then $f : A \to B$ is an injection.

7. Prove Part (2) of Theorem 6.19

Let A, B, and C be nonempty sets and let $f : A \to B$ and $g : B \to C$. If $g \circ f : A \to C$ is a surjection, then $g : B \to C$ is a surjection.

8. For each of the following, give an example of functions $f : A \to B$ and $g : B \to C$ that satisfy the stated conditions, or explain why no such example exists.

   (a) The function f is a surjection, but the function $g \circ f$ is not a surjection.

   (b) The function f is an injection, but the function $g \circ f$ is not an injection.

   (c) The function g is a surjection, but the function $g \circ f$ is not a surjection.

   (d) The function g is an injection, but the function $g \circ f$ is not an injection.

   (e) The function f is not a surjection, but the function $g \circ f$ is a surjection.

   (f) The function f is not an injection, but the function $g \circ f$ is an injection.

   (g) The function f is not an injection, but the function $g \circ f$ is an injection.

   (h) The function g is not an injection, but the function $g \circ f$ is an injection.

6.5   Inverse Functions

Preview Activity 1 (A Function as a Set of Ordered Pairs).
   Any function $f : A \to B$ can be thought of as a set of ordered pairs that is a subset of $A \times B$. This subset is

$$f = \{(a, f(a)) \mid a \in A\}, \qquad \text{or} \qquad f = \{(a,b) \in A \times B \mid b = f(a)\}.$$

Note: Since f is the name of the function, it is customary to use f as the name for the set of ordered pairs.

Let $A = \{1, 2, 3\}$ and let $B = \{a, b\}$.

1. Define the function $F : A \to B$ by $F(1) = a$, $F(2) = b$, and $F(3) = b$.

   Write the function F as a set of three ordered pairs in $A \times B$.

On the other hand, if we started with $A = \{1, 2, 3\}$, $B = \{a, b\}$, and

$$G = \{(1, a), (2, a), (3, b)\} \subseteq A \times B,$$

then we could think of G as a function from A to B with $G(1) = a$, $G(2) = a$, and $G(3) = b$. The idea is to use the first coordinate of each ordered pair as the input, and the second coordinate as the output. However, not every subset of $A \times B$ can be used to define a function from A to B.

2. Could we use the following subset of $A \times B$ to define a function from A to B? Explain.

$$f = \{(1, a), (2, a), (3, a), (1, b)\}$$

3. Could we use the following subset of $A \times B$ to define a function from A to B? Explain.

$$g = \{(1, a), (2, b), (3, a)\}$$

4. Could we use the following subset of $A \times B$ to define a function from A to B? Explain.

$$h = \{(1, a), (2, b)\}$$

Preview Activity 2 (A Composition of Two Specific Functions).
Let $A = \{a, b, c, d\}$ and let $B = \{p, q, r, s\}$.

1. Construct an example of a function $f : A \to B$ that is a bijection. Draw an arrow diagram for this function.

2. On your arrow diagram, draw an arrow from each element of B back to its corresponding element in A. Explain why this defines a function from B to A.

3. If the name of the function in Part (2) is g, so that $g : B \to A$, what are $g(p)$, $g(q)$, $g(r)$, and $g(s)$?

4. Complete the following tables for $g \circ f : A \to A$ and $f \circ g : B \to B$.

x	$(g \circ f)(x)$
a	
b	
c	
d	

y	$(f \circ g)(y)$
p	
q	
r	
s	

5. What do you observe about the tables of values in Part (4)?

Preview Activity 3 (Cubes and Cube Roots).

Let $f : \mathbb{R} \to \mathbb{R}$ be defined by $f(x) = x^3$. Let $g : \mathbb{R} \to \mathbb{R}$ be defined by $g(x) = \sqrt[3]{x}$.

Complete the following tables:

x	$f(x)$
0	
1	
2	
3	
-1	
-3	

x	$g(x)$
0	
1	
8	
27	
-1	
-27	

1. What is happening? In particular, how are the ordered pairs of the functions f and g related?

2. Solve the equation $(2t - 1)^3 = 20$. Explain how the cube root function, g, was used in solving this equation.

3. Solve the equation $\sqrt[3]{t - 3} = 2$. Explain how the cubing function, f, was used in solving this equation.

4. For each $x \in \mathbb{R}$, determine $(g \circ f)(x)$ and $(f \circ g)(x)$.

The Ordered Pair Representation of a Function

Up to this point, we have not really used the ordered pair representation of a function. Since it does provide a convenient way to introduce the concept of the inverse of a function, we will use it here. If we have a function $f : A \to B$, we can generate a set of ordered pairs f that is a subset of $A \times B$ as follows:

$$f = \{(a, f(a)) \mid a \in A\}, \text{ or}$$
$$f = \{(a, b) \in A \times B \mid b = f(a)\}.$$

Since, $\text{dom}\,(f) = A$, we know that

$$\text{For every } a \in A, \text{ there exists a } b \in B \text{ such that } (a, b) \in f. \tag{1}$$

Specifically, we use $b = f(a)$. This says that every element of A can be used as an input. In addition, to be a function, each input can produce only one output. In terms of ordered pairs, this means that there will never be two ordered pairs (a, b) and (a, c) in the function f where $a \in A$, $b, c \in B$, and $b \neq c$. We can formulate this as a conditional statement as follows:

$$\text{For every } a \in A \text{ and every } b, c \in B,$$
$$\text{if } (a, b) \in f \text{ and } (a, c) \in f, \text{ then } b = c. \tag{2}$$

This also means that if we start with a subset f of $A \times B$ that satisfies Conditions (1) and (2), then we can consider f to be a function from A to B by using $b = f(a)$ whenever (a, b) is an ordered pair in f. We have established the following theorem.

Theorem 6.20. *Let A and B be nonempty sets and let f be a subset of $A \times B$ that satisfies the following two properties:*

- *For every $a \in A$, there exists $b \in B$ such that $(a, b) \in f$; and*

- *For every $a \in A$ and every $b, c \in B$, if $(a,\ b) \in f$ and $(a, c) \in f$, then $b = c$.*

If we use $f(a) = b$ whenever $(a, b) \in f$, then f is a function from A to B.

Note: The first condition in Theorem 6.20 means that every element of A is an input, and the second condition insures that every input has exactly one output. Many texts will use Theorem 6.20 as the definition of a function.

Example 6.21. Let $A = \{1, 2, 3\}$ and let $B = \{a, b\}$. Then

- The set of ordered pairs $F = \{(1, a), (2, a)\}$ cannot be used to define a function from A to B since F does not satisfy the first condition of Theorem 6.20.

- Also, the set of ordered pairs $G = \{(1, a), (2, b), (3, c), (2, c)\}$ is not a function since G does not satisfy the second condition of Theorem 6.20. [The ordered pairs $(2, b)$ and $(2, c)$ are in G.]

The Inverse of a Function

In previous mathematics courses, we learned that the exponential function (with base e) and the natural logarithm functions are inverses of each other. This was often expressed using the following two properties:

- For each $x \in \mathbb{R}$, if $y = e^x$, then $\ln y = x$.

- For each $x \in \mathbb{R}$ such that $x > 0$, if $y = \ln x$, then $e^y = x$.

In each case, when x is the input for one of the functions, y is the output. If y is used as the input for the other function, then this produces x as the output. In essence, the inverse function reverses the action of the original function. In terms of ordered pairs (input-output pairs), this means that if (x, y) is an ordered pair for a function, then (y, x) is an ordered pair for its inverse. This idea of reversing the roles of the first and second coordinates is the basis for our definition of the inverse of a function.

Definition. Let $f : A \to B$ be a function. The **inverse of f**, denoted by f^{-1}, is the set of ordered pairs $\{(b, a) \in B \times A \mid f(a) = b\}$. That is,

$$f^{-1} = \{(b, a) \in B \times A \mid f(a) = b\}.$$

If we use the ordered pair representation for f, we could also write

$$f^{-1} = \{(b, a) \in B \times A \mid (a, b) \in f\}.$$

Notice that this definition does not state that f^{-1} is a function. It is simply a subset of $B \times A$. Later in the text, we will say that this means that f^{-1} is a **relation** from B to A. This fact, however, is not important to us now. We are mainly interested in the following question:

Under what conditions will the inverse of the function $f : A \to B$ be a function from B to A?

Activity 6.22 (Exploring the Inverse of a Function).

Let $A = \{a, b, c\}$, $B = \{a, b, c, d\}$, and $C = \{p, q, r\}$. Define

$f : A \to C$ by	$g : A \to C$ by	$h : B \to C$ by
$f(a) = r$	$g(a) = p$	$h(a) = p$
$f(b) = p$	$g(b) = q$	$h(b) = q$
$f(c) = q$	$g(c) = p$	$h(c) = r$
		$h(d) = q$

1. Draw an arrow diagram for each function.

2. Determine the inverse of each function as a set of ordered pairs.

3. (a) Is f^{-1} a function from C to A? Explain.

   (b) Is g^{-1} a function from C to A? Explain.

   (c) Is h^{-1} a function from C to B? Explain.

4. Draw an arrow diagram for each inverse from Part (3) that is a function. Use your existing arrow diagram from Part (1) to draw this arrow diagram.

5. Make a conjecture about what conditions on a function $F : S \to T$ will ensure that its inverse is a function from T to S.

By our definition, if $f : A \to B$ is a function, then f^{-1} is defined as a subset of $B \times A$. However, f^{-1} may or may not be a function from B to A. For example, suppose that $s, t \in A$ with $s \neq t$ and $f(s) = f(t)$. This is represented in the following arrow diagram:

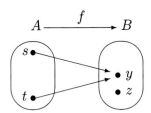

In this case, if we try to reverse the arrows, we will not get a function from B to A. This is because $(y, s) \in f^{-1}$ and $(y, t) \in f^{-1}$ with $s \neq t$. Consequently, f^{-1} is not a function. This suggests that when f is not an injection, then f^{-1} is not a function.

Also, if f is not a surjection, then there exists a $z \in B$ such that $f(a) \neq z$ for all $a \in A$, such as in the diagram above. In other words, there is no ordered pair in f with z as the second coordinate. This means that there would be no ordered pair in f^{-1} with z as a first coordinate. Consequently, f^{-1} cannot be a function from B to A.

This motivates the statement in Theorem 6.23. In the proof of this theorem, we will frequently change back and forth from the input-output representation of a function and the ordered pair representation of a function. The idea is that if $G : S \to T$ is a function, then for $s \in S$ and

$t \in T$,
$$G(s) = t \text{ if and only if } (s,t) \in G.$$

When we use the ordered pair representation of a function, we will also use the ordered pair representation of its inverse. In this case, we know that

$$(s,t) \in G \text{ if and only if } (t,s) \in G^{-1}.$$

Theorem 6.23. *Let A and B be nonempty sets and let $f : A \to B$. The inverse of f is a function from B to A if and only if f is a bijection.*

Proof. Let A and B be nonempty sets and let $f : A \to B$. We will first assume that f is a bijection and prove that f^{-1} is a function from B to A. To do this, we will show that f^{-1} satisfies the two conditions of Theorem 6.20.

We first choose $b \in B$. Since the function f is a surjection, there exists an $a \in A$ such that $f(a) = b$. This implies that $(a,b) \in f$ and hence that $(b,a) \in f^{-1}$. Thus, each element of B is the first coordinate of an ordered pair in f^{-1}, and hence f^{-1} satisfies the first condition of Theorem 6.20.

To prove that f^{-1} satisfies the second condition of Theorem 6.20, we must show that each element of B is the first coordinate of exactly one ordered pair in f^{-1}. So let $b \in B$, $a_1, a_2 \in A$ and assume that

$$(b,a_1) \in f^{-1} \text{ and } (b,a_2) \in f^{-1}.$$

This means that $(a_1,b) \in f$ and $(a_2,b) \in f$. We can then conclude that

$$f(a_1) = b \text{ and } f(a_2) = b.$$

But this means that $f(a_1) = f(a_2)$. Since f is a bijection, it is an injection, and we can conclude that $a_1 = a_2$. This proves that b is the first element of only one ordered pair in f^{-1}. Consequently, we have proven that f^{-1} satisfies both conditions of Theorem 6.20, and hence that f^{-1} is a function from B to A.

We now assume that f^{-1} is a function from B to A and prove that f is a bijection. First, to prove that f is an injection, we assume that $a_1, a_2 \in A$ and that $f(a_1) = f(a_2)$. We wish to show that $a_1 = a_2$. If we let $b = f(a_1) = f(a_2)$, we can conclude that

$$(a_1,b) \in f \text{ and } (a_2,b) \in f.$$

But this means that

$$(b,a_1) \in f^{-1} \text{ and } (b,a_2) \in f^{-1}.$$

Since we have assumed that f^{-1} is a function, we can conclude that $a_1 = a_2$. Hence, f is an injection.

Now, to prove that f is a surjection, we choose $b \in B$ and will show that there exists an $a \in A$ such that $f(a) = b$. Since f^{-1} is a function, b must be the first coordinate of some ordered pair in f^{-1}. Consequently, there exists an $a \in A$ such that

$$(b, a) \in f^{-1}.$$

Now, this implies that $(a, b) \in f$ and hence that $f(a) = b$. This proves that f is a surjection. Since we have also proven that f is an injection, we conclude that f is a bijection. ∎

Inverse Function Notation

In the situation where $f : A \to B$ is a bijection and f^{-1} is a function from B to A, we can write $f^{-1} : B \to A$. In this case, we frequently say that f is an **invertible function**, and we usually do not use the ordered pair representation for either f or f^{-1}. Instead of writing $(a, b) \in f$, we write $f(a) = b$, and instead of writing $(b, a) \in f^{-1}$, we write $f^{-1}(b) = a$. Using the fact that $(a, b) \in f$ if and only if $(b, a) \in f^{-1}$, we can now write $f(a) = b$ if and only if $f^{-1}(b) = a$. We summarize this in Theorem 6.24.

Theorem 6.24. *Let A and B be nonempty sets and let $f : A \to B$ be a bijection. Then $f^{-1} : B \to A$ is a function, and for every $a \in A$ and $b \in B$,*

$$f(a) = b \text{ if and only if } f^{-1}(b) = a.$$

Example 6.25. Let $\mathbb{R}^+ = \{x \in \mathbb{R} \mid x > 0\}$. Define

$f : \mathbb{R} \to \mathbb{R}$ by $f(x) = x^3$; and

$g : \mathbb{R} \to \mathbb{R}^+$ by $g(x) = e^x$.

Notice that $\mathbb{R}^+$ is the codomain of g. We can then say that both f and g are bijections. Consequently, the inverses of these functions are also functions. In fact,

$f^{-1} : \mathbb{R} \to \mathbb{R}$ by $f^{-1}(y) = \sqrt[3]{y}$; and

$g^{-1} : \mathbb{R}^+ \to \mathbb{R}$ by $g^{-1}(y) = \ln y$.

For each function (and its inverse), we can write the result of Theorem 6.24 as follows:

Theorem 6.24	**Translates to:**
For $x, y \in \mathbb{R}$, $f(x) = y$ if and only if $f^{-1}(y) = x$.	For $x, y \in \mathbb{R}$, $x^3 = y$ if and only if $\sqrt[3]{y} = x$.
For $x \in \mathbb{R}$, $y \in \mathbb{R}^+$, $g(x) = y$ if and only if $g^{-1}(y) = x$.	For $x \in \mathbb{R}$, $y \in \mathbb{R}^+$, $e^x = y$ if and only if $\ln y = x$.

Theorems about Inverse Functions

The next two results in this section are two important theorems about inverse functions. The first is actually a corollary of Theorem 6.24.

Corollary 6.26. *Let A and B be nonempty sets and let $f : A \to B$ be a bijection. Then*

$$f^{-1} \circ f = I_A \text{ and } f \circ f^{-1} = I_B$$

where I_A is the identity function on A, and I_B is the identity function on B.

Proof. Let A and B be nonempty sets and assume that $f : A \to B$ is a bijection. Recall that for every $x \in A$, $I_A(x) = x$. So, let $x \in A$ and let $f(x) = y$. By Theorem 6.24, we can conclude that $f^{-1}(y) = x$. Therefore,

$$\begin{aligned}
\left(f^{-1} \circ f\right)(x) &= f^{-1}(f(x)) \\
&= f^{-1}(y) \\
&= x \\
&= I_A(x).
\end{aligned}$$

Hence, for each $x \in A$, $\left(f^{-1} \circ f\right)(x) = I_A(x)$, and since $f^{-1} \circ f : A \to A$ and $I_A : A \to A$, we conclude that $f^{-1} \circ f = I_A$.

The proof that $f \circ f^{-1} = I_B$ is left as an exercise. ∎

Example 6.25. (continued) For the cubing function and the cube root function, we have seen that

$$\text{For } x, y \in \mathbb{R}, \; x^3 = y \text{ if and only if } \sqrt[3]{y} = x.$$

Notice that

- If we substitute $x^3 = y$ into the equation $\sqrt[3]{y} = x$, we obtain

$$\sqrt[3]{x^3} = x.$$

- If we substitute $\sqrt[3]{y} = x$ into the equation $x^3 = y$, we obtain

$$\left(\sqrt[3]{y}\right)^3 = y.$$

This is an illustration of Corollary 6.26. We can see this by using $f : \mathbb{R} \to \mathbb{R}$ defined by $f(x) = x^3$ and $f^{-1} : \mathbb{R} \to \mathbb{R}$ defined by $f^{-1}(x) = \sqrt[3]{x}$. Then $f^{-1} \circ f : \mathbb{R} \to \mathbb{R}$ and $f^{-1} \circ f = I_{\mathbb{R}}$. So, for $x \in \mathbb{R}$,

$$\left(f^{-1} \circ f\right)(x) = I_{\mathbb{R}}(x)$$
$$f^{-1}(f(x)) = x$$
$$f^{-1}\left(x^3\right) = x$$
$$\sqrt[3]{x^3} = x.$$

Similarly, the equation $f \circ f^{-1} = I_{\mathbb{R}}$ can be used to obtain the equation $\left(\sqrt[3]{y}\right)^3 = y$ for each $y \in \mathbb{R}$.

We will now consider the case where $f : A \to B$ and $g : B \to C$ are both bijections. In this case, $f^{-1} : B \to A$ and $g^{-1} : C \to B$. We will use the following diagram to illustrate this situation.

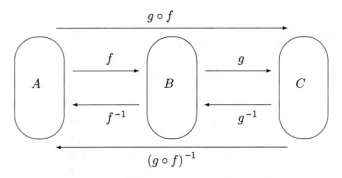

By Theorem 6.18, $g \circ f : A \to C$ is also a bijection. Hence, by Theorem 6.23, $(g \circ f)^{-1}$ is a function and in fact, $(g \circ f)^{-1} : C \to A$. Notice that we can also form the composition of g^{-1} followed by f^{-1} to get $f^{-1} \circ g^{-1} : C \to A$. This diagram helps illustrate the result of the next theorem.

Theorem 6.27. *Let $f : A \to B$ and $g : B \to C$ be bijections. Then $g \circ f$ is a bijection and $(g \circ f)^{-1} = f^{-1} \circ g^{-1}$.*

Proof. Let $f : A \to B$ and $g : B \to C$ be bijections. Then $f^{-1} : B \to A$ and $g^{-1} : C \to B$. Hence, $f^{-1} \circ g^{-1} : C \to A$. Also, by Theorem 6.18, $g \circ f : A \to C$ is a bijection, and hence $(g \circ f)^{-1} : C \to A$. We will now prove that for each $z \in C$, $(g \circ f)^{-1}(z) = (f^{-1} \circ g^{-1})(z)$.

Let $z \in C$. Since the function g is a surjection, there exists a $y \in B$ such that

$$g(y) = z. \tag{1}$$

Also, since f is a surjection, there exists an $x \in A$ such that

$$f(x) = y. \tag{2}$$

Now, Equations (1) and (2) can be written in terms of the respective inverse functions as

$$g^{-1}(z) = y; \text{ and} \tag{3}$$
$$f^{-1}(y) = x. \tag{4}$$

Using Equations (3) and (4), we see that

$$\begin{aligned}
(f^{-1} \circ g^{-1})(z) &= f^{-1}(g^{-1}(z)) \\
&= f^{-1}(y) \\
&= x. \tag{5}
\end{aligned}$$

Using Equations (1) and (2) again, we see that $(g \circ f)(x) = z$. However, in terms of the inverse function, this means that

$$(g \circ f)^{-1}(z) = x \tag{6}$$

Comparing Equations (5) and (6), we have shown that for all $z \in C$, $(g \circ f)^{-1}(z) = (f^{-1} \circ g^{-1})(z)$. This proves that $(g \circ f)^{-1} = f^{-1} \circ g^{-1}$. $\blacksquare$

Activity 6.28 (Construction of Inverse Functions). If $f : A \to B$ is a bijection, then we know that its inverse is a function. If we are given a formula for the function f, it may be desirable to determine a formula for the function f^{-1}. This can sometimes be done, while at other times it is very difficult or even impossible.

Construction of an Inverse Function

Let $f : \mathbb{R} \to \mathbb{R}$ be defined by $f(x) = 2x^3 - 7$.

1. Sketch a graph of this function and use the graph to explain why the function is a bijection (one-to-one and onto). A formal proof is not required.

One way to attempt to find a formula for the inverse of this function is to set $y = f(x)$ and solve for x. In this case, we are using x for the input of f and y for the output of f. By solving for x in terms of y, we are attempting to write a formula where y is the input and x is the output. This formula represents the inverse function.

2. Solve the equation $y = 2x^3 - 7$ for x. Use this to write a formula for $f^{-1}(y)$ where $f^{-1} : \mathbb{R} \to \mathbb{R}$.

3. Use the result of Part (2) to verify that for each $x \in \mathbb{R}$, $f^{-1}(f(x)) = x$ and for each $y \in \mathbb{R}$, $f(f^{-1}(y)) = y$.

<u>Note</u>: If we write $y = f(x)$ for the function f, we are using x as the independent variable (input) and y as the dependent variable (output). In the case where f^{-1} is a function, we write $x = f^{-1}(y)$, and y is the independent variable for f^{-1}. However, when we are using real functions, we frequently want to graph them. So we traditionally write each function as a function of x. Even if we have

$$f^{-1}(y) = \sqrt[3]{\frac{y+7}{2}},$$

we will frequently write

$$f^{-1}(x) = \sqrt[3]{\frac{x+7}{2}}.$$

Existence of an Inverse Function

In calculus we learned that a function G is an **increasing function** provided that $G(x_2) > G(x_1)$ whenever $x_2 > x_1$. We also learned that a differentiable function G is increasing on an interval provided that $G'(x) > 0$ for all x in that interval.

1. Explain why an increasing function is a one-to-one function or an injection.

Let $g : \mathbb{R} \to \mathbb{R}$ be defined by $g(x) = x^3 + x - 5$.

2. Use $g'(x)$ to show that g is an injection, and use a graph of g to show that g is a surjection. Explain why this proves the existence of an inverse function $g^{-1} : \mathbb{R} \to \mathbb{R}$.

3. If $y = g(x)$ and hence $y = x^3 + x - 5$, do you think it is possible to solve this equation for x to obtain a formula for $g^{-1}(y)$? Explain.

Activity 6.29 (The Inverse Sine Function). In order to obtain an inverse function, it is sometimes necessary to restrict the domain (or the codomain) of a function.

1. Let $f : \mathbb{R} \to \mathbb{R}$ be defined by $f(x) = \sin x$. Explain why the inverse of the function f is not a function. (A graph may be helpful.)

Notice that if we use the ordered pair representation, then the sine function can be represented as

$$f = \{(x,\ y) \in \mathbb{R} \times \mathbb{R} \mid y = \sin x\}.$$

If we denote the inverse of the sine function by $\sin^{-1}$, then

$$f^{-1} = \{(y, x) \in \mathbb{R} \times \mathbb{R} \mid y = \sin x\}.$$

Part (1) proves that f^{-1} is not a function. However, in previous mathematics courses, we frequently used the "inverse sine function." This is not really the inverse of the sine function as defined in Part (1) but rather, it is the inverse of the sine function **restricted to the domain** $\left[-\dfrac{\pi}{2}, \dfrac{\pi}{2}\right]$.

2. Explain why the function $F : \left[-\dfrac{\pi}{2}, \dfrac{\pi}{2}\right] \to [-1, 1]$ defined by $F(x) = \sin x$ is a bijection.

The inverse of the function in Part (2) is itself a function and is called the **inverse sine function** (or sometimes the **arcsine function**).

3. What is the domain of the inverse sine function? What are the range and codomain of the inverse sine function?

Let us now use $F(x) = \text{Sin}(x)$ to represent the restricted sine function in Part (2). Therefore, $F^{-1}(x) = \text{Sin}^{-1}(x)$ can be used to represent the inverse sine function. Observe that:

$$F : \left[-\dfrac{\pi}{2}, \dfrac{\pi}{2}\right] \to [-1, 1] \ \text{ and } \ F^{-1} : [-1, 1] \to \left[-\dfrac{\pi}{2}, \dfrac{\pi}{2}\right]$$

4. Using this notation, explain why

$$\text{Sin}^{-1} y = x \text{ if and only if } \left[y = \sin x \text{ and } -\dfrac{\pi}{2} \leq x \leq \dfrac{\pi}{2}\right];$$

$$\text{Sin}\left(\text{Sin}^{-1}(y)\right) = y \text{ for all } y \in [-1, 1]; \text{ and}$$

$$\text{Sin}^{-1}\left(\text{Sin}(x)\right) = x \text{ for all } x \in \left[-\dfrac{\pi}{2}, \dfrac{\pi}{2}\right].$$

Exercises 6.5

1. Let $A = \{1, 2, 3\}$ and $B = \{a, b, c\}$.

    (a) Construct an example of a function $f : A \to B$ that is not a bijection. Write the inverse of this function as a set of ordered pairs. Is the inverse of f a function? Explain. If so, draw an arrow diagram for f and f^{-1}.

    (b) Construct an example of a function $g : A \to B$ that is a bijection. Write the inverse of this function as a set of ordered pairs. Is the inverse of g a function? Explain. If so, draw an arrow diagram for g and g^{-1}.

2. Let $S = \{a, b, c, d\}$. Define $f : S \to S$ by defining f to be the following set of ordered pairs.

$$f = \{(a, c), (b, b), (c, d), (d, a)\}.$$

    (a) Draw an arrow diagram to represent the function f. Is the function f a bijection?

    (b) Write the inverse of f, f^{-1}, as a set of ordered pairs. Is f^{-1} a function? Explain.

    (c) Draw an arrow diagram for f^{-1} using the arrow diagram from Exercise (2a).

    (d) Compute $(f^{-1} \circ f)(x)$ and $(f \circ f^{-1})(x)$ for each x in S. What theorem does this illustrate?

3. This exercise uses ideas in Preview Activity 3.

    (a) Inverse functions can be used to help solve certain equations. The idea is to use an inverse function to undo the function. For example, we can often use the cube root function to help solve an equation involving a cube. For example, the main step in solving the equation

$$(2t - 1)^3 = 20$$

is to take the cube root of each side of the equation. This gives

$$\sqrt[3]{(2t - 1)^3} = \sqrt[3]{20}$$
$$2t - 1 = \sqrt[3]{20}.$$

Explain how this step in solving the equation is a use of Corollary 6.26.

(b) A main step in solving the equation $e^{2t-1} = 20$ is to take the natural logarithm of both sides of this equation. Explain how this step is a use of Corollary 6.26, and then solve the resulting equation to obtain a solution for t in terms of the natural logarithm function.

(c) How are the methods of solving the equations in Exercise (3a) and Exercise (3b) similar?

4. Let $f : A \to B$ and $g : B \to A$. Let I_A and I_B be the identity functions on the sets A and B, respectively. Prove each of the following:

(a) If $g \circ f = I_A$, then f is an injection.

(b) If $f \circ g = I_B$, then f is a surjection.

(c) If $g \circ f = I_A$ and $f \circ g = I_B$, then f and g are bijections and $g = f^{-1}$.

5. This exercise uses ideas that are contained in Activity 6.28. Let $\mathbb{R}^+ = \{y \in \mathbb{R} \mid y > 0\}$. Define $f : \mathbb{R} \to \mathbb{R}^+$ by $f(x) = e^{2x-1}$

(a) Set $y = e^{2x-1}$ and solve for x in terms of y.

(b) Use your work in Exercise (5a) to define a function $g : \mathbb{R}^+ \to \mathbb{R}$.

(c) For each $x \in \mathbb{R}$, determine $(g \circ f)(x)$ and for each $y \in \mathbb{R}^+$, determine $(f \circ g)(y)$.

(d) Use Exercise (4) to explain why $g = f^{-1}$.

6. (a) Define $f : \mathbb{R} \to \mathbb{R}$ by $f(x) = e^{-x^2}$. Is the inverse of f a function? Justify your conclusion.

(b) Let $\mathbb{R}^* = \{x \in \mathbb{R} \mid x \geq 0\}$. Define $g : \mathbb{R}^* \to (0, 1]$ by $g(x) = e^{-x^2}$. Is the inverse of g a function? Justify your conclusion.

7. This exercise uses ideas discussed in Activity 6.29.

(a) Let $f : \mathbb{R} \to \mathbb{R}$ be defined by $f(x) = x^2$. Explain why the inverse of f is not a function.

(b) Restrict the domain and codomain of the squaring function from Exercise (7a) so that the resulting (restricted squaring) function is a bijection.

(c) Explain how to define the square root function as the inverse of the function in Exercise (7b).

(d) True or false: $(\sqrt{x})^2 = x$ for all $x \in \mathbb{R}$ such that $x \geq 0$.

(e) True or false: $\sqrt{x^2} = x$ for all $x \in \mathbb{R}$.

8. Prove the following:

  If $f : A \to B$ is a bijection, then $f^{-1} : B \to A$ is also a bijection.

9. For each natural number k, let A_k be a set, and for each natural number n, let $f_n : A_n \to A_{n+1}$.

  For example, $f_1 : A_1 \to A_2$, $f_2 : A_2 \to A_3$, $f_3 : A_3 \to A_4$, and so on.

  Use mathematical induction to prove that for each natural number n with $n \geq 2$, if $f_1, f_2, \ldots, f_n$ are all bijections, then $f_n \circ f_{n-1} \circ \cdots \circ f_2 \circ f_1$ is a bijection and

  $$(f_n \circ f_{n-1} \circ \cdots \circ f_2 \circ f_1)^{-1} = f_1^{-1} \circ f_2^{-1} \circ \cdots \circ f_{n-1}^{-1} \circ f_n^{-1}.$$

  Note: This is an extension of Theorem 6.27. In fact, Theorem 6.27 is the basis step of this proof for $n = 2$.

Chapter 7

Equivalence Relations

7.1 Relations

Preview Activity 1 (The Telephone Keypad).

The following diagram represents the keypad for a telephone.

1	A B C 2	D E F 3
G H I 4	J K L 5	M N O 6
P Q R S 7	T U V 8	W X Y Z 9
*	0	#

Let $\Gamma = \{0, 1, 2, \ldots, 8, 9\}$, and let Δ be the set of all upper case letters in the English alphabet. (Note: We are using the uppercase Greek letters gamma, Γ, and delta, Δ, to avoid confusion with the letters on the keypad.) Define Ω, the uppercase Greek letter omega, to be the following subset of $\Gamma \times \Delta$:

$$\Omega = \{(x, y) \in \Gamma \times \Delta \mid x \text{ and } y \text{ are on the same button}\}.$$

1. Determine at least five different ordered pairs in $\Gamma \times \Delta$ that are elements of Ω.

2. Determine at least five different ordered pairs in $\Gamma \times \Delta$ that are not elements of Ω.

3. Determine all elements of the set $\{ x \in \Gamma \mid (\exists y \in \Delta)\, ((x, y) \in \Omega)\}$.

4. Determine all elements of the set $\{y \in \Delta \mid (\exists x \in \Gamma)\, ((x, y) \in \Omega)\}$.

Preview Activity 2 (An Equation with Two Variables).
    In Section 2.1, we introduced the concept of the **truth set of a predicate with one variable**. This was defined to be the set of all elements in the universal set that can be substituted for the variable to make the predicate a true proposition.
    Assume that x and y represent real numbers. Then

$$x^2 + y^2 = 25$$

is a predicate with two variables. An element of the truth set of this predicate (also called a solution of the equation) is an ordered pair (a, b) of real numbers so that when a is substituted for x and b is substituted for y, the predicate becomes a true proposition (a true equation in this case).

1. List five different elements of the truth set (often called the **solution set**) of this predicate with two variables.

2. Sketch the graph of the equation $x^2 + y^2 = 25$ in the xy-coordinate plane. What does the graph of this equation show?

3. Write a description of the solution set S of the equation $x^2 + y^2 = 25$ using set builder notation.

Preview Activity 3 (The United States of America).
    Let A be the set of all states in the United States. Consider the following subset of $A \times A$:

$$R = \{(x, y) \in A \times A \mid x \text{ and } y \text{ have a land border in common}\}.$$

1. Use the roster method for defining a set to list all the elements in the following set:
$$\{y \in A \mid (\text{Michigan}, y) \in R\}.$$

2. Use the roster method for defining a set to list all the elements in the following set:

$$\{x \in A \,|\, (x, \text{Michigan}) \in R\}.$$

3. Find three different examples of two ordered pairs, (x, y) and (y, z) such that $(x, y) \in R$, $(y, z) \in R$, but $(x, z) \notin R$, or explain why no such example exists.

4. In Sections 6.2 and 6.5, we learned how to represent a function as a set of ordered pairs, and we learned under what conditions a set of ordered pairs can be used to define a function.

   Can the set R be used to define a function from the set A to the set A? Explain.

In Section 6.1, we introduced the formal definition of a function from one set to another set. The notion of a function can be thought of as one way of relating the elements of one set with those of another set (or the same set). A function is a special type of **relation** in the sense that each element of the first set, the domain, is "related" to exactly one element of the second set, the codomain.

This idea of relating the elements of one set to those of another set is not restricted to functions. For example, we may say that one integer, a, is related to another integer, b, provided that a is congruent to b modulo 3. Notice that this relation of congruence modulo 3 also provides a way of relating one integer to another integer. However, in this case, an integer a is related to more than one other integer. For example,

$$
\begin{array}{lll}
5 \text{ is related to } 5 & \text{since} & 5 \equiv 5 \pmod{3} \\
5 \text{ is related to } 2 & \text{since} & 5 \equiv 2 \pmod{3} \\
5 \text{ is related to } 8 & \text{since} & 5 \equiv 8 \pmod{3} \\
5 \text{ is related to } \text{-}1 & \text{since} & 5 \equiv -1 \pmod{3}
\end{array}
$$

Notice that, as with functions, each relation of the form $a \equiv b \pmod{3}$, involves two integers a and b, and hence involves an ordered pair (a, b) which is an element of $\mathbb{Z} \times \mathbb{Z}$. We begin our study of relations by repeating the definition of the Cartesian product of two sets from Section 4.4.

Definition. The **Cartesian product** of two sets A and B, written $A \times B$, is the set of all ordered pairs (a, b), where $a \in A$ and $b \in B$. That is,

$$A \times B = \{(a, b) \mid a \in A \text{ and } b \in B\}.$$

We frequently read $A \times B$ as "A cross B."

In the case where the two sets are the same, we will write A^2 for $A \times A$. That is,

$$A^2 = A \times A = \{(a, b) \mid a \in A \text{ and } b \in A\}$$

Recall from Section 6.5 that a function $f : A \to B$ can be thought of as a set of ordered pairs that is a special type of subset of $A \times B$. We generalize this idea to make a formal definition of a relation from the set A to the set B.

Definition. A **relation from the set A to the set B** is a subset of $A \times B$. That is, R is a collection of ordered pairs where for each ordered pair in R, the first coordinate is an element of A and the second coordinate is an element of B.

If $A = B$ and R is a subset of $A \times A$, we say that R is a **relation on A**.

In Section 6.1, we defined the domain and range of a function. We make similar definitions for a relation.

Definition. If R is a relation from the set A to the set B, then the subset of A consisting of all the first coordinates of the ordered pairs in R is called the **domain** of R and is abbreviated as $\operatorname{dom}(R)$. The subset of B consisting of all the second coordinates of the ordered pairs in R is called the **range** of R, and is denoted by $\operatorname{range}(R)$. That is,

$$\operatorname{dom}(R) = \{u \in A \mid (u, y) \in R \text{ for at least one } y \in B\}$$
$$\operatorname{range}(R) = \{v \in B \mid (x, v) \in R \text{ for at least one } x \in A\}.$$

Example 7.1. A relation was studied in each of the three Preview Activities for this section.

For Preview Activity 1, we let $\Gamma = \{0, 1, 2, \ldots, 8, 9\}$, and let Δ be the set of all upper case letters in the English alphabet. Since

$$\Omega = \{(x, y) \in \Gamma \times \Delta \mid x \text{ and } y \text{ are on the same button}\},$$

Ω is a relation from the set Γ to the set Δ. In Problem (3) of Preview Activity 1, we actually determined the domain of this relation, and in Problem (4), we determined the range of this relation.

$$\begin{aligned}
\text{dom}\,(\Omega) &= \{x \in \Gamma \mid (\exists y \in \Delta)\,((x, y) \in \Omega)\} \\
&= \{2, 3, 4, 5, 6, 7, 8, 9\}.
\end{aligned}$$

$$\text{range}\,(\Omega) = \{y \in \Delta \mid (\exists x \in \Gamma)\,((x, y) \in \Omega)\} = \Delta$$

Activity 7.2 (Examples of Relations).

1. Let $S = \{(x, y) \in \mathbb{R} \times \mathbb{R} \mid x^2 + y^2 = 64\}$.

   (a) Explain why S is a relation on $\mathbb{R}$.

   (b) Find all values of x such that $(x, 4) \in S$. Find all values of x such that $(x, 9) \in S$.

   (c) What is the domain of the relation S? What is the range of S?

   (d) Since S is a relation on $\mathbb{R}$, its elements can be graphed in the coordinate plane. Describe the graph of the relation S.

2. Let A be the set of all states in the United States, and let

   $$R = \{(x, y) \in A \times A \mid x \text{ and } y \text{ have a border in common}\}.$$

   (a) Explain why R is a relation on A.

   (b) What is the domain of the relation R? What is the range of the relation R?

   (c) Is the following statement true or false? Explain.
   For all $x, y \in A$, if $(x, y) \in R$, then $(y, x) \in R$.

   (d) Is the following statement true or false? Explain.
   For all $x, y \in A$, if $(x, y) \in R$ and $(y, z) \in R$, then $(x, z) \in R$.

Some Standard Mathematical Relations

There are many different relations in mathematics. For example, two real numbers can be considered to be related if one number is less than the other number. We call this the "less than" relation on $\mathbb{R}$. If $x, y \in \mathbb{R}$ and x is less than y, we often write $x < y$. As a set of ordered pairs, this relation is $R_<$, where

$$R_< = \{(x, y) \in \mathbb{R} \times \mathbb{R} \mid x < y\}.$$

With many mathematical relations, we do not write the relation as a set of ordered pairs even though, technically, it is a set of ordered pairs. Table 7.1 describes some standard mathematical relations.

Name	Open Sentence	Relation as a Set of Ordered Pairs
The "less than" relation on $\mathbb{R}$	$x < y$	$\{(x, y) \in \mathbb{R} \times \mathbb{R} \mid x < y\}$
The "equality" relation on $\mathbb{R}$	$x = y$	$\{(x, y) \in \mathbb{R} \times \mathbb{R} \mid x = y\}$
The "divides" relation on $\mathbb{R}$	$m \mid n$	$\{(m, n) \in \mathbb{Z} \times \mathbb{Z} \mid m \text{ divides } n\}$
The "subset" relation on $\mathcal{P}(U)$	$S \subseteq T$	$\{(S, T) \in P(U) \times P(U) \mid S \subseteq T\}$
The "element of" relation from U to $\mathcal{P}(U)$	$x \in S$	$\{(x, S) \in U \times P(U) \mid x \in S\}$
The "congruence modulo n" relation on $\mathbb{Z}$	$a \equiv b \pmod{n}$	$\{(a, b) \in \mathbb{Z} \times \mathbb{Z} \mid a \equiv b \pmod{n}\}$

Table 7.1: Standard Mathematical Relations

Notation for Relations

The mathematical relations in Table 7.1 all used a relation symbol between the two elements that form the ordered pair in $A \times B$. For this reason, we often do the same thing for a general relation from the set A to the set B. So if R is a relation from A to B, and $x \in A$ and $y \in B$, we use the notation

$$x \, R \, y \quad \text{to mean} \quad (x, y) \in R; \text{ and}$$
$$x \, \cancel{R} \, y \quad \text{to mean} \quad (x, y) \notin R.$$

In some cases, we will even use a generic relation symbol for defining a new relation or speaking about relations in a general context. Perhaps the most commonly used symbol is "$\sim$", read "tilde" or "squiggle" or "is related to." When we do this, we will write

$$x \sim y \quad \text{means the same thing as} \quad (x, y) \in R; \text{ and}$$
$$x \not\sim y \quad \text{means the same thing as} \quad (x, y) \notin R$$

Activity 7.3 (The Divides Relation). Whenever we have spoken about one integer dividing another integer, we were working with the "divides" relation on $\mathbb{Z}$. In particular, we can write

$$D = \{(m, n) \in \mathbb{Z} \times \mathbb{Z} \mid m \text{ divides } n\}.$$

In this case, we have a specific notation for "divides", and we write

$$m \mid n \quad \text{if and only if} \quad (m, n) \in D.$$

1. What is the domain of the "divides" relation?

2. What is the range of the "divides" relation?

3. Are the following statements true or false? Explain.

   (a) For every $a \in \mathbb{Z}$, $a \mid a$.

   (b) For all $a, b \in \mathbb{Z}$, if $a \mid b$, then $b \mid a$.

   (c) For all $a, b, c \in \mathbb{Z}$, if $a \mid b$ and $b \mid c$, then $a \mid c$.

Functions as Relations

If we have a function $f : A \to B$, we can generate a set of ordered pairs f that is a subset of $A \times B$ as follows:

$$f = \{(a, f(a)) \mid a \in A\}, \text{ or}$$
$$f = \{(a, b) \in A \times B \mid b = f(a)\}.$$

This means that f is a relation from A to B. Since, $\text{dom}(f) = A$, we know that

$$\text{For every } a \in A, \text{ there exists a } b \in B \text{ such that } (a, b) \in f. \tag{1}$$

When $(a, b) \in f$, we write $b = f(a)$. This says that every element of A can be used as an input. In addition, to be a function, each input can produce only one output. In terms of ordered pairs, this means that there will never be two ordered pairs (a, b) and (a, c) in the function f where $a \in A$, $b, c \in B$, and $b \neq c$. We can formulate this as a conditional statement as follows:

For every $a \in A$ and every $b, c \in B$, if $(a, b) \in f$ and $(a, c) \in f$,

then $b = c$. $\qquad$ (2)

This means that a function f from A to B is a relation from A to B that satisfies Conditions (1) and (2). (See Theorem 6.20 in Section 6.5.)

The Inverse of a Relation

In Section 6.5, we introduced the **inverse of a function**. If $f : A \rightarrow B$ is a function, then the inverse of f, denoted by f^{-1}, is defined as

$$f^{-1} = \{(b, a) \in B \times A \mid f(a) = b\}.$$

If we use the ordered pair representation for f, we could also write

$$f^{-1} = \{(b, a) \in B \times A \mid (a, b) \in f\}.$$

Now that we know about relations, we see that f^{-1} is always a relation from B to A. The concept of the inverse of a function is actually a special case of the more general concept of the inverse of a relation, which we now define.

Definition. Let R be a relation from the set A to the set B. The **inverse of R**, written R^{-1} and read "R inverse," is the relation from B to A defined by

$$R^{-1} = \{(y, x) \in B \times A \mid (x, y) \in R\}, \text{ or}$$
$$R^{-1} = \{(y, x) \in B \times A \mid x \, R \, y\}.$$

That is, R^{-1} is the subset of $B \times A$ consisting of all ordered pairs (y, x) such that $x \, R \, y$.

An Example of an Inverse Relation

Let D be the "divides" relation on $\mathbb{Z}$. See Activity 7.3. So

$$D = \{(m, n) \in \mathbb{Z} \times \mathbb{Z} \mid m \text{ divides } n\}.$$

This means that we can write

$$m \mid n \quad \text{if and only if} \quad (m, n) \in D.$$

So, in this case,

$$\begin{aligned} D^{-1} &= \{(n, m) \in \mathbb{Z} \times \mathbb{Z} \mid (m, n) \in D\} \\ &= \{(n, m) \in \mathbb{Z} \times \mathbb{Z} \mid m \text{ divides } n\}. \end{aligned}$$

Now, if we would like to focus on the first coordinate instead of the second coordinate in D^{-1}, we know that "m divides n" means the same thing as "n is a multiple of m." Hence,

$$D^{-1} = \{(n, m) \in \mathbb{Z} \times \mathbb{Z} \mid n \text{ is a multiple of } m\}.$$

We can say that the inverse of the "divides" relation on $\mathbb{Z}$ is the "is a multiple of" relation on $\mathbb{Z}$.

Theorem 7.4, which follows, contains some elementary facts about inverse relations. The proofs of these results are included in Activity 7.5.

Theorem 7.4. *Let R be a relation from the set A to the set B. Then*

1. *The domain of R^{-1} is the range of R. That is, $dom\left(R^{-1}\right) = range\left(R\right)$.*

2. *The range of R^{-1} is the domain of R. That is, $range\left(R^{-1}\right) = dom\left(R\right)$.*

3. *The inverse of R^{-1} is R. That is, $\left(R^{-1}\right)^{-1} = R$.*

Activity 7.5 (Proving Theorem 7.4). To prove Part (1) of Theorem 7.4, observe that the goal is to prove that two sets are equal,

$$\text{dom}\left(R^{-1}\right) = \text{range}\left(R\right).$$

One way to do this is to prove that each is a subset of the other. Another way is to use a sequence of "if and only if" statements.

To prove that dom $(R^{-1}) \subseteq$ range (R), we can start by choosing an arbitrary element of dom (R^{-1}). So let $y \in$ dom (R^{-1}). The goal now is to prove that $y \in$ range (R). What does it mean to say that $y \in$ dom (R^{-1})? It means that there exists an $x \in A$ such that

$$(y, x) \in R^{-1}.$$

Now, what does it mean to say that $(y, x) \in R^{-1}$? It means that $(x, y) \in R$. What does this tell us about y?

Complete the proof of Part (1) of Theorem 7.4. Then, complete the proofs of Part (2) and Part (3) of Theorem 7.4.

Exercises 7.1

1. Let $A = \{a, b, c\}$, $B = \{p, q, r\}$, and $R = \{(a, p), (b, q), (c, p), (a, q)\}$.

   (a) Use the roster method to list all the elements of $A \times B$. Explain why $A \times B$ can be considered to be a relation from A to B.

   (b) Explain why R is a relation from A to B.

   (c) What is the domain of R? What is the range of R?

   (d) What is R^{-1}, the inverse relation of R?

2. Let $A = \{a, b, c\}$ and let $R = \{(a, a), (a, c), (b, b), (b, c), (c, a), (c, b)\}$.
   Are the following statements true or false? Explain.

   (a) For each $x \in A$, $x \, R \, x$.

   (b) For every $x, y \in A$, if $x \, R \, y$, then $y \, R \, x$.

   (c) For every $x, y, z \in A$, if $x \, R \, y$ and $y \, R \, z$, then $x \, R \, z$.

   (d) R is a function from A to A.

3. Let A be the set of all females citizens of the United States. Let D be the relation on A defined by

$$D = \{(x, y) \in A \times A \mid x \text{ is a daughter of } y\}.$$

   That is, $x \, D \, y$ means that x is a daughter of y.

   (a) Describe those elements of A that are in the domain of D.

(b) Describe those elements of A that are in the range of D.

(c) Is the relation D a function from A to A? Explain.

(d) Use set builder notation to define the inverse relation of D.

4. Let U be a nonempty set, and let R be the "subset relation" on $\mathcal{P}(U)$. That is,

$$R = \{(S,T) \in \mathcal{P}(U) \times \mathcal{P}(U) \mid S \subseteq T\}.$$

(a) Write the predicate $(S,T) \in R$ using standard subset notation.

(b) What is the domain of this subset relation, R?

(c) What is the range of this subset relation, R?

(d) Use set builder notation to define the inverse relation, R^{-1}.

(e) Is R a function from $\mathcal{P}(U)$ to $\mathcal{P}(U)$? Explain.

5. Let U be a nonempty set, and let R be the "element of" relation from U to $\mathcal{P}(U)$. That is,

$$R = \{(x,S) \in U \times \mathcal{P}(U) \mid x \in S\}.$$

(a) What is the domain of this "element of" relation, R?

(b) What is the range of this "element of" relation, R?

(c) Use set builder notation to define the inverse relation, R^{-1}.

(d) Is R a function from U to $\mathcal{P}(U)$? Explain.

6. Let $S = \{(x,y) \in \mathbb{R} \times \mathbb{R} \mid x^2 + y^2 = 100\}$. Note that S is a relation on $\mathbb{R}$.

(a) Find all values of x such that $(x,6) \in S$. Find all values of x such that $(x,9) \in S$.

(b) What is the domain of the relation S?

(c) What is the range of the relation S?

(d) What is the inverse of the relation S?

(e) Is the relation S a function from $\mathbb{R}$ to $\mathbb{R}$? Explain.

(f) Since S is a relation on $\mathbb{R}$, its elements can be graphed in the co-ordinate plane. Describe the graph of the relation S. Is the graph consistent with your answers in Exercises (6a) through (6e)? Explain.

7. Repeat Exercise(6) using the relation

$$S = \left\{ (x, y) \in \mathbb{R} \times \mathbb{R} \mid y = \sqrt{100 - x^2} \right\}.$$

What is the connection between this relation and the relation in Exercise (6)?

8. Let $R_<$ be the "less than" relation on $\mathbb{R}$. That is,

$$R_< = \{(x, y) \in \mathbb{R} \times \mathbb{R} \mid x < y\}.$$

(a) What is the domain of the relation $R_<$?

(b) What is the range of the relation $R_<$?

(c) What is the inverse of the relation $R_<$?

(d) Is the relation $R_<$ a function from $\mathbb{R}$ to $\mathbb{R}$? Explain.

9. Prove the following proposition:

> Let A and B be nonempty sets, and let R and S be relations from A to B. If $R \subseteq S$, then $R^{-1} \subseteq S^{-1}$.

<u>Note</u>: Remember that a relation is a set. Consequently, we can talk about one relation being a subset of another relation. Another thing to remember is that the elements of a relation are ordered pairs.

7.2   Equivalence Relations

Preview Activity 1 (Directed Graphs and Relations).
If A is a (small) finite set, a relation R on A can be specified by simply listing all the ordered pairs in R. For example, if $A = \{1, 2, 3, 4\}$, then

$$R = \{(1, 1), (4, 4), (1, 3), (3, 2), (1, 2), (2, 1)\}$$

is a relation on A. A convenient way to represent such a relation is to draw a point in the plane for each of the elements of A and then for each $(x, y) \in R$ (or $x \, R \, y$), we draw an arrow starting at point x and pointing to point y. If $(x, x) \in R$ (or $x \, R \, x$), we draw a loop at the point x. The resulting diagram is called a **directed graph** or a **digraph**. The diagram in Figure 7.1 is a digraph for the relation R.

In a directed graph, the points are called the **vertices**. So each element of A corresponds to a **vertex**. The arrows, including the loops, are called the **directed edges** of the directed graph.

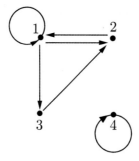

Figure 7.1: Directed Graph for a Relation

1. Let $A = \{1, 2, 3, 4\}$ and $R = \{(1, 1), (2, 2), (3, 3), (4, 4), (1, 3), (3, 2)\}$. Draw a digraph for this relation and then determine whether the following statements are true or false.

   (a) For each $x \in A$, $x \, R \, x$.

   (b) For every $x, y \in A$, if $x \, R \, y$, then $y \, R \, x$.

   (c) For every $x, y, z \in A$, if $x \, R \, y$ and $y \, R \, z$, then $x \, R \, z$.

2. Let $A = \{1, 2, 3, 4\}$ and $S = \{(1, 1), (1, 4), (2, 4), (4, 1), (4, 2)\}$. Draw a digraph for this relation and then determine whether the following statements are true or false.

   (a) For each $x \in A$, $x \, S \, x$.

   (b) For every $x, y \in A$, if $x \, S \, y$, then $y \, S \, x$.

   (c) For every $x, y, z \in A$, if $x \, S \, y$ and $y \, S \, z$, then $x \, R \, z$.

Preview Activity 2 (Properties of Relations).

In Preview Activity (1), the same three questions were asked about two different relations on the set $A = \{1, 2, 3, 4\}$. These were questions about certain properties of relations, and these properties occur frequently enough to warrant names.

Definition. Let A be a nonempty set and let R be a relation on A.

- The relation R is **reflexive on A** provided that for each $x \in A$, $x \mathrel{R} x$ or, equivalently, $(x, x) \in R$.

- The relation R is **symmetric** provided that for every $x, y \in A$, if $x \mathrel{R} y$, then $y \mathrel{R} x$ or, equivalently, for every $x, y \in A$, if $(x, y) \in R$, then $(y, x) \in R$.

- The relation R is **transitive** provided that for every $x, y, z \in A$, if $x \mathrel{R} y$ and $y \mathrel{R} z$, then $x \mathrel{R} z$ or, equivalently, for every $x, y, z \in A$, if $(x, y) \in R$ and $(y, z) \in R$, then $(x, z) \in R$.

Let U be a finite, nonempty set and let $\mathcal{P}(U)$ be the power set of U. Define the relations $\sim$ and $\approx$ on $\mathcal{P}(U)$ as follows:

For $A, B \in P(U)$,

- $A \sim B$ if and only if $A \cap B = \emptyset$. That is, the ordered pair (A, B) is in the relation $\sim$ if and only if A and B are disjoint.

- $A \approx B$ if and only if $|A| = |B|$. That is, the ordered pair (A, B) is in the relation $\approx$ if and only if A and B have the same cardinality (same number of elements).

1. Is the relation $\sim$ a reflexive relation on $\mathcal{P}(U)$? Is it a symmetric relation? Is it a transitive relation?

2. Is the relation $\approx$ a reflexive relation on $\mathcal{P}(U)$? Is it a symmetric relation? Is it a transitive relation?

Preview Activity 3 (Review of Congruence Modulo n).

1. Let $a, b \in \mathbb{Z}$ and let $n \in \mathbb{N}$. On page 77 of Section 3.1, we defined what it means to say that a is congruent to b modulo n. Write this definition and state two different conditions that are equivalent to the definition.

2. Let $a, b \in \mathbb{Z}$ and let $n \in \mathbb{N}$. We use $a \equiv b \pmod{n}$ as a notation for "a is congruent to b modulo n." Explain why congruence modulo n is a relation on $\mathbb{Z}$.

3. Carefully review Theorem 3.5 and the proofs given on page 78 of Section 3.1. In terms of the properties of relations introduced in Preview Activity 2, what does this theorem say about the relation of congruence modulo n on the integers?

4. Write a complete statement of Theorem 3.21 and Corollary 3.22.

5. Write a proof of the symmetric property for congruence modulo n. That is, prove the following:

> Let $a, b \in \mathbb{Z}$ and let $n \in \mathbb{N}$. If $a \equiv b \pmod{n}$, then $b \equiv a \pmod{n}$.

In mathematics, as in real life, it is often convenient to think of two different things as being essentially the same. For example, when you go to a store to buy a cold soft drink, the cans of soft drinks in the cooler are often sorted by brand and type of soft drink. The Coca Colas are grouped together, the Pepsi Colas are grouped together, the Dr. Peppers are grouped together, and so on. When we choose a particular can of one type of soft drink, we are assuming that all the cans are essentially the same. Even though the specific cans of one type of soft drink are physically different, it makes no difference which can we choose. In doing this, we are saying that the cans of one type of soft drink are equivalent, and we are using the mathematical notion of an equivalence relation.

Directed Graphs and Properties of Relations

An equivalence relation on a set is a relation with a certain combination of properties that allow us to sort the elements of the set into certain classes. In this section, we will focus on the properties that define an equivalence relation, and in the next section, we will see how these properties allow us to sort or partition the elements of the set into certain classes. The three properties used to define an equivalence relation are the reflexive, symmetric, and transitive properties. These properties were introduced in Preview Activity 2 and will be repeated in the following descriptions of how these three properties can be visualized on a directed graph. In Preview Activity 1, we used directed graphs, or digraphs, to represent relations on finite sets.

- Let A be a nonempty set and let R be a relation on A. The relation R is **reflexive on A** provided that for each $x \in A$, $x \, R \, x$, or equivalently, $(x, x) \in R$.

This means that if a reflexive relation is represented on a digraph, there would have to be a loop at each vertex, as is shown in the following figure.

- Let A be a nonempty set and let R be a relation on A. The relation R is **symmetric** provided that for every $x, y \in A$, if $x \ R \ y$, then $y \ R \ x$ or, equivalently, for every $x, y \in A$, if $(x, y) \in R$, then $(y, x) \in R$.

  This means that if a symmetric relation is represented on a digraph, then anytime there is a directed edge from one vertex to a second vertex, there would be a directed edge from the second vertex to the first vertex, as is shown in the following figure.

- Let A be a nonempty set and let R be a relation on A. The relation R is **transitive** provided that for every $x, y, z \in A$, if $x \ R \ y$ and $y \ R \ z$, then $x \ R \ z$ or, equivalently, for every $x, y, z \in A$, if $(x, y) \in R$ and $(y, z) \in R$, then $(x, z) \in R$. So if a transitive relation is represented by a digraph, then anytime there is a directed edge from a vertex x to a vertex y and a directed edge from y to a vertex z, there would be a directed edge from x to z.

  In addition, if a transitive relation is represented by a digraph, then anytime there is a directed edge from a vertex x to a vertex y and a directed edge from y to the vertex x, there would be loops at x and y. These two situations are illustrated as follows:

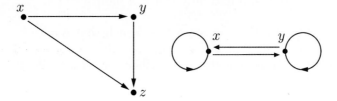

There are other properties of relations that can be studied, but we will restrict ourselves to these three, as they are the ones pertinent to the study of equivalence relations.

Definition. Let A be a nonempty set. A relation $\sim$ on the set A is an **equivalence relation** provided that $\sim$ is reflexive, symmetric, and transitive. For $a, b \in A$, if $\sim$ is an equivalence relation on A and $a \sim b$, we say that **a is equivalent to b**.

Activity 7.6 (Negations of the Properties of Relations). Let A be a nonempty set and let R be a relation on A.

1. Carefully explain what it means to say that the relation R is not reflexive on the set A.

2. Carefully explain what it means to say that the relation R is not symmetric.

3. Carefully explain what it means to say that the relation R is not transitive.

There is a subtle difference between the reflexive property and the other two properties. The reflexive property states that some ordered pairs actually belong to the relation R, or some elements of A are related. The reflexive property has a universal quantifier and hence, we must prove that for all $x \in A$, $x\,R\,x$. Symmetry and transitivity, on the other hand, are defined by conditional sentences. We often use a direct proof for these properties, and so we start by assuming the hypothesis and then showing that the conclusion must follow from the hypothesis.

Example 7.7. Let M be the relation on $\mathbb{Z}$ defined as follows:

For $a, b \in \mathbb{Z}$, $a\,M\,b$ if and only if a is a multiple of b.

So $a\,M\,b$ if and only if there exists a $k \in \mathbb{Z}$ such that $a = b \cdot k$.

- The relation M is reflexive on $\mathbb{Z}$ since for all $x \in \mathbb{Z}$, $x = x \cdot 1$ and hence, $x\,M\,x$.

- Notice that $4\,M\,2$, but $2\,\cancel{M}\,4$. So there exist integers x and y such that $x\,M\,y$ but $y\,\cancel{M}\,x$. Hence, the relation M is not symmetric.

- Now assume that $x \ M \ y$ and $y \ M \ z$. Then there exist integers p and q such that
$$x = y \cdot p \text{ and } y = z \cdot q.$$
Substituting the second equation into the first equation, we see that $x = z \cdot (p \cdot q)$. Since $p \cdot q \in \mathbb{Z}$, we have shown that x is a multiple of z and hence $x \ M \ z$. Therefore, M is a transitive relation.

Activity 7.8 (Examples of Relations with Certain Properties).
Let $A = \{1, 2, 3, 4\}$. We can define relations on the set A by simply listing the ordered pairs in the relation, and we can use digraphs to represent the relations. For example,
A relation on A that is reflexive on A but not symmetric and not transitive is the relation in Part (1) of Preview Activity 1. This relation is

$$R = \{(1,1), (2,2), (3,3), (4,4), (1,3), (3,2)\}.$$

- This relation is reflexive on A because $(x, x) \in R$ for every $x \in A$.

- This relation is not symmetric since $(1,3) \in R$ but $(3,1) \notin R$.

- This relation is not transitive since $(1,3) \in R$ and $(3,2) \in R$, but $(1,2) \notin R$.

1. Explain why $R = \{(1,2), (2,1)\}$ is a relation on A that is symmetric but not reflexive on A and not transitive .

2. Find a relation on A that is transitive but neither reflexive on A nor symmetric.

3. Find a relation on A that is reflexive on A and symmetric but not transitive.

4. Find a relation on A that is reflexive on A and transitive but not symmetric.

5. Find a relation on A that is symmetric and transitive but not reflexive on A.

The relations in this activity show that the three properties (reflexivity, symmetry, and transitivity) are independent of each other in the sense that it is possible to construct a relation that satisfies any one of the properties but not the other two. It is also possible to construct a relation that satisfies any two of the properties but not the third.

Congruence Modulo n

One of the important equivalence relations we will study in detail in this text is that of congruence modulo n. We reviewed this relation in Preview Activity 3.

Let $n \in \mathbb{N}$. For $a, b \in \mathbb{Z}$, we have defined a to be congruent to b modulo n, denoted by $a \equiv b \pmod{n}$, as follows:

$$a \equiv b \pmod{n} \text{ provided that } n \mid (a - b).$$

This is equivalent to saying that $a \equiv b \pmod{n}$ provided that there exists an integer k such that $a - b = nk$.

Theorem 3.5 on page 78 tells us that this is an equivalence relation on $\mathbb{Z}$. Recall that by the Division Algorithm, if $a \in \mathbb{Z}$, then there exist unique integers q and r such that

$$a = nq + r \text{ and } 0 \le r < n.$$

Theorem 3.21 and Corollary 3.22 then tell us that $a \equiv r \pmod{n}$. That is, a is congruent modulo n, to its remainder r when it is divided by n. When we use the term "remainder" in this context, we always mean the remainder r with $0 \le r < n$ that is guaranteed by the Division Algorithm. We can use this idea to prove the following theorem.

Theorem 7.9. *Let $n \in \mathbb{N}$ and let $a, b \in \mathbb{Z}$. Then $a \equiv b \pmod{n}$ if and only if a and b have the same remainder when divided by n.*

Proof. Let $n \in \mathbb{N}$ and let $a, b \in \mathbb{Z}$. We will first prove that if a and b have the same remainder when divided by n, then $a \equiv b \pmod{n}$.

Assume that a and b have the same remainder when divided by n, and let r be this common remainder. Then, by Theorem 3.21,

$$a \equiv r \pmod{n} \text{ and } b \equiv r \pmod{n}.$$

Since congruence modulo n is an equivalence relation, it is a symmetric relation. Hence, since $b \equiv r \pmod{n}$, we can conclude that $r \equiv b \pmod{n}$. Combining this with the fact that $a \equiv r \pmod{n}$, we now have

$$a \equiv r \pmod{n} \text{ and } r \equiv b \pmod{n}.$$

We can now use the transitive property to conclude that $a \equiv b \pmod{n}$. This proves that if a and b have the same remainder when divided by n, then $a \equiv b \pmod{n}$.

We will now prove that if $a \equiv b \pmod{n}$, then a and b have the same remainder when divided by n. Assume that $a \equiv b \pmod{n}$, and let r be the least nonnegative remainder when b is divided by n. Then $0 \leq r < n$ and by Theorem 3.21,

$$b \equiv r \pmod{n}.$$

Now, using the facts that $a \equiv b \pmod{n}$ and $b \equiv r \pmod{n}$, we can use the transitive property to conclude that

$$a \equiv r \pmod{n}.$$

This means that there exists an integer q such that $a - r = nq$ or that

$$a = nq + r.$$

Since we already know that $0 \leq r < n$, the last equation tells us that r is the least nonnegative remainder when a is divided by n. Hence we have proven that if $a \equiv b \pmod{n}$, then a and b have the same remainder when divided by n. ∎

Examples of Other Equivalence Relations

1. The first relation in Preview Activity 2 was not reflexive and was not transitive. Hence, it is not an equivalence relation.

   However, the second relation in Preview Activity 2 is an equivalence relation. It is repeated here.

   Let U be a finite, nonempty set and let $P(U)$ be the power set of U. Define the relation $\approx$ on $P(U)$ as follows:

   For $A, B \in P(U)$, $A \approx B$ if and only if $|A| = |B|$.

   This relation states that two subsets of U are equivalent provided that they have the same number of elements.

2. Let A be a nonempty set. The **equality relation on A** is an equivalence relation. This relation is also called the **identity relation on A** and is denoted by I_A, where

$$I_A = \{(x, x) \mid x \in A\}.$$

3. Define the relation $\sim$ on $\mathbb{R}$ as follows:

For $a, b \in \mathbb{R}$, $a \sim b$ if and only if there exists an integer k such that $a - b = 2k\pi$.

We will prove that the relation $\sim$ is an equivalence relation on $\mathbb{R}$. The relation $\sim$ is reflexive on $\mathbb{R}$ since for each $a \in \mathbb{R}$, $a - a = 0 = 2 \cdot 0 \cdot \pi$.

Now, let $a, b \in \mathbb{R}$ and assume that $a \sim b$. We will prove that $b \sim a$. Since $a \sim b$, there exists an integer k such that

$$a - b = 2k\pi.$$

By multiplying both sides of this equation by -1, we obtain

$$(-1)(a - b) = (-1)(2k\pi)$$
$$b - a = 2(-k)\pi.$$

Since $-k \in \mathbb{Z}$, the last equation proves that $b \sim a$. Hence, we have proven that if $a \sim b$, then $b \sim a$, and therefore, the relation $\sim$ is symmetric.

To prove transitivity, let $a, b, c \in \mathbb{R}$ and assume that $a \sim b$ and $b \sim c$. We will prove that $a \sim c$. Now, there exist integers k and n such that

$$a - b = 2k\pi \text{ and } b - c = 2n\pi.$$

By adding these two equations, we see that

$$(a - b) + (b - c) = 2k\pi + 2n\pi$$
$$a - c = 2(k + n)\pi.$$

By the closure properties of the integers, $k + n \in \mathbb{Z}$. So this proves that $a \sim c$, and hence the relation $\sim$ is transitive.

We have now proven that $\sim$ is an equivalence relation on $\mathbb{R}$. This equivalence relation is important in trigonometry. If $a \sim b$, then there exists an integer k such that $a - b = 2k\pi$ and hence $a = b + k(2\pi)$. Since the sine and cosine functions are periodic with a period of 2π, we see that

$$\sin a = \sin(b + k(2\pi)) = \sin b, \text{ and}$$
$$\cos a = \cos(b + k(2\pi)) = \cos b.$$

Therefore, when $a \sim b$, each of the trigonometric functions have the same value at a and b.

4. For an example from Euclidean geometry, we define a relation P on the set $\mathcal{L}$ of all lines in the plane as follows:

   For $l_1, l_2 \in \mathcal{L}$, $l_1 \, P \, l_2$ if and only if l_1 is parallel to l_2 or $l_1 = l_2$.

   We added the second condition to the definition of P to ensure that P is reflexive on $\mathcal{L}$. Theorems from Euclidean geometry tell us that if l_1 is parallel to l_2, then l_2 is parallel to l_1, and if l_1 is parallel to l_2 and l_2 is parallel to l_3, then l_1 is parallel to l_3. (Drawing pictures will help visualize these properties.) This tells us that the relation P is reflexive, symmetric, and transitive, and hence an equivalence relation on $\mathcal{L}$.

Exercises 7.2

1. Let $A = \{a, b\}$ and let $R = \{(a, b)\}$. Justify your answers to each of the following:

   (a) Is R a reflexive relation on A?

   (b) Is R a symmetric relation?

   (c) Is R a transitive relation?

   (d) Is R an equivalence relation on A?

2. Let $R = \{(x, y) \in \mathbb{R} \times \mathbb{R} \mid |x| + |y| = 4\}$. Then R is a relation on $\mathbb{R}$. Justify your answers to each of the following:

   (a) Is R a reflexive relation on $\mathbb{R}$?

   (b) Is R a symmetric relation?

   (c) Is R a transitive relation?

   (d) Is R an equivalence relation on $\mathbb{R}$?

3. Let $A = \{1, 2, 3, 4, 5\}$. The identity relation on A is

   $$I_A = \{(1, 1), (2, 2), (3, 3), (4, 4), (5, 5)\}.$$

   Determine an equivalence relation on A that is different from I_A or explain why this is not possible.

4. Let $A = \{a, b, c\}$. For each of the following, draw a directed graph that represents a relation with the specified properties.

(a) A relation on A that is symmetric but not transitive

(b) A relation on A that is transitive but not symmetric

(c) A relation on A that is symmetric and transitive but not reflexive on A

(d) A relation on A that is not reflexive on A, is not symmetric, and is not transitive

(e) A relation on A, other than the identity relation, that is an equivalence relation on A

5. Let $f : \mathbb{R} \to \mathbb{R}$ be defined by $f(x) = x^2 - 4$ for each $x \in \mathbb{R}$. Define a relation $\sim$ on $\mathbb{R}$ as follows:

For $a, b \in \mathbb{R}$, $a \sim b$ if and only if $f(a) = f(b)$.

(a) Is the relation $\sim$ an equivalence relation on $\mathbb{R}$? Justify your conclusion.

(b) Determine all real numbers in the following set:

$$C = \{x \in \mathbb{R} \mid x \sim 5\}.$$

6. Repeat Exercise (5) using the function $f : \mathbb{R} \to \mathbb{R}$ that is defined by $f(x) = x^2 - 3x - 7$ for each $x \in \mathbb{R}$.

7. (a) Repeat Exercise (5a) using the function $f : \mathbb{R} \to \mathbb{R}$ that is defined by $f(x) = \sin x$ for each $x \in \mathbb{R}$.

(b) Determine all real numbers in the following set:

$$C = \{x \in \mathbb{R} \mid x \sim \pi\}.$$

8. Define the relation $\sim$ on $\mathbb{R}$ as follows:

For $x, y \in \mathbb{R}$, $x \sim y$ if and only if $y - x$ is an integer.

(a) Is the relation $\sim$ an equivalence relation on $\mathbb{R}$? Justify your conclusion.

(b) List four different elements of the set $C = \{x \in \mathbb{R} \mid x \sim \pi\}$.

(c) Describe all the elements in the set C without using the symbol $\sim$ for the equivalence relation.

9. Let X be a nonempty set and let $\mathcal{P}(X)$ be the power set of X. That is, $\mathcal{P}(X)$ is the set of all subsets of X.

For A and B in $\mathcal{P}(X)$, define $A \sim B$ to mean that there exists a bijection $f : A \to B$. Prove that $\sim$ is an equivalence relation on $\mathcal{P}(X)$.

Hint: Use results from Sections 6.4 and 6.5.

10. Define the relation $\approx$ on $\mathbb{R} \times \mathbb{R}$ as follows:

For $(a, b), (c, d) \in \mathbb{R} \times \mathbb{R}$, $(a, b) \approx (c, d)$ if and only if $a^2 + b^2 = c^2 + d^2$.

   (a) Prove that $\approx$ is an equivalence relation on $\mathbb{R} \times \mathbb{R}$.

   (b) List four different elements of the set
$$C = \{(x, y) \in \mathbb{R} \times \mathbb{R} \mid (x, y) \approx (4, 3)\}.$$

   (c) Give a geometric description of set C.

7.3   Equivalence Classes

Preview Activity 1 (Sets Associated with a Relation).
Let $A = \{a, b, c, d, e\}$, and let

$$R = \{(a, a), (b, b), (c, c), (d, d), (e, e), (a, b), (b, a), (c, d), (d, c)\};$$
$$S = \{(b, b), (c, c), (d, d), (e, e), (a, b), (b, c), (a, d), (c, d), (d, c)\}.$$

1. Draw a digraph that represents the relation R on A. Explain why R is an equivalence relation on A.

2. Draw a digraph that represents the relation S on A. Explain why S is not an equivalence relation on A.

For each $y \in A$, define the following subsets of A:

$$R[y] = \{x \in A \mid x \, R \, y\} = \{x \in A \mid (x, y) \in R\};$$
$$S[y] = \{x \in A \mid x \, S \, y\} = \{x \in A \mid (x, y) \in S\}.$$

3. **(a)** Determine $R[a]$, $R[b]$, $R[c]$, $R[d]$, and $R[e]$.

   (b) Which of the sets $R[a]$, $R[b]$, $R[c]$, $R[d]$, and $R[e]$ are equal?

   (c) Which of the sets $R[a]$, $R[b]$, $R[c]$, $R[d]$, and $R[e]$ are disjoint?

4. (a) Determine $S[a]$, $S[b]$, $S[c]$, $S[d]$, and $S[e]$.

(b) Which of the sets $S[a]$, $S[b]$, $S[c]$, $S[d]$, and $S[e]$ are equal?

(c) Which of the sets $S[a]$, $S[b]$, $S[c]$, $S[d]$, and $S[e]$ are disjoint?

Preview Activity 2 (Congruence Modulo 3).

1. Use the roster method to list the elements of each of the following sets:

(a) The set $C[0]$ of all integers a that are congruent to 0 modulo 3. That is, $C[0] = \{a \in \mathbb{Z} \mid a \equiv 0 \pmod 3\}$.

(b) The set $C[1]$ of all integers a that are congruent to 1 modulo 3. That is, $C[1] = \{a \in \mathbb{Z} \mid a \equiv 1 \pmod 3\}$.

(c) The set $C[2]$ of all integers a that are congruent to 2 modulo 3. That is, $C[2] = \{a \in \mathbb{Z} \mid a \equiv 2 \pmod 3\}$.

(d) The set $C[3]$ of all integers a that are congruent to 3 modulo 3. That is, $C[3] = \{a \in \mathbb{Z} \mid a \equiv 3 \pmod 3\}$.

2. Consider the three sets, $C[0]$, $C[1]$, and $C[2]$ in Parts (1a), (1b), and (1c).

(a) Determine the intersection of any two of these sets. That is, determine $C[0] \cap C[1]$, $C[0] \cap C[2]$, and $C[1] \cap C[2]$.

(b) Let $n = 734$. What is the remainder when n is divided by 3? Which of the three sets, if any, contains $n = 734$?

(c) Repeat Part (2b) for $n = 79$.

(d) Repeat Part (2b) for $n = -79$.

(e) Do you think that $C[0] \cup C[1] \cup C[2] = \mathbb{Z}$? Explain.

As was indicated at the beginning of Section 7.2, an equivalence relation on a set is a relation with a certain combination of properties (reflexive, symmetric, and transitive) that allow us to sort the elements of the set into certain classes. We saw this happen in the Preview Activities. Specifically, in Preview Activity 2, we used the equivalence relation of congruence modulo 3 on $\mathbb{Z}$ to construct the following three sets:

$$C[0] = \{a \in \mathbb{Z} \mid a \equiv 0 \pmod 3\},$$
$$C[1] = \{a \in \mathbb{Z} \mid a \equiv 1 \pmod 3\}, \text{ and}$$
$$C[2] = \{a \in \mathbb{Z} \mid a \equiv 2 \pmod 3\}.$$

The main results that we want to use now are Theorem 3.21 and Corollary 3.22 on page 112. This corollary tells us that for any $a \in \mathbb{Z}$, a is congruent to precisely one of the integers 0, 1, or 2. Consequently, the integer a must be congruent to 0, 1, or 2, and it cannot be congruent to two of these numbers. Thus

1. For each $a \in \mathbb{Z}$, $a \in C[0]$, $a \in C[1]$, or $a \in C[2]$; and

2. $C[0] \cap C[1] = \emptyset$, $C[0] \cap C[2] = \emptyset$, and $C[1] \cap C[2] = \emptyset$.

This means that the relation of congruence modulo 3 sorts the integers into three distinct sets, or classes, and that each pair of these sets have no elements in common. Theorem 3.21 and the Division Algorithm tell us that each integer is congruent, modulo 3, to its remainder when it is divided by 3, and Theorem 7.9 tells us that two integers are congruent modulo 3 if and only if they both have the same remainder when divided by 3. If we use a rectangle to represent $\mathbb{Z}$, we can divide that rectangle into three smaller rectangles, corresponding to $C[0]$, $C[1]$, and $C[2]$, and we might picture this situation as follows:

The Integers

$C[0]$ consisting of all integers with a remainder of 0 when divided by 3	$C[1]$ consisting of all integers with a remainder of 1 when divided by 3	$C[2]$ consisting of all integers with a remainder of 2 when divided by 3

Each integer is in exactly one of the three sets $C[0]$, $C[1]$, or $C[2]$, and two integers are congruent modulo 3 if and only if they are in the same set.

We will see that, in a similar manner, if n is any natural number, then the relation of congruence modulo n can be used to sort the integers into n classes. We will also see that in general, if we have an equivalence relation R on a set A, we can sort the elements of the set A into classes in a similar manner.

Definition. Let $\sim$ be an equivalence relation on a nonempty set A. For each $a \in A$, the **equivalence class of a** determined by $\sim$ is the subset of A, denoted by $[a]$, consisting of all the elements of A that are equivalent to a. That is,

$$[a] = \{x \in A \mid x \sim a\}.$$

We read $[a]$ as "the equivalence class of a" or as "bracket a."

<u>Note</u>: We use the notation $[a]$ when only one equivalence relation is being used. If there is more than one equivalence relation, then we need to distinguish between the equivalence classes for each relation. We often use something like $[a]_\sim$, or if R is the name of the relation, we can use $R[a]$ for the equivalence class of a determined by R.

In any case, always remember that when we are working with any equivalence relation on a set A and if $a \in A$, then *the equivalence class $[a]$ is a subset of A*.

Example 7.10 (Congruence Modulo n and Congruence Classes).

In Preview Activity 2, we used the notation $C[k]$ for the set of all integers that are congruent to k modulo 3. We could have used a similar notation for equivalence classes, and this would have been perfectly acceptable. However, the notation $[a]$ is probably the most common notation for the equivalence class of a. We will now use this same notation when dealing with congruence modulo n.

Let $n \in \mathbb{N}$. Congruence modulo n is an equivalence relation on $\mathbb{Z}$. So, for $a \in \mathbb{Z}$,

$$[a] = \{x \in \mathbb{Z} \mid x \equiv a \pmod{n}\}.$$

In this case, $[a]$ is called the **congruence class of a modulo n**.

We have seen that congruence modulo 3 divides the integers into three distinct congruence classes. Each congruence class consists of those integers with the same remainder when divided by 3. In a similar manner, if we use congruence modulo 2, we simply divide the integers into two classes. One class will consist of all the integers that have a remainder of 0 when divided by 2, and the other class will consist of all the integers that have a remainder of 1 when divided by 2. That is, congruence modulo 2 simply divides the integers into the even and odd integers.

We will soon prove that if $\sim$ is an equivalence relation on the set A, then we can "sort" the elements of A into distinct equivalence classes.

As another example, in Preview Activity 1, the relation R was an equivalence relation. For that Preview Activity, we used $R[x]$ to denote the equivalence class of $x \in A$, and these equivalence classes were either equal or disjoint.

However, in Preview Activity 1, the relation S was not an equivalence relation, and hence we do not use the term "equivalence class." We should note, however, that the sets $S[x]$ were not equal and were not disjoint.

This exhibits one of the main distinctions between equivalence relations and relations that are not equivalence relations.

For an equivalence relation on a set A, every element of A is in its own equivalence class, two elements are equivalent if and only if their equivalence classes are equal, and two equivalence classes are either identical or they are disjoint. These properties of equivalence classes will be proven in Theorem 7.11.

Theorem 7.11. *Let A be a nonempty set and let $\sim$ be an equivalence relation on A.*

1. *For each $a \in A$, $a \in [a]$.*

2. *For each $a, b \in A$, $a \sim b$ if and only if $[a] = [b]$.*

3. *For each $a, b \in A$, $[a] = [b]$ or $[a] \cap [b] = \emptyset$.*

Proof. Let A be a nonempty set and assume that $\sim$ is an equivalence relation on A. To prove the first part of the theorem, let $a \in A$. Since $\sim$ is an equivalence relation on A, it is reflexive on A. Thus, $a \sim a$, and we can conclude that $a \in [a]$.

The second part of this theorem is a biconditional sentence. We will prove it by proving two conditional statements. We will first prove that if $a \sim b$, then $[a] = [b]$. So, let $a, b \in A$ and assume that $a \sim b$. We will now prove that the two sets $[a]$ and $[b]$ are equal. We will do this by proving that each is a subset of the other.

First, assume that $x \in [a]$. Then, by definition, $x \sim a$. Since we have assumed that $a \sim b$, we can use the transitive property of $\sim$ to conclude that $x \sim b$, and this means that $x \in [b]$. This proves that $[a] \subseteq [b]$.

We now assume that $y \in [b]$. This means that $y \sim b$, and hence by the symmetric property, that $b \sim y$. Again, we are assuming that $a \sim b$. So we have

$$a \sim b \text{ and } b \sim y.$$

We use the transitive property to conclude that $a \sim y$, and then using the symmetric property, we conclude that $y \sim a$. This proves that $y \in [a]$, and hence, that $[b] \subseteq [a]$. This means that we can conclude that if $a \sim b$, then $[a] = [b]$.

We must now prove that if $[a] = [b]$, then $a \sim b$. Let $a, b \in A$ and assume that $[a] = [b]$. Using the first part of the theorem, we know that $a \in [a]$

and since the two sets are equal, this tells us that $a \in [b]$. Hence by the definition of $[b]$, we conclude that $a \sim b$. This completes the proof of the second part of the theorem.

For the third part of the theorem, let $a, b \in A$. Since this part of the theorem is a disjunction, we will consider two cases: either

- $[a] \cap [b] = \emptyset$, or

- $[a] \cap [b] \neq \emptyset$.

In the case where $[a] \cap [b] = \emptyset$, the first part of the disjunction is true, and hence there is nothing to prove. So we assume that $[a] \cap [b] \neq \emptyset$ and will show that $[a] = [b]$. Since $[a] \cap [b] \neq \emptyset$, there is an element x in A such that

$$x \in [a] \cap [b].$$

This means that $x \in [a]$ and $x \in [b]$. Consequently, $x \sim a$ and $x \sim b$, and so the first part of the theorem tells us that $[x] = [a]$ and $[x] = [b]$. Hence, we may conclude that $[a] = [b]$. Therefore, we have proven that $[a] = [b]$ or $[a] \cap [b] = \emptyset$. ∎

Theorem 7.11 gives the primary properties of equivalence classes. Consequences of these properties will be explored in the exercises. The following table restates the properties in Theorem 7.11 and a verbal description of each one.

Formal Statement from Theorem 7.11	Verbal Description
For each $a \in A$, $a \in [a]$.	Every element of A is in its own equivalence class.
For each $a, b \in A$, $a \sim b$ if and only if $[a] = [b]$.	Two elements of A are equivalent if and only if their equivalence classes are equal.
For each $a, b \in A$, $[a] = [b]$ or $[a] \cap [b] = \emptyset$.	Any two equivalence classes are either equal or they are disjoint. This means that if two equivalence classes are not disjoint then they must be equal.

The results of Theorem 7.11 are consistent with all the equivalence relations studied in the Preview Activities. Since this theorem applies to all equivalence relations, it applies to the relation of congruence modulo n on the integers. These results for congruence modulo n are given in the following corollary.

Corollary 7.12. *Let $n \in \mathbb{N}$. For each $a \in \mathbb{Z}$, let $[a]$ represent the congruence class of a modulo n.*

1. *For each $a \in \mathbb{Z}$, $a \in [a]$.*

2. *For each $a, b \in \mathbb{Z}$, $a \equiv b \pmod{n}$ if and only if $[a] = [b]$.*

3. *For each $a, b \in \mathbb{Z}$, $[a] = [b]$ or $[a] \cap [b] = \emptyset$.*

For the equivalence relation of congruence modulo n, Theorem 3.21 and Corollary 3.22 tell us that each integer is congruent to its remainder when divided by n, and that each integer is congruent modulo n to precisely one of one of the integers $0, 1, 2, \ldots, n-1$. This means that each integer is in precisely one of the congruence classes $[0]$, $[1]$, $[2]$, $\ldots, [n-1]$. Hence, Corollary 7.12 gives us the following result.

Corollary 7.13. *Let $n \in \mathbb{N}$. For each $a \in \mathbb{Z}$, let $[a]$ represent the congruence class of a modulo n.*

1. $\mathbb{Z} = [0] \cup [1] \cup [2] \cup \cdots \cup [n-1]$.

2. *For $j, k \in \{0, 1, 2, \ldots, n-1\}$, if $j \neq k$, then $[j] \cap [k] = \emptyset$.*

Example 7.14. Let $f : \mathbb{R} \to \mathbb{R}$ be defined by $f(x) = x^2 - 4$ for each $x \in \mathbb{R}$. Define a relation $\sim$ on $\mathbb{R}$ as follows:

$$\text{For } a, b \in \mathbb{R}, \ a \sim b \text{ if and only if } f(a) = f(b).$$

In Exercise (5) of Section 7.2, we proved that $\sim$ is an equivalence relation on $\mathbb{R}$. Consequently, each real number has an equivalence class. For example,

$$\begin{aligned}
[5] &= \{x \in \mathbb{R} \mid x \sim 5\} \\
&= \{x \in \mathbb{R} \mid f(x) = f(5)\} \\
&= \{x \in \mathbb{R} \mid x^2 - 4 = 21\} \\
&= \{x \in \mathbb{R} \mid x^2 = 25\} \\
&= \{-5, 5\}.
\end{aligned}$$

Similarly, $[-5] = \{-5, 5\}$, $[10] = \{-10, 10\}$, and $[-10] = \{-10, 10\}$.

For this equivalence relation, there are infinitely many equivalence classes, and so it is not possible to list all of them. However, if $a \in \mathbb{R}$, we see that

$$
\begin{aligned}
[a] &= \{x \in \mathbb{R} \mid x \sim a\} \\
&= \{x \in \mathbb{R} \mid f(x) = f(a)\} \\
&= \{x \in \mathbb{R} \mid x^2 - 4 = a^2 - 4\} \\
&= \{x \in \mathbb{R} \mid x^2 = a^2\} \\
&= \{-a, a\}.
\end{aligned}
$$

Partitions and Equivalence Relations

A partition of a set A is a collection of subsets of A that "breaks up" the set A into disjoint subsets. Technically, each pair of subsets in the collection must be disjoint. We then say that the collection of subsets is **pairwise disjoint**. We introduce the following formal definition.

Definition. Let A be a nonempty set, and let $\mathcal{C}$ be a collection of subsets of A. That is, $\mathcal{C} \subseteq \mathcal{P}(A)$. The collection of subsets $\mathcal{C}$ is a **partition of A** provided that

1. For each $V \in \mathcal{C}$, $V \neq \emptyset$.

2. For each $x \in A$, there exists a $V \in \mathcal{C}$ such that $x \in V$.

3. For every $V, W \in \mathcal{C}$, $V = W$ or $V \cap W = \emptyset$.

There is a close relation between partitions and equivalence classes since the equivalence classes of an equivalence relation form a partition of the underlying set, as will be proven in Theorem 7.15. The proof of this theorem relies on the results in Theorem 7.11.

Theorem 7.15. *Let $\sim$ be an equivalence relation on the nonempty set A. Then the collection $\mathcal{C}$ of all equivalence classes determined by $\sim$ is a partition of the set A.*

Proof. Let $\sim$ be an equivalence relation on the nonempty set A, and let $\mathcal{C}$ be the collection of all equivalence classes determined by $\sim$. That is,

$$
\mathcal{C} = \{[a] \mid a \in A\}.
$$

We will use Theorem 7.11 to prove that $\mathcal{C}$ is a partition of A.

Part (1) of Theorem 7.11 states that for each $a \in A$, $a \in [a]$. In terms of the equivalence classes, this means that each equivalence class is nonempty since each element of A is in its own equivalence class. Consequently, $\mathcal{C}$, the collection of all equivalence classes determined by $\sim$, satisfies the first two conditions of the definition of a partition.

We must now show that the collection $\mathcal{C}$ of all equivalence classes determined by $\sim$ satisfies the third condition for being a partition. That is, we need to show that any two equivalence classes are either equal or are disjoint. However, this is exactly the result in Part (3) of Theorem 7.11.

Hence, we have proven that the collection $\mathcal{C}$ of all equivalence classes determined by $\sim$ is a partition of the set A. ∎

Note: Theorem 7.15 has shown us that if $\sim$ is an equivalence relation on a nonempty set A, then the collection of the equivalence classes determined by $\sim$ form a partition of the set A.

This process can be reversed. This means that given a partition $\mathcal{C}$ of a nonempty set A, we can define an equivalence relation on A whose equivalence classes are precisely the subsets of A that form the partition. This will be explored in Activity 7.16.

Activity 7.16 (A Partition Defines an Equivalence Relation).

1. Let $A = \{a, b, c, d, e\}$ and let $\mathcal{C} = \{\{a, b, c\}, \{d, e\}\}$.

   (**a**) Explain why $\mathcal{C}$ is a partition of A.

   Define a relation $\sim$ on A as follows: For $x, y \in A$, $x \sim y$ if and only if there exists a set U in $\mathcal{C}$ such that $x \in U$ and $y \in U$.

   (**b**) Prove that $\sim$ is an equivalence relation on the set A, and then determine all the equivalence classes for $\sim$. How does the collection of all equivalence classes compare to $\mathcal{C}$?

2. What we did for the specific partition in Part (1) can be done for any partition of a set. So, to generalize Part (1), we let A be a nonempty set and let $\mathcal{C}$ be a partition of A. We then define a relation $\sim$ on A as follows:

   For $x, y \in A$, $x \sim y$ if and only if there exists a set U in $\mathcal{C}$ such that $x \in U$ and $y \in U$.

(a) Prove that $\sim$ is an equivalence relation on the set A.

(b) Let $a \in A$ and let $U \in C$ such that $a \in U$. Prove that $[a] = U$.

Exercises 7.3

1. Let $A = \{a, b, c, d, e\}$ and let $\sim$ be the relation on A that is represented by the following directed graph.

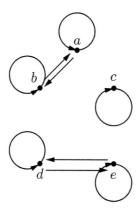

(a) Prove that $\sim$ is an equivalence relation on the set A.

(b) Determine each of the equivalence classes determined by this equivalence relation.

2. Let $A = \{0, 1, 2, 3, \ldots, 999, 1000\}$. Define the relation R on A as follows:

   For $x, y \in A$, $x \, R \, y$ if and only if x and y have the same number of digits.

   Prove that R is an equivalence relation on the set A and determine all of the distinct equivalence classes determined by R.

3. Determine all of the congruence classes for the relation of congruence modulo 5 on the set of integers.

4. Let $A = \mathbb{Z} \times (\mathbb{Z} - \{0\})$. That is, $A = \{(a, b) \in \mathbb{Z} \times \mathbb{Z} \mid b \neq 0\}$. Define the relation $\approx$ on A as follows:

   For $(a, b), (c, d) \in A$, $(a, b) \approx (c, d)$ if and only if $ad = bc$.

(a) Prove that $\approx$ is an equivalence relation on A.

(b) Why was it necessary to include the restriction that $b \neq 0$ in the definition of the set A?

(c) Determine an equation that gives a relation between a and b if $(a, b) \in A$ and $(a, b) \approx (2, 3)$.

(d) Determine at least four different elements in $[(2, 3)]$, the equivalence class of $(2, 3)$.

(e) Use set builder notation to describe $[(2, 3)]$, the equivalence class of $(2, 3)$.

5. Let A be a nonempty set and let $\sim$ be an equivalence relation on A. Prove each of the following:

(a) For each $a, b \in A$, $a \not\sim b$ if and only if $[a] \cap [b] = \emptyset$.

(b) For each $a, b \in A$, if $[a] \neq [b]$, then $[a] \cap [b] = \emptyset$.

(c) For each $a, b \in A$, if $[a] \cap [b] \neq \emptyset$, then $[a] = [b]$.

7.4   Modular Arithmetic

Preview Activity 1 (Congruence Modulo 5).

For this preview activity, we will only use the relation of congruence modulo 5 on the set of integers.

1. Find five different integers a such that $a \equiv 4 \pmod 5$. That is, find five different integers in the congruence class of 4.

2. Find five different integers b such that $b \equiv 2 \pmod 5$.

3. Calculate $s = (a + b)$ using several values of a from Part (1) and several values of b from Part (2). For each sum s that is calculated, find r so that $0 \leq r < 5$ and $s \equiv r \pmod 5$.

4. Calculate $p = (a \cdot b)$ using several values of a from Part (1) and several values of b from Part (2). For each product p that is calculated, find r so that $0 \leq r < 5$ and $p \equiv r \pmod 5$.

Preview Activity 2 (Congruence Modulo 6).

For this preview activity, we will only use the relation of congruence modulo 6 on the set of integers.

1. Find five different integers a such that $a \equiv 4 \pmod 6$. That is, find five different integers in the congruence class of 4.

2. Find five different integers b such that $b \equiv 3 \pmod 6$.

3. Calculate $s = (a + b)$ using several values of a from Part (1) and several values of b from Part (2). For each sum s that is calculated, find r so that $0 \leq r < 6$ and $s \equiv r \pmod 6$.

4. Calculate $p = (a \cdot b)$ using several values of a from Part (1) and several values of b from Part (2). For each product p that is calculated, find r so that $0 \leq r < 6$ and $p \equiv r \pmod 6$.

Preview Activity 3 (The Remainder When Dividing by 9).

If a and b are integers with $b > 0$, then from the Division Algorithm, we know that there exist unique integers q and r such that

$$a = bq + r \text{ and } 0 \leq r < b.$$

In this activity, we are interested in the remainder r. Notice that $r = a - bq$. So, given a and b, if we can calculate q, then we can calculate r.

We can use the "int" function on a calculator to calculate q. [The "int" function is the "greatest integer function." If x is a real number, then int (x) is the greatest integer that is less than or equal to x.]

So, in the context of the Division Algorithm, $q = $ int $\left(\dfrac{a}{b}\right)$. Consequently,

$$r = a - b \cdot \text{int} \left(\frac{a}{b}\right).$$

If n is a positive integer, we will let $s(n)$ denote the sum of the digits of n. For example, if $n = 731$, then

$$s(731) = 7 + 3 + 1 = 11.$$

For each of the following values of n, calculate:

> The remainder when n is divided by 9;
> The value of $s(n)$; and
> The remainder when $s(n)$ is divided by 9.

1. $n = 49$ **3.** $n = 4672$ **5.** $n = 51381$

2. $n = 731$ **4.** $n = 9845$ **6.** $n = 305877$

The Integers Modulo n

Let $n \in \mathbb{N}$. Since the relation of congruence modulo n is an equivalence relation on $\mathbb{Z}$, we can discuss its equivalence classes. Recall that in this situation, we refer to the equivalence classes as congruence classes.

> **Definition.** Let $n \in \mathbb{N}$. The set of congruence classes for the relation of congruence modulo n on $\mathbb{Z}$ is the set of **integers modulo n**, or the set of integers mod n. We will denote this set of congruence classes by $\mathbb{Z}_n$.

Corollary 7.13 tells us that

$$\mathbb{Z} = [0] \cup [1] \cup [2] \cup \cdots \cup [n-1].$$

In addition, we know that each integer is congruent to precisely one of the integers $0, 1, 2, \ldots, n-1$. This tells us that one way to represent $\mathbb{Z}_n$ is

$$\mathbb{Z}_n = \{[0], [1], [2], \ldots [n-1]\}.$$

Consequently, the set $\mathbb{Z}_n$ has n distinct elements. Now, the finite set $\mathbb{Z}_n$ is closely related to the infinite set $\mathbb{Z}$. In fact, each integer has a congruence class, but many of these congruence classes are identical. In fact, we know that there are only n distinct congruence classes, namely, $[0]$, $[1]$, $[2], \ldots [n-1]$.

Now, the set of integers $\mathbb{Z}$ is a set that has two operations: addition and multiplication. Moreover, we know that $\mathbb{Z}$ is closed under addition and multiplication. Is it possible to define operations of addition and multiplication on $\mathbb{Z}_n$?

Indeed, one of the basic problems dealt with in modern algebra is to determine if the arithmetic operations on one set "transfer" to a related set. In this case, the related set is $\mathbb{Z}_n$. For example, in the integers modulo 5, $\mathbb{Z}_5$, is it possible to add the congruence classes $[4]$ and $[2]$ as follows?

$$
\begin{aligned}
[4] \oplus [2] &= [4+2] \\
&= [6] \\
&= [1].
\end{aligned}
$$

We have used the symbol $\oplus$ to denote addition in $\mathbb{Z}_5$ so that we do not confuse it with addition in $\mathbb{Z}$. This looks simple enough, but there is a problem. The congruence classes [4] and [2] are not numbers, *they are infinite sets*. We have to make sure that we get the same answer no matter what element of [4] we use and no matter what element of [2] we use.

For example,

$$9 \equiv 4 \pmod 5 \quad \text{and so} \quad 9 \in [4]. \text{ Also,}$$

$$7 \equiv 2 \pmod 5 \quad \text{and so,} \quad 7 \in [2].$$

Thus, we know that [9] = [4] and [7] = [2]. Do we get the same result if we add [9] and [7] in the way we did when we added [4] and [2]? The following computation confirms that we do:

$$[9] \oplus [7] = [9 + 7]$$
$$= [16]$$
$$= [1].$$

This is one of the ideas that was explored in Preview Activities 1 and 2. The main difference is that in the Preview Activities, we used the relation of congruence, and here we are using congruence classes. All of the examples in Preview Activity 1 should have illustrated the properties of congruence modulo 5 in the following table. The left side shows the properties in terms of the congruence relation and the right side shows the properties in terms of the congruence classes.

If $a \equiv 4 \pmod 5$ and $b \equiv 2 \pmod 5$, then	If $[a] = [4]$ and $[b] = [2]$, then
• $(a + b) \equiv (4 + 2) \pmod 5$.	• $[a + b] = [4 + 2]$;
• $(a \cdot b) \equiv (4 \cdot 2) \pmod 5$.	• $[a \cdot b] = [4 \cdot 2]$.

Preview Activity 2 illustrated similar properties for congruence modulo 6. These are illustrations of general properties that we have already proven in Theorem 3.24. We repeat the statement of the theorem here because it is so important for defining the operations of addition and multiplication in $\mathbb{Z}_n$.

Theorem 3.24 *Let n be a natural number and let $a, b, c,$ and d be integers. Then*

    1. *If $a \equiv b \pmod{n}$ and $c \equiv d \pmod{n}$, then $(a + c) \equiv (b + d) \pmod{n}$.*

    2. *If $a \equiv b \pmod{n}$ and $c \equiv d \pmod{n}$, then $ac \equiv bd \pmod{n}$.*

    3. *If $a \equiv b \pmod{n}$ and $m \in \mathbb{N}$, then $a^m \equiv b^m \pmod{n}$.*

Since $x \equiv y \pmod{n}$ if and only if $[x] = [y]$, we can restate the result of this Theorem 3.24 in terms of congruence classes in $\mathbb{Z}_n$.

Corollary 7.17. *Let n be a natural number and let a, b, c, and d be integers. Then, in $\mathbb{Z}_n$,*

    1. *If $[a] = [b]$ and $[c] = [d]$, then $[a + c] = [b + d]$.*

    2. *If $[a] = [b]$ and $[c] = [d]$, then $[a \cdot c] = [b \cdot d]$.*

    3. *If $[a] = [b]$ and $m \in \mathbb{N}$, then $[a]^m = [b]^m$.*

    Because of Corollary 7.17, we know that the following formal definition of addition and multiplication of congruence classes in $\mathbb{Z}_n$ is independent of the choice of the elements we choose from each class. We say that these definitions of addition and multiplication are **well defined**.

> **Definition.** Let $n \in \mathbb{N}$. **Addition and multiplication** in $\mathbb{Z}_n$ are defined as follows: For $[a], [c] \in \mathbb{Z}_n$,
>
> $$[a] \oplus [c] = [a + c] \text{ and } [a] \odot [c] = [ac].$$
>
> The term **modular arithmetic** is used to refer to the operations of addition and multiplication of congruence classes in the integers modulo n.

Activity 7.18 (Something that Is Not Well Defined). Define the following
subsets of $\mathbb{Z}$:

$$A = \{\ldots, -9, -5, -1, 0, 4, 8, \ldots\},$$
$$B = \{\ldots, -12, -8, -4, 1, 5, 9 \ldots\},$$
$$C = \{\ldots -11, -7, -3, 3, 7, 11, \ldots\},$$
$$D = \{\ldots -10, -6, -2, 2, 6, 10, \ldots\}.$$

1. Explain why $\mathcal{U} = \{A, B, C, D\}$ is a partition of $\mathbb{Z}$. We will call A, B, C, and D the classes of this partition.

2. Find an example of integers a and b in A and c and d in B such that $a + c$ and $b + d$ are in different classes of the partition $\mathcal{U}$.

3. Use Part (2) to explain why it is not possible to define $A \oplus B$ as the class containing $a + c$ if $a \in A$ and $c \in B$.

4. Find an example of integers a and b in A and c and d in B such that $a \cdot c$ and $b \cdot d$ are in different sets contained in the partition $\mathcal{U}$.

5. Use Part (4) to explain why it is not possible to define $A \odot B$ as the class containing $a \cdot c$ if $a \in A$ and $c \in B$.

Let $n \in \mathbb{N}$. We now have an addition and multiplication defined on $\mathbb{Z}_n$, the integers modulo n. These operations are

$$[a] \oplus [c] = [a + c] \text{ and } [a] \odot [c] = [ac].$$

Always remember that for each of these equations, the operations on the left, $\oplus$ and $\odot$, are the new operations that are being defined. The operations on the right side of the equations, $+$ and $\cdot$, are the known operations of addition and multiplication in $\mathbb{Z}$.

Since $\mathbb{Z}_n$ is a finite set, it is possible to construct complete addition and multiplication tables for $\mathbb{Z}_n$.

Example 7.19 (Addition and Multiplication Tables for $\mathbb{Z}_5$).

Verify that the following addition and multiplication tables for $\mathbb{Z}_5$ are correct.

$\oplus$	$[0]$	$[1]$	$[2]$	$[3]$	$[4]$
$[0]$	$[0]$	$[1]$	$[2]$	$[3]$	$[4]$
$[1]$	$[1]$	$[2]$	$[3]$	$[4]$	$[0]$
$[2]$	$[2]$	$[3]$	$[4]$	$[0]$	$[1]$
$[3]$	$[3]$	$[4]$	$[0]$	$[1]$	$[2]$
$[4]$	$[4]$	$[0]$	$[1]$	$[2]$	$[3]$

$\odot$	$[0]$	$[1]$	$[2]$	$[3]$	$[4]$
$[0]$	$[0]$	$[0]$	$[0]$	$[0]$	$[0]$
$[1]$	$[0]$	$[1]$	$[2]$	$[3]$	$[4]$
$[2]$	$[0]$	$[2]$	$[4]$	$[1]$	$[3]$
$[3]$	$[0]$	$[3]$	$[1]$	$[4]$	$[2]$
$[4]$	$[0]$	$[4]$	$[3]$	$[2]$	$[1]$

Activity 7.20 (Modular Arithmetic in $\mathbb{Z}_4$ and $\mathbb{Z}_6$).

1. Construct complete addition and multiplication tables for $\mathbb{Z}_4$ and $\mathbb{Z}_6$.

2. Recall that the zero product property is true in the real numbers.

For all $a, b \in \mathbb{R}$, if $a \cdot b = 0$, then $a = 0$ or $b = 0$.

Write the contrapositive of the conditional statement in this property.

3. Is the following statement true or false? Justify your conclusion.

For all $[a], [b] \in \mathbb{Z}_6$, if $[a] \odot [b] = [0]$, then $[a] = [0]$ or $[b] = [0]$.

4. Is the following statement true or false? Justify your conclusion.

For all $[a], [b] \in \mathbb{Z}_5$, if $[a] \odot [b] = [0]$, then $[a] = [0]$ or $[b] = [0]$.

Divisibility Tests

Let $n \in \mathbb{N}$ and let $s(n)$ denote the sum of the digits of n. For example, if $n = 731$, then $s(731) = 7 + 3 + 1 = 11$. In Preview Activity 3, we saw that

$$731 \equiv 2 \pmod 9 \text{ and } 11 \equiv 2 \pmod 9.$$

In fact, for every example in Preview Activity 3, we saw that n and $s(n)$ were congruent modulo 9 since they both had the same remainder when divided by 9. The concepts of congruence and congruence classes can help prove that this is always true.

We will use the case of $n = 731$ to illustrate the general process. We must use our standard place value system. By this, we mean that we will write 731 as follows:

$$731 = \left(7 \times 10^2\right) + \left(3 \times 10^1\right) + \left(1 \times 10^0\right). \tag{1}$$

The idea is to now use the definition of addition and multiplication in $\mathbb{Z}_9$ to convert Equation (1) to an equation in $\mathbb{Z}_9$. We do this as follows:

$$
\begin{aligned}
[731] &= \left[\left(7 \times 10^2\right) + \left(3 \times 10^1\right) + \left(1 \times 10^0\right)\right] \\
&= \left[7 \times 10^2\right] \oplus \left[3 \times 10^1\right] \oplus \left[1 \times 10^0\right] \\
&= \left([7] \odot \left[10^2\right]\right) \oplus \left([3] \odot [10]\right) \oplus \left([1] \odot [1]\right). \tag{2}
\end{aligned}
$$

We now use the facts that $10^2 \equiv 1 \pmod 9$ and $10 \equiv 1 \pmod 9$ to conclude that $\left[10^2\right] = [1]$ and $[10] = [1]$. Hence, we can use these facts and

Equation (2) to obtain

$$
\begin{aligned}
[731] &= \left([7] \odot \left[10^2\right]\right) \oplus \left([3] \odot [10]\right) \oplus \left([1] \odot [1]\right) \\
&= \left([7] \odot [1]\right) \oplus \left([3] \odot [1]\right) \oplus \left([1] \odot [1]\right) \\
&= [7] \oplus [3] \oplus [1] \\
&= [7 + 3 + 1].
\end{aligned}
\tag{3}
$$

Equation (3) tells us that 731 has the same remainder when divided by 9 as the sum of its digits. It is easy to check that the sum of the digits is 11 and hence has a remainder of 2. This means that when 731 is divided by 9, the remainder is 2.

To prove that any natural number has the same remainder when divided by 9 as the sum of its digits, it is helpful to introduce notation for the decimal representation of a natural number. The notation we will use is similar to the notation for the number 731 in Equation (1).

In general, if $n \in \mathbb{N}$, and $n = a_k a_{k-1} \cdots a_1 a_0$ is the decimal representation of n, then

$$
n = \left(a_k \times 10^k\right) + \left(a_{k-1} \times 10^{k-1}\right) + \cdots + \left(a_1 \times 10^1\right) + \left(a_0 \times 10^0\right).
$$

This can also be written using summation notation as follows:

$$
n = \sum_{j=0}^{k} \left(a_j \times 10^j\right).
$$

Using congruence classes for congruence modulo 9, we have

$$
\begin{aligned}
[n] &= \left[\left(a_k \times 10^k\right) + \left(a_{k-1} \times 10^{k-1}\right) + \cdots + \left(a_1 \times 10^1\right) + \left(a_0 \times 10^0\right)\right] \\
&= \left[a_k \times 10^k\right] \oplus \left[a_{k-1} \times 10^{k-1}\right] \oplus \cdots \oplus \left[a_1 \times 10^1\right] \oplus \left[a_0 \times 10^0\right] \\
&= \left([a_k] \odot \left[10^k\right]\right) \oplus \left([a_{k-1}] \odot \left[10^{k-1}\right]\right) \oplus \cdots \\
&\qquad\qquad \oplus \left([a_1] \odot \left[10^1\right]\right) \oplus \left([a_0] \odot \left[10^0\right]\right).
\end{aligned}
\tag{4}
$$

One last detail is needed. It is given in Proposition 7.21. The proof by mathematical induction is left as an exercise.

Proposition 7.21. *If n is a nonnegative integer, then $10^n \equiv 1 \pmod{9}$, and hence for the equivalence relation of congruence modulo 9, $[10^n] = [1]$.*

Now, using Equation (4) and Proposition 7.21, we obtain

$$[n] = ([a_k] \odot [1]) \oplus ([a_{k-1}] \odot [1]) \oplus \cdots \oplus ([a_1] \odot [1]) \oplus ([a_0] \odot [1])$$
$$= [a_k] \oplus [a_{k-1}] \oplus \cdots \oplus [a_1] \oplus [a_0]$$
$$= [a_k + a_{k-1} + \cdots + a_1 + a_0]. \tag{5}$$

If we let $s(n)$ denote the sum of the digits of n, then

$$s(n) = a_k + a_{k-1} + \cdots + a_1 + a_0,$$

and Equation (5) shows that $[n] = [s(n)]$. This completes the proof of Theorem 7.22.

Theorem 7.22. *Let $n \in \mathbb{N}$ and let $s(n)$ denote the sum of the digits of n. Then*

1. *$[n] = [s(n)]$, using congruence classes modulo 9.*

2. *$n \equiv s(n) \pmod 9$.*

3. *$9 \mid n$ if and only if $9 \mid s(n)$.*

Part (3) of Theorem 7.22 is called a **divisibility test**. If gives a necessary and sufficient condition for a natural number to be divisible by 9. Other divisibility tests will be explored in Activity 7.23 and the exercises. Most of these divisibility tests can be proven in a manner similar to the proof of the divisibility test for 9.

Activity 7.23 (Other Divisibility Tests). Let $n \in \mathbb{N}$ and let $s(n)$ denote the sum of the digits of n. So if we write

$$n = \left(a_k \times 10^k\right) + \left(a_{k-1} \times 10^{k-1}\right) + \cdots + \left(a_1 \times 10^1\right) + \left(a_0 \times 10^0\right),$$

then $s(n) = a_k + a_{k-1} + \cdots + a_1 + a_0$.

1. Use mathematical induction to prove that if n is an integer and $n \geq 2$, then $10^n \equiv 0 \pmod 4$. Hence, for congruence classes modulo 4, if n is an integer and $n \geq 2$, then $[10^n] = [0]$.

2. Prove each of the following:

(a) $[n] = [10a_1 + a_0]$, using congruence classes modulo 4.

(b) $n \equiv (10a_1 + a_0) \pmod 4$.

(c) $4 \mid n$ if and only if $4 \mid (10a_1 + a_0)$.

3. Use mathematical induction to prove that if n is a nonnegative integer, then $10^n \equiv 1 \pmod 3$. Hence, for congruence classes modulo 3, if n is a nonnegative integer, then $[10^n] = [1]$.

4. Prove each of the following:

   (a) $[n] = [s(n)]$, using congruence classes modulo 3.

   (b) $n \equiv s(n) \pmod 3$.

   (c) $3 \mid n$ if and only if $3 \mid s(n)$.

Activity 7.24 (Using Congruence Modulo 4). The set $\mathbb{Z}_n$ is a finite set, and hence one way to prove things about $\mathbb{Z}_n$ is to simply use the n elements in $\mathbb{Z}_n$ as the n cases for a proof using cases. For example, if $n \in \mathbb{Z}$, then in $\mathbb{Z}_4$, $[n] = [0]$, $[n] = [1]$, $[n] = [2]$, or $[n] = [3]$.

1. Prove that if $n \in \mathbb{Z}$, then in $\mathbb{Z}_4$, $[n]^2 = [0]$ or $[n]^2 = [1]$. Use this to conclude that in $\mathbb{Z}_4$, $[n^2] = [0]$ or $[n^2] = [1]$.

2. Translate the equations $[n^2] = [0]$ and $[n^2] = [1]$ in $\mathbb{Z}_4$ into congruences modulo 4.

3. Use the result of Part (2) in Activity 7.23 to determine the value of r so that $r \in \mathbb{Z}$, $0 \le r < 3$, and

$$104\ 257\ 833\ 259 \equiv r \pmod 4.$$

   That is, $[104\ 257\ 833\ 259] = [r]$ in $\mathbb{Z}_4$.

4. Is the natural number $104\ 257\ 833\ 259$ a perfect square? Justify your conclusion.

Exercises 7.4

1. **(a)** Complete the addition and multiplication tables for $\mathbb{Z}_2$.

   (b) Complete the addition and multiplication tables for $\mathbb{Z}_3$.

2. **(a)** Complete the addition and multiplication tables for $\mathbb{Z}_7$.

   (b) Complete the addition and multiplication tables for $\mathbb{Z}_8$.

3. The set $\mathbb{Z}_n$ contains n elements. One way to solve an equation in $\mathbb{Z}_n$ is to substitute each of these n elements in the equation to check which ones are solutions. In $\mathbb{Z}_n$, when parentheses are not used, we follow the usual order of operations, which means that multiplications are done first and then additions. Solve each of the following equations:

   (a) $[x]^2 = [1]$ in $\mathbb{Z}_4$

   (b) $[x]^2 = [1]$ in $\mathbb{Z}_8$

   (c) $[x]^4 = [1]$ in $\mathbb{Z}_5$

   (d) $[x]^2 \oplus [3] \odot [x] = [3]$ in $\mathbb{Z}_6$

   (e) $[x]^2 \oplus [1] = [0]$ in $\mathbb{Z}_5$

   (f) $[3] \odot [x] \oplus [2] = [0]$ in $\mathbb{Z}_5$

   (g) $[3] \odot [x] \oplus [2] = [0]$ in $\mathbb{Z}_6$

   (h) $[3] \odot [x] \oplus [2] = [0]$ in $\mathbb{Z}_9$

4. In each case, determine if the statement is true or false.

   (a) For all $[a] \in \mathbb{Z}_6$, if $[a] \neq [0]$, then there exists a $[b] \in \mathbb{Z}_6$ such that $[a] \odot [b] = [1]$.

   (b) For all $[a] \in \mathbb{Z}_5$, if $[a] \neq [0]$, then there exists a $[b] \in \mathbb{Z}_5$ such that $[a] \odot [b] = [1]$.

5. In each case, determine if the statement is true or false.

   (a) For all $[a], [b] \in \mathbb{Z}_6$, if $[a] \neq [0]$ and $[b] \neq [0]$, then $[a] \odot [b] \neq [0]$.

   (b) For all $[a], [b] \in \mathbb{Z}_5$, if $[a] \neq [0]$ and $[b] \neq [0]$, then $[a] \odot [b] \neq [0]$.

6. Use mathematical induction to prove Proposition 7.21.

   If n is a nonnegative integer, then $10^n \equiv 1 \pmod 9$, and hence for the equivalence relation of congruence modulo 9, $[10^n] = [1]$.

7. Use mathematical induction to prove that if n is an integer and $n \geq 1$, then $10^n \equiv 0 \pmod 5$. Hence, for congruence classes modulo 5, if n is an integer and $n \geq 1$, then $[10^n] = [0]$.

8. Let $n \in \mathbb{N}$ and assume

$$n = \left(a_k \times 10^k\right) + \left(a_{k-1} \times 10^{k-1}\right) + \cdots + \left(a_1 \times 10^1\right) + \left(a_0 \times 10^0\right).$$

Use the result in Exercise (7) to help prove each of the following:

    (a) $[n] = [a_0]$, using congruence classes modulo 5.

    (b) $n \equiv a_0 \pmod 5$.

    (c) $5 \mid n$ if and only if $5 \mid a_0$.

9. Use mathematical induction to prove that if n is an integer and $n \geq 3$, then $10^n \equiv 0 \pmod 8$. Hence, for congruence classes modulo 8, if n is an integer and $n \geq 3$, then $[10^n] = [0]$.

10. Let $n \in \mathbb{N}$ and assume

$$n = \left(a_k \times 10^k\right) + \left(a_{k-1} \times 10^{k-1}\right) + \cdots + \left(a_1 \times 10^1\right) + \left(a_0 \times 10^0\right).$$

Use the result in Exercise (9) to help develop a divisibility test for 8. Prove that your divisibility test is correct.

11. Use mathematical induction to prove that if n is a nonnegative integer then $10^n \equiv (-1)^n \pmod{11}$. Hence, for congruence classes modulo 11, if n is a non-negative integer, then $[10^n] = [(-1)^n]$.

12. Let $n \in \mathbb{N}$ and assume

$$n = \left(a_k \times 10^k\right) + \left(a_{k-1} \times 10^{k-1}\right) + \cdots + \left(a_1 \times 10^1\right) + \left(a_0 \times 10^0\right).$$

Use the result in Exercise (11) to help prove each of the following:

    (a) $n \equiv \sum_{j=0}^{k} (-1)^j a_j \pmod{11}$.

    (b) $[n] = \left[\sum_{j=0}^{k} (-1)^j a_j\right]$, using congruence classes modulo 11.

    (c) $11 \mid n$ if and only if $11 \mid \sum_{j=0}^{k} (-1)^j a_j$.

13. **(a)** Prove the following proposition:

    Let $[a], [b] \in \mathbb{Z}_3$. In $\mathbb{Z}_3$, if $[a]^2 + [b]^2 = [0]$, then $[a] = 0$ and $[b] = [0]$.

    (b) Use Exercise (13a) to prove the following proposition:

    Let $a, b \in \mathbb{Z}$. If $(a^2 + b^2) \equiv 0 \pmod 3$, then $a \equiv 0 \pmod 3$ and $b \equiv 0 \pmod 3$.

    (c) Use Exercise (13b) to prove the following proposition:

    Let $a, b \in \mathbb{Z}$. If 3 divides $(a^2 + b^2)$, then 3 divides a and 3 divides b.

14. Prove the following proposition:

    Let $a \in \mathbb{Z}$. If there exist integers b and c such that $a = b^4 + c^4$, then the units digit of a must be 0, 1, 2, 5, 6, or 7.

15. Is the following proposition true or false? Justify your conclusion.

    Let $n \in \mathbb{Z}$. If n is odd, then $8 \mid (n^2 - 1)$. [Hint: What are the possible values of $n \pmod 8$?]

16. Prove the following proposition:

    Let $n \in \mathbb{N}$. If $n \equiv 7 \pmod 8$, then n is not the sum of three squares. That is, there do not exist natural numbers a, b, and c such that $n = a^2 + b^2 + c^2$.

Chapter 8

Topics in Number Theory

8.1 The Greatest Common Divisor

Preview Activity 1 (A Review of Divisibility).

1. Let m and n be integers. Explain what it means to say that m divides n. Recall that we use the notation $m \mid n$ to indicate that the integer m divides the integer n.

2. Give two different examples of a pair of integers where the first integer divides the second integer.

3. Give two different examples of a pair of integers where the first integer does not divide the second integer.

4. Let m and n be integers. Explain what it means to say that m does not divide n.

5. According to the definition of "divides," does the integer 0 divide the integer 10? Explain.

6. According to the definition of "divides," does the integer 10 divide the integer 0? Explain.

7. Is the following proposition true or false? Justify your conclusion.

Let $a, b, c \in \mathbb{Z}$. If $a \mid (b \cdot c)$, then $a \mid b$ or $a \mid c$.

Preview Activity 2 (The Greatest Common Divisor).

Let a and b be integers, not both 0. A **common divisor** of a and b is any integer that divides both a and b. The *largest* natural number that divides both a and b is called the **greatest common divisor** of a and b. The greatest common divisor of a and b is denoted by $\gcd(a, b)$.

1. Use the roster method to list the elements of the set that contains all the natural numbers that are divisors of 48.

2. Use the roster method to list the elements of the set that contains all the natural numbers that are divisors of 84.

3. Determine the intersection of the two sets in Parts (1) and (2). This set contains all the natural numbers that are common divisors of 48 and 84.

4. What is the greatest common divisor of 48 and 84?

5. Use the method suggested in Parts (1) through (4) to determine each of the following: $\gcd(8, -12)$, $\gcd(0, 5)$, $\gcd(8, 27)$, and $\gcd(14, 28)$.

6. Make a conjecture about how the common divisors of a and b are related to the greatest common divisor of a and b.

Preview Activity 3 (The GCD and the Division Algorithm).

1. Write a complete statement of the Division Algorithm for integers. (See Section 3.4.)

When we speak of the quotient and the remainder when we "divide an integer a by the positive integer b," we will always mean the quotient q and the remainder r guaranteed by the Division Algorithm.

2. Each row in the following table contains values for the integers a and b. In this table, the value of r is the remainder (from the Division Algorithm) when a is divided by b. Complete each row in this table by determining $\gcd(a, b)$, r, and $\gcd(b, r)$.

a	b	$\gcd(a,b)$	Remainder r	$\gcd(b,r)$
44	12			
75	21			
50	33			

3. Formulate a conjecture based on the results of the table in Part (2).

The System of Integers

Number theory is a study of the system of integers, which consists of the set of integers, $\mathbb{Z} = \{\ldots, -3, -2, -1, 0, 1, 2, 3, \ldots\}$ and the various properties of this set under the usual operations of addition and multiplication and under the usual ordering relation of "less than." The properties of the integers in Table 8.1 will be considered axioms in this text.

For all integers a, b, and c:

Closure Properties for Addition and Multiplication	$a + b \in \mathbb{Z}$ and $ab \in \mathbb{Z}$
Commutative Properties for Addition and Multiplication	$a + b = b + a$, and $ab = ba$
Associative Properties for Addition and Multiplication	$(a + b) + c = a + (b + c)$ and $(ab)\,c = a\,(bc)$
Distributive Properties of Multiplication over Addition	$a\,(b + c) = ab + ac$, and $(b + c)\,a = ba + ca$
Additive and Multiplicative Identity Properties	$a + 0 = 0 + a = a$, and $a \cdot 1 = 1 \cdot a = a$
Additive Inverse Property	$a + (-a) = (-a) + a = 0.$

Table 8.1: Axioms for the Integers

We will also assume the properties of the integers shown in Table 8.2. These properties can be proven from the preceding properties. (However, we will not do so here.)

Zero Property of Multiplication	If $a \in \mathbb{Z}$, then $a \cdot 0 = 0 \cdot a = 0.$
Cancellation Properties of Addition and Multiplication	If $a, b, c \in \mathbb{Z}$ and $a + b = a + c$, then $b = c$. If $a, b, c \in \mathbb{Z}$, $a \neq 0$ and $ac = bc$, then $b = c$.
Trichotomy Law	If $a \in \mathbb{Z}$, then exactly one of the following statements is true: $a < 0$, $a = 0$, or $a > 0$.

Table 8.2: Properties of the Integers

We have already studied a good deal of number theory in this course in our discussion of proof methods. In particular, we have studied even and odd integers, divisibility of integers, congruence, and the Division Algorithm.

See Theorems 3.10 and 3.11 at the end of Section 3.2 for a summary of results concerning even and odd integers as well as some results concerning properties of divisors. We reviewed some of these properties and the Division Algorithm in the Preview Activities.

The Greatest Common Divisor

Recall that we use the notation $m \mid n$ to indicate that the integer m divides the integer n. This also means that m is a factor of n, that n is a multiple of m, and that n is divisible by m.

In this section, we will study the greatest common divisor of two integers. We repeat the definition given in Preview Activity 2.

Definition. Let a and b be integers, not both 0. A **common divisor** of a and b is any integer that divides both a and b. The largest natural number that divides both a and b is called the **greatest common divisor** of a and b. The greatest common divisor of a and b is denoted by $\gcd(a, b)$.

Remarks about the Definition of Greatest Common Divisor

1. If $a, b \in \mathbb{Z}$, and a and b are not both 0, and if $d \in \mathbb{N}$, then $d = \gcd(a, b)$ provided that it satisfies all of the following properties:

- $d \mid a$ and $d \mid b$. That is, d is a common divisor of a and b.
- If k is a natural number such that $k \mid a$ and $k \mid b$, then $k \leq d$. That is, any other common divisor of a and b is less than or equal to d.

2. Consequently, a natural number d is not the greatest common divisor of a and b provided that it does not satisfy at least one of these properties. That is, d is not equal to $\gcd(a, b)$ provided that

- d does not divide a or d does not divide b; or
- There exists a natural number k such that $k \mid a$ and $k \mid b$ and $k > d$.

This means that d is not the greatest common divisor of a and b provided that it is not a common divisor of a and b or that there exists a common divisor of a and b that is greater than d.

In the Preview Activities, we determined the greatest common divisors for several pairs of integers. The process was to list all the divisors of both integers, then list all the common divisors of both integers, and finally, from the list of all common divisors, find the greatest (largest) common divisor. This method works reasonably well for small integers but can get quite cumbersome if the integers are large. Before we develop an efficient method for determining the greatest common divisor of two integers, we need to establish some properties of greatest common divisors.

One property was suggested in Preview Activity 2. If we look at the results in Part (5) of that preview activity, we should observe that any common divisor of a and b will divide $\gcd(a, b)$. In fact, the primary goals of the remainder of this section are

1. To find an efficient method for determining $\gcd(a, b)$, where a and b are integers.

2. To prove that the natural number $\gcd(a, b)$ is the only natural number d that satisfies the following properties:

   - d divides a,
   - d divides b, and
   - if k is a natural number such that $k \mid a$ and $k \mid b$, then $k \mid d$.

The second goal is only slightly different from the definition of the greatest common divisor. The only difference is in the third condition where $k \leq d$ was replaced by $k \mid d$.

We will first consider the case where a and b are integers with $a \neq 0$ and $b > 0$. The proof of the result stated in the second goal contains a method (called the Euclidean Algorithm) for determining the greatest common divisors of the two integers a and b. The main idea of the method is to keep replacing the pair of integers (a, b) with another pair of integers (b, r), where $0 \leq r < b$ and $\gcd(b, r) = \gcd(a, b)$. This idea was explored in Preview Activity 3. Lemma 8.1 is a conjecture that could have been formulated in Preview Activity 3.

Lemma 8.1. *Let c and d be integers, not both equal to zero. If q and r are integers such that $c = d \cdot q + r$, then $\gcd(c, d) = \gcd(d, r)$.*

Proof. Let c and d be integers, not both equal to zero. Assume that q and r are integers such that $c = d \cdot q + r$. For ease of notation, we will let

$$m = \gcd(c, d) \text{ and } n = \gcd(d, r).$$

Now, m divides c and m divides d. Consequently, there exist integers x and y such that $c = mx$ and $d = my$. Hence,

$$r = c - d \cdot q$$
$$r = mx - (my) q$$
$$r = m (x - yq).$$

But this means that m divides r. Since m divides d and m divides r, m is less than or equal to $\gcd(d, r)$. Thus, $m \leq n$.

Using a similar argument, we see that n divides d and n divides r. Since $c = d \cdot q + r$, we get that n divides c. Hence, n divides c and n divides d. Thus, $n \leq \gcd(c, d)$ or $n \leq m$. We now have $m \leq n$ and $n \leq m$. Hence, $m = n$ and $\gcd(c, d) = \gcd(d, r)$. ∎

Example 8.2 (Illustrations of Lemma 8.1). We completed several examples illustrating Lemma 8.1 in Preview Activity 3. For another example, let $c = 54$ and $d = 8$. We can verify that $\gcd(54, 8) = 2$.

Now, if we use the Division Algorithm for dividing 54 by 8, we obtain $54 = 6 \cdot 8 + 6$. So the remainder (r) when 54 is divided by 8 is 6. Notice that $\gcd(8, 6) = 2$. Hence

$$\gcd(54, 8) = \gcd(8, 6).$$

The key to finding the greatest common divisor (in more complicated cases) is to use the Division Algorithm again, this time with 8 and r. (In this case, $r = 6$.) We now find integers q_2 and r_2 such that

$$8 = r \cdot q_2 + r_2$$
$$8 = 6q_2 + r_2.$$

(The result should be $r_2 = 2$.) Notice that $\gcd(6, 2) = 2 = \gcd(54, 8)$.

This example illustrates the main idea of the **Euclidean Algorithm** for finding $\gcd(a, b)$, which is explained in the proof of the following theorem.

Theorem 8.3. *Let a and b be integers with $a \neq 0$ and $b > 0$. Then $\gcd(a, b)$ is the only natural number d such that*

(a) *d divides a,*

(b) *d divides b, and*

(c) *if k is an integer that divides both a and b, then k divides d.*

Proof. Let a and b be integers with $a \neq 0$ and $b > 0$, and let $d = \gcd(a, b)$. By the Division Algorithm, there exist integers q_1 and r_1 such that

$$a = b \cdot q_1 + r_1, \text{ and } 0 \leq r_1 < b. \tag{1}$$

If $r_1 = 0$, then Equation (1) implies that b divides a. Hence, $b = d = \gcd(a, b)$ and this number satisfies Conditions (a), (b), and (c).

If $r_1 > 0$, then by Lemma 8.1, $\gcd(a, b) = \gcd(b, r_1)$. We use the Division Algorithm again to obtain integers q_2 and r_2 such that

$$b = r_1 \cdot q_2 + r_2, \text{ and } 0 \leq r_2 < r_1. \tag{2}$$

If $r_2 = 0$, then Equation (2) implies that r_1 divides b. This means that $r_1 = \gcd(b, r_1)$. But we have already seen that $\gcd(a, b) = \gcd(b, r_1)$. Hence, $r_1 = \gcd(a, b)$. In addition, if k is an integer that divides both a and b, then using Equation (1), we see that $r_1 = a - b \cdot q_1$, and hence k divides r_1. This shows that $r_1 = \gcd(a, b)$ since r_1 satisfies Conditions (a), (b), and (c).

If $r_2 > 0$, then by Lemma 8.1, $\gcd(b, r_1) = \gcd(r_1, r_2)$. But we have already seen that $\gcd(a, b) = \gcd(b, r_1)$. Hence, $\gcd(a, b) = \gcd(r_1, r_2)$. We now continue to apply the Division Algorithm to produce a sequence of pairs of integers (all of which have the same greatest common divisor). This is summarized in the following table:

Original Pair	Equation from Division Algorithm	Inequality from Division Algorithm	New Pair
(a, b)	$a = b \cdot q_1 + r_1$	$0 \leq r_1 < b$	(b, r_1)
(b, r_1)	$b = r_1 \cdot q_2 + r_2$	$0 \leq r_2 < r_1$	(r_1, r_2)
(r_1, r_2)	$r_1 = r_2 \cdot q_3 + r_3$	$0 \leq r_3 < r_2$	(r_2, r_3)
(r_2, r_3)	$r_2 = r_3 \cdot q_4 + r_4$	$0 \leq r_4 < r_3$	(r_3, r_4)
(r_3, r_4)	$r_3 = r_4 \cdot q_5 + r_5$	$0 \leq r_5 < r_4$	(r_4, r_5)
$\vdots$	$\vdots$	$\vdots$	$\vdots$

From the inequalities in the third column of this table, we have a strictly decreasing sequence of nonnegative integers $(b > r_1 > r_2 > r_3 > r_4 \cdots)$. Consequently, a term in this sequence must eventually be equal to zero. Let p be the smallest natural number such that $r_{p+1} = 0$. This means that the last two rows in the table above will be

Original Pair	Equation from Division Algorithm	Inequality from Division Algorithm	New Pair
(r_{p-2}, r_{p-1})	$r_{p-2} = r_{p-1} \cdot q_p + r_p$	$0 \leq r_p < r_{p-1}$	(r_{p-1}, r_p)
(r_{p-1}, r_p)	$r_{p-1} = r_p \cdot q_{p+1} + 0$		

Remember that this table was constructed by repeated use of Lemma 8.1 and that the greatest common divisor of each pair of integers produced equals $\gcd(a, b)$. Also, the last row in the table indicates that r_p divides r_{p-1}. This means that $\gcd(r_{p-1}, r_p) = r_p$ and hence $r_p = \gcd(a, b)$.

This proves that $r_p = \gcd(a, b)$ satisfies Conditions (a) and (b) of this theorem. Now assume that k is an integer such that k divides a and k divides b. We proceed through the table row by row. First, since $r_1 = a - b \cdot q$, we see that

$$k \text{ must divide } r_1.$$

The second row tells us that $r_2 = b - r_1 \cdot q_2$. Since k divides b and k divides r_1, we conclude that

$$k \text{ divides } r_2.$$

Continuing with each row, we see that k divides each of the remainders r_1, r_2, r_3, ..., r_p. This means that $r_p = \gcd(a, b)$ satisfies Condition (c) of the theorem. ∎

Activity 8.4 (Using the Euclidean Algorithm).

1. Use the Euclidean Algorithm to determine $\gcd(180, 126)$. Notice that we have deleted the third column (Inequality from Division Algorithm) from the table. It is not needed in the computations.

Original Pair	Equation from Division Algorithm	New Pair
$(180, 126)$	$180 = 126 \cdot 1 + 54$	$(126, 54)$
$(126, 54)$	$126 =$	

Consequently, $\gcd(180, 126) = $ _____.

2. Use the Euclidean algorithm to determine $\gcd(4208, 288)$.

Original Pair	Equation from Division Algorithm	New Pair
$(4208, 288)$	$4208 = 288 \cdot 14 + 176$	$(288, \quad)$

Consequently, $\gcd(4208, 288) = $ _____.

Some Remarks about Theorem 8.3

Theorem 8.3 was proven with the assumptions that $a, b \in \mathbb{Z}$ with $a \neq 0$ and $b > 0$. A more general version of this theorem can be proven with $a, b \in \mathbb{Z}$ and $b \neq 0$. This can be proven using Theorem 8.3 and the results in the following lemma.

Lemma 8.5. *Let $a, b \in \mathbb{Z}$ with $b \neq 0$. Then*

    1. $\gcd(0, b) = |b|$.

    2. *If $\gcd(a, b) = d$, then $\gcd(a, -b) = d$.*

The proofs of these results are included in the exercises. An application of this result is given in the next example.

Example 8.6. Let $a = 234$ and $b = -42$. We will use the Euclidean algorithm to determine $\gcd(234, 42)$.

Step	Original Pair	Equation from Division Algorithm	New Pair
1	$(234, 42)$	$234 = 42 \cdot 5 + 24$	$(42, 24)$
2	$(42, 24)$	$42 = 24 \cdot 1 + 18$	$(24, 18)$
3	$(24, 18)$	$24 = 18 \cdot 1 + 6$	$(18, 6)$
4	$(18, 6)$	$18 = 6 \cdot 3$	

So $\gcd(234, 42) = 6$ and hence $\gcd(234, -42) = 6$.

We will use Example 8.6 to illustrate another use of the Euclidean algorithm. It is possible to use the steps of the Euclidean algorithm in reverse order to write $\gcd(a, b)$ in terms of a and b. We will use these steps in reverse order to find integers m and n such that $\gcd(234, 42) = 234m + 42n$. The idea is to start with the row with the last nonzero remainder and work backward as shown in the following table:

Explanation	Result
First, use the equation in Step 3 to write 6 in terms of 24 and 18.	$6 = 24 - 18 \cdot 1$
Use the equation in Step 2 to write $18 = 42 - 24 \cdot 1$. Substitute this into the preceding result and simplify.	$6 = 24 - 18 \cdot 1$ $= 24 - (42 - 24 \cdot 1)$ $= 42 \cdot (-1) + 24 \cdot 2$
We now have written 6 in terms of 42 and 24. Use the equation in Step 1 to write $24 = 234 - 42 \cdot 5$. Substitute this into the preceding result and simplify.	$6 = 42 \cdot (-1) + 24 \cdot 2$ $= 42 \cdot (-1) + (234 - 42 \cdot 5) \cdot 2$ $= 234 \cdot 2 + 42 \cdot (-11)$

Hence, we can write

$$\gcd(234, 42) = 234 \cdot 2 + 42 \cdot (-11).$$

(Check this with a calculator.) In this case, we say that we have written $\gcd(234, 42)$ as a linear combination of 234 and 42. More generally, we have the following definition.

Definition. Let a and b be integers. A **linear combination** of a and b is an integer of the form $ax + by$, where x and y are integers.

Activity 8.7 (Writing the gcd as a linear combination). Use the results from Activity 8.4 to

1. Write $\gcd(180, 126)$ as a linear combination of 180 and 126.

2. Write $\gcd(4208, 288)$ as a linear combination of 4208 and 288.

The previous example and activity illustrate the following important result in number theory, which will be used in the next section to help prove some other significant results.

Theorem 8.8. *Let a and b be integers, not both 0. Then $\gcd(a, b)$ can be written as a linear combination of a and b. That is, there exist integers x and y such that $\gcd(a, b) = ax + by$.*

We will not give a formal proof of this theorem. Hopefully, the examples and activities provide evidence for its validity. The idea is to use the steps of the Euclidean algorithm in reverse order to write $\gcd(a, b)$ as a linear combination of a and b. For example, assume the completed table for the Euclidean Algorithm is

Step	Original Pair	Equation from Division Algorithm	New Pair
1	(a, b)	$a = b \cdot q_1 + r_1$	(b, r_1)
2	(b, r_1)	$b = r_1 \cdot q_2 + r_2$	(r_1, r_2)
3	(r_1, r_2)	$r_1 = r_2 \cdot q_3 + r_3$	(r_2, r_3)
4	(r_2, r_3)	$r_2 = r_3 \cdot q_4 + 0$	(r_3, r_4)

We can use Step 3 to write $r_3 = \gcd(a, b)$ as a linear combination of r_1 and r_2. We can then solve the equation in Step 2 for r_2, and use this to write $r_3 = \gcd(a, b)$ as a linear combination of r_1 and b. Finally, we can use the equation in Step 1 to solve for r_1 and use this to write $r_3 = \gcd(a, b)$ as a linear combination of a and b.

In general, if we can write $r_p = \gcd(a, b)$ as a linear combination of a pair in a given row, then we can use the equation in the preceding step to write $r_p = \gcd(a, b)$ as a linear combination of the pair in this preceding row.

The notational details of this induction argument get quite involved. Many mathematicians prefer to prove Theorem 8.8 using a property of the natural numbers called the Well-Ordering Principle. **The Well-Ordering Principle** for the natural numbers states that any non-empty set of natural numbers must contain a least element. It can be proven that the Well-Ordering Principle is equivalent to the Principle of Mathematical Induction.

Exercises 8.1

1. Find each of the following greatest common divisors by first listing all of the common divisors of each pair of integers.

    (a) $\gcd(21, 28)$ (c) $\gcd(58, 63)$ (e) $\gcd(110, 215)$

    (b) $\gcd(-21, 28)$ (d) $\gcd(0, 12)$ (f) $\gcd(110, -215)$

2. (a) Let $a \in \mathbb{Z}$ and let $d \in \mathbb{N}$. Prove that if $d \mid a$ and $d \mid (a + 1)$, then $d \mid 1$, and hence $d = 1$.

    (b) Let $a \in \mathbb{Z}$. Find the greatest common divisor of the consecutive integers a and $a + 1$. That is, determine $\gcd(a, a + 1)$.

3. (a) Let $a \in \mathbb{Z}$ and let $d \in \mathbb{N}$. Prove that if $d \mid a$ and $d \mid (a + 2)$, then $d \mid 2$.

    (b) Let $a \in \mathbb{Z}$. Find the greatest common divisor of a and $a + 2$. That is, determine $\gcd(a, a + 2)$.

4. For each of the following pairs of integers, use the Euclidean algorithm to find $\gcd(a, b)$ and to write $\gcd(a, b)$ as a linear combination of a and b. That is, find integers m and n such that $d = am + bn$.

   (a) $a = 36, b = 60$ (c) $a = 12628, b = 21361$

   (b) $a = 901, b = 935$ (d) $a = 72, b = 714$

5. Let $a, b \in \mathbb{Z}$ with $b \neq 0$. Prove each of the following:

   (a) $\gcd(0, b) = |b|$.

   (b) If $\gcd(a, b) = d$, then $\gcd(a, -b) = d$. That is, $\gcd(a, -b) = \gcd(a, b)$.

6. (a) Notice that $\gcd(11, 17) = 1$. Find integers x and y such that $11x + 17y = 1$.

   (b) Let $m, n \in \mathbb{Z}$. Write the sum $\dfrac{m}{11} + \dfrac{n}{17}$ as a single fraction.

   (c) Find two rational numbers with denominators of 11 and 17, respectively, whose sum is equal to $\dfrac{10}{187}$. Hint: Write the rational numbers in the form $\dfrac{m}{11}$ and $\dfrac{n}{17}$, where $m, n \in \mathbb{Z}$. Then write

   $$\frac{m}{11} + \frac{n}{17} = \frac{10}{187}.$$

   Use Exercises (6a) and (6b) to determine m and n.

7. Visit the Euclidean algorithm Web site at

   http://bigcheese.math.sc.edu/~ sumner/numbertheory/euclidean/euclidean.html.

   (a) Use this Web site to determine $\gcd(40608, 151280)$ and to write this greatest common divisor as a linear combination of 40608 and 151280.

   (b) Use this Web site to determine $\gcd(380488, 6251740)$ and to write this greatest common divisor as a linear combination of 380488 and 6251740.

8.2   Prime Numbers and Prime Factorizations

Preview Activity 1 (Linear Combinations of Integers).
Let $a = 20$ and $b = 12$.

1. What is the greatest common divisor of 20 and 12?

2. Let $d = \gcd(20, 12)$. Write d as a linear combination of 20 and 12.

3. Generate at least six different linear combinations of 20 and 12. Does $\gcd(20, 12)$ divide each of these linear combinations?

4. Generate at least six different linear combinations of 21 and -6. Does $\gcd(21, -6)$ divide each of these linear combinations?

5. Complete the proof of Proposition 4.10 in Section 4.2.

   Proposition 4.10 *Let a, b, and d be integers. If d divides a and d divides b, then for all integers x and y, d divides ax + by.*

   Proof. Let a, b, and d be integers, and assume that d divides a and d divides b. We will prove that for all integers x and y, d divides $ax + by$.

   So let $x \in \mathbb{Z}$ and let $y \in \mathbb{Z}$. Since d divides a, there exists an integer m such that ....

   ∎

Preview Activity 2 (Exploring Examples where a divides $b \cdot c$).

1. Find at least three different examples of integers a, b, and c such that $a \mid (bc)$ but a does not divide b and a does not divide c. In each case, compute $\gcd(a, b)$ and $\gcd(a, c)$.

2. Find at least three different examples of integers a, b, and c such that $\gcd(a, b) = 1$ and $a \mid (bc)$. In each example, is there any relation between the integers a and c?

3. Formulate a conjecture based on your work in Parts (1) and (2).

Preview Activity 3 (Prime Factorizations).

Recall that a natural number p is a **prime number** provided that it is greater than 1 and the only natural numbers that divide p are 1 and p. A natural number other than 1 that is not a prime number is a **composite number**. The number 1 is neither prime nor composite.

1. Give examples of four natural numbers that are prime and four natural numbers that are composite.

Theorem 5.12 in Section 5.2 states that every natural number greater than 1 is either a prime number or a product of prime numbers.

When a composite number is written as a product of prime numbers, we say that we have obtained a **prime factorization** of that composite number.

2. Write the number 40 as a product of prime numbers by first writing $40 = 2 \cdot 20$ and then factoring 20 into a product of primes. Next, write the number 40 as a product of prime numbers by first writing $40 = 5 \cdot 8$ and then factoring 8 into a product of primes.

3. In Part (2), we used two different methods to obtain a prime factorization of 40. Did these methods produce the same prime factorization or different prime factorizations? Explain.

4. Repeat Parts (2) and (3) with 150. First, start with $150 = 3 \cdot 50$, and then start with $150 = 5 \cdot 30$.

In Section 8.1, we introduced the concept of the greatest common divisor of two integers. We showed how the Euclidean algorithm can be used to find the greatest common divisor of two integers, a and b, and also showed how to use the results of the Euclidean algorithm to write the greatest common divisor of a and b as a linear combination of a and b.

In this section, we will use these results to help prove the so-called Fundamental Theorem of Arithmetic, which states that any natural number greater than 1 that is not prime can be written as product of primes in "essentially" only one way. This means that given two prime factorizations, the prime factors are exactly the same, and the only difference may be in the order in which the prime factors are written. We start with more results concerning greatest common divisors.

More Results about Greatest Common Divisors

Let $a, b \in \mathbb{Z}$, not both 0, and let $d = \gcd(a, b)$. Theorem 8.8 states that d can be written as a linear combination of a and b. Now, since $d \mid a$ and $d \mid b$, we can use the result of Proposition 4.10 to conclude that for all $x, y \in \mathbb{Z}$, $d \mid (ax + by)$. This means that d divides every linear combination of a and b. We summarize these results in Theorem 8.9.

Theorem 8.9. *Let $a, b \in \mathbb{Z}$, not both 0, and let $d = \gcd(a, b)$.*

1. *The greatest common divisor, d, is a linear combination of a and b. That is, there exist integers m and n such that $d = am + bn$.*

2. *The greatest common divisor, d, divides every linear combination of a and b. That is, for all $x, y \in \mathbb{Z}$, $d \mid (ax + by)$.*

In Preview Activity 2, we constructed several examples of integers a, b, and c such that $a \mid (bc)$ but a does not divide b and a does not divide c. For each example, we observed that $\gcd(a, b) \neq 1$ and $\gcd(a, c) \neq 1$.

We also constructed several examples where $a \mid (bc)$ and $\gcd(a, b) = 1$. In all of these cases, we noted that a divides c. Integers whose greatest common divisor is equal to 1 are given a special name.

Definition. Two nonzero integers a and b are **relatively prime** provided that $\gcd(a, b) = 1$.

Activity 8.10 (Relatively Prime Integers).

1. Construct at least three different examples where p is a prime number, $a \in \mathbb{Z}$, and $p \mid a$. In each example, what is $\gcd(a, p)$? Based on these examples, formulate a conjecture about $\gcd(a, p)$ when $p \mid a$.

2. Construct at least three different examples where p is a prime number, $a \in \mathbb{Z}$, and p does not divide a. In each example, what is $\gcd(a, p)$? Based on these examples, formulate a conjecture about $\gcd(a, p)$ when p does not divide a.

3. Give at least three different examples of integers a and b where a is not prime, b is not prime, and $\gcd(a, b) = 1$, or explain why it is not possible to construct such examples.

Theorem 8.11. *Let a and b be nonzero integers, and let p be a prime number.*

1. *If a and b are relatively prime, then there exist integers m and n such that $am + bn = 1$. That is, 1 can be written as linear combination of a and b.*

2. *If $p \mid a$, then $\gcd(a, p) = p$.*

3. *If p does not divide a, then $\gcd(a, p) = 1$.*

Part (1) of Theorem 8.11 is actually a corollary of Theorem 8.9. Parts (2) and (3) could have been the conjectures you formulated in Activity 8.10. The proofs will be included in the exercises.

Given nonzero integers a and b, we have seen that it is possible to use the Euclidean algorithm to write their greatest common divisor as a linear combination of a and b. We have also seen that this can sometimes be a tedious, time-consuming process, which is why people have programmed computers to do this. Fortunately, in many proofs of number theory results, we do not actually have to construct this linear combination since simply knowing that it exists can be useful in proving results. This will be illustrated in the proof of Theorem 8.12, which is based on work in Preview Activity 2.

Theorem 8.12. *Let a, b, and c be integers. If a and b are relatively prime and $a \mid (bc)$, then $a \mid c$.*

The explorations in Preview Activity 2 were related to this theorem. Before we prove this theorem, we will explore the forward-backward process for the proof.

The goal is to prove that $a \mid c$. A standard way to do this is to prove that there exists an integer q such that

$$c = aq. \tag{1}$$

Now, we are given that $a \mid (bc)$. This means that there exists an integer k such that

$$bc = ak. \tag{2}$$

It may seem tempting to divide both sides of Equation (2) by b, but if we do so, we run into problems with the fact that the integers are not closed with respect to division. Instead, we look at the other part of the hypothesis, which is that a and b are relatively prime. This means that $\gcd(a, b) = 1$.

How can we use this? This means that a and b have no common factors except for 1. In light of Equation (2), it seems reasonable that any factor of a, then, must also be a factor of c. But how do we formalize this?

One conclusion that we can use is that since $\gcd(a, b) = 1$, by Theorem 8.11, we know that there exist integers m and n such that

$$am + bn = 1. \tag{3}$$

We may consider solving Equation (3) for b and substituting this into Equation (2). The problem, again, is that in order to solve Equation (3) for b, we need to divide by n.

Before doing anything else, we should look at the goal in Equation (1). We need to introduce c into Equation (3). One way to do this is to multiply both sides of Equation (3) by c. (This keeps us in the system of integers since the integers are closed under multiplication.) This gives:

$$(am + bn)\,c = 1 \cdot c$$
$$acm + bcn = c. \tag{4}$$

Notice that the left side of Equation (4) contains a term, bcn, that contains bc. This means that we can use Equation (2) and substitute $bc = ak$ in Equation (4). After doing this, we can factor the left side of the equation to prove that $a \mid c$.

Activity 8.13 (Completing the Proof of Theorem 8.12).

Write a complete proof of Theorem 8.12.

Corollary 8.14.

1. Let $a, b \in \mathbb{Z}$, and let p be a prime number. If $p \mid (ab)$, then $p \mid a$ or $p \mid b$.

2. Let p be a prime number, let $n \in \mathbb{N}$, and let $a_1, a_2, \ldots, a_n \in \mathbb{Z}$. If $p \mid (a_1 a_2 \cdots a_n)$, then there exists a natural number k with $1 \leq k \leq n$ such that $p \mid a_k$.

Part (1) of Corollary 8.14 is a corollary of Theorem 8.12. Part (2) is proven using mathematical induction. The basis step is the case where $n = 1$, and Part (1) is the case where $n = 2$. The proofs of these two results are included in the exercises.

Historical Note

Part (1) of Corollary 8.14 is known as **Euclid's Lemma**. Most people associate geometry with *Euclid's Elements*, but these books also contain many basic results in number theory. Many of the results that are contained in this section appeared in *Euclid's Elements*.

We are now ready to prove the Fundamental Theorem of Arithmetic. The first part of this theorem was proven in Theorem 5.12 in Section 5.2. This theorem states that each natural number greater than 1 is either a prime number or is a product of prime numbers. Before we state the Fundamental Theorem of Arithmetic, we will discuss some notational conventions that will help us with the proof. We start with an example.

We will use $n = 120$. Since $5 \mid 120$, we can write $120 = 5 \cdot 24$. In addition, we can factor 24 as $24 = 2 \cdot 2 \cdot 2 \cdot 3$. So we can write

$$
\begin{aligned}
120 &= 5 \cdot 24 \\
&= 5 \left(2 \cdot 2 \cdot 2 \cdot 3 \right).
\end{aligned}
$$

This is a prime factorization of 120, but it is not the way we usually write this factorization. Most often, we will write the prime number factors in ascending order. So we write

$$
120 = 2 \cdot 2 \cdot 2 \cdot 3 \cdot 5 \text{ or } 120 = 2^3 \cdot 3 \cdot 5.
$$

Now, let $n \in \mathbb{N}$. To write the prime factorization of n with the prime factors in ascending order requires that if we write $n = p_1 p_2 \cdots p_r$, where $p_1, p_2, \cdots p_r$ are prime numbers, we will have $p_1 \leq p_2 \leq \cdots \leq p_r$.

Theorem 8.15 (The Fundamental Theorem of Arithmetic).

1. *Each natural number greater than 1 is either a prime number or is a product of prime numbers.*

2. *Let $n \in \mathbb{N}$ with $n > 1$. Assume that*

$$
n = p_1 p_2 \cdots p_r \text{ and that } n = q_1 q_2 \cdots q_s
$$

where $p_1, p_2, \cdots p_r$ and $q_1, q_2, \cdots q_s$ are primes with $p_1 \leq p_2 \leq \cdots \leq p_r$ and $q_1 \leq q_2 \leq \cdots \leq q_s$. Then, $r = s$, and for each j from 1 to r, $p_j = q_j$.

Proof. The first part of this theorem was proven in Theorem 5.12. We will prove the second part of the theorem by induction on n using the Second Principle of Mathematical Induction. (See Section 5.2.) For each natural number n with $n > 1$, let $P(n)$ be

> If $n = p_1 p_2 \cdots p_r$ and $n = q_1 q_2 \cdots q_s$, where $p_1, p_2, \cdots p_r$ and $q_1, q_2, \cdots q_s$ are primes with $p_1 \le p_2 \le \cdots \le p_r$ and $q_1 \le q_2 \le \cdots \le q_s$, then, $r = s$, and for each j from 1 to r, $p_j = q_j$.

For the basis step, we notice that since 2 is a prime number, its only factorization is $2 = 1 \cdot 2$. This means that the only equation of the form $2 = p_1 p_2 \cdots p_r$, where $p_1, p_2, \cdots p_r$ are prime numbers, is the case where $r = 1$ and $p_1 = 2$. This proves that $P(2)$ is true.

For the inductive step, let $k \in \mathbb{N}$ with $k \ge 2$. We will assume that $P(2), P(3), \ldots, P(k)$ are true. The goal now is to prove that $P(k+1)$ is true. To prove this, we assume that $k + 1$ has two prime factorizations and then prove that these prime factorizations are the same. So we assume that

> $k + 1 = p_1 p_2 \cdots p_r$ and that $k + 1 = q_1 q_2 \cdots q_s$, where $p_1, p_2, \cdots p_r$ and $q_1, q_2, \cdots q_s$ are primes with $p_1 \le p_2 \le \cdots \le p_r$ and $q_1 \le q_2 \le \cdots \le q_s$.

We must now prove that $r = s$, and for each j from 1 to r, $p_j = q_j$. We can break our proof into two cases: (1) $p_1 \le q_1$; and (2) $q_1 \le p_1$. Since one of these must be true, and since the proofs will be similar, we can assume, without loss of generality, that $p_1 \le q_1$.

Since $k + 1 = p_1 p_2 \cdots p_r$, we know that $p_1 \mid (k + 1)$, and hence we may conclude that $p_1 \mid (q_1 q_2 \cdots q_s)$. We now use Corollary 8.14 to conclude that there exists a j with $1 \le j \le s$ such that $p_1 \mid q_j$. Since p_1 and q_j are primes, we conclude that

$$p_1 = q_j.$$

Now, we have also assumed that $q_1 \le q_j$ for all j, and hence, we know that $q_1 \le p_1$. However, we have also assumed that $p_1 \le q_1$. Hence,

$$p_1 = q_1.$$

We now use this and the fact that $k + 1 = p_1 p_2 \cdots p_r = q_1 q_2 \cdots q_s$ to conclude that

$$p_2 \cdots p_r = q_2 \cdots q_s.$$

The product in the previous equation is less that $k + 1$. Hence, we can apply our induction hypothesis to these factorizations and conclude that $r = s$, and for each j from 2 to r, $p_j = q_j$.

This completes the proof that if $P(2), P(3), \ldots, P(k)$ are true, then $P(k+1)$ is true. Hence, by the Second Principle of Mathematical Induction, we conclude that $P(n)$ is true for all $n \in \mathbb{N}$ with $n \geq 2$. This completes the proof of the theorem. ∎

Note: We often shorten the result of the Fundamental Theorem of Arithmetic by simply saying that each natural number greater than one that is not a prime has a **unique factorization** as a product of primes. This simply means that if $n \in \mathbb{N}$, $n > 1$, and n is not prime, then no matter how we choose to factor n into a product of primes, we will always have the same prime factors. The only difference may be in the order in which we write the prime factors.

Prime numbers have fascinated mathematicians for centuries. For example, we can easily start writing a list of prime numbers in ascending order. Following is a list of the prime numbers less than 100.

2, 3, 5, 7, 11, 13, 17, 19, 23, 29, 31, 37, 41, 43, 47, 53, 59, 61, 67, 71, 73, 79, 83, 89, 97.

This list contains the first 25 prime numbers. Does this list ever stop? The question was answered in *Euclid's Elements*, and the result is stated in the next theorem. This proof is one of the classical proofs by contradiction.

Theorem 8.16. *There are infinitely many prime numbers.*

Proof. We will use a proof by contradiction. We assume that there are only finitely many primes, and let

$$p_1, p_2, \ldots, p_m$$

be the list of all the primes. Let

$$M = p_1 p_2 \cdots p_m + 1. \tag{1}$$

Notice that $M \neq 1$. So M is either a prime number or, by the Fundamental Theorem of Arithmetic, M is a product of prime numbers. In either case, M has a prime factor. Since we have listed all the prime numbers, this means that there exists a j with $1 \leq j \leq m$ such that $p_j \mid M$. Now, we can rewrite Equation (1) as follows:

$$1 = M - p_1 p_2 \cdots p_m. \tag{2}$$

We have proven $p_j \mid M$, and since p_j is one of the prime factors of $p_1 p_2 \cdots p_m$, we can also conclude that $p_j \mid (p_1 p_2 \cdots p_m)$. Since p_j divides both of the terms on the right side of Equation (2), we can use this equation to conclude that $p_j \mid 1$. This is a contradiction. Hence, our assumption that there are only finitely many primes is false, and so there must be infinitely many primes.   ■

Comments about Prime Numbers

1. The Distribution of Prime Numbers

There are infinitely many primes, but when we write a list of the prime numbers, we can see some long sequences of consecutive natural numbers that contain no prime numbers. For example, there are no prime numbers between 113 and 127. The following proposition shows that there exist arbitrarily long sequences of consecutive natural numbers containing no prime numbers. A guided proof of this proposition is included in Exercise (15).

Proposition 8.17. *For any natural number n, there exist at least n consecutive natural numbers that are composite numbers.*

There are many unanswered questions about prime numbers, two of which will now be discussed.

2. The Twin Prime Conjecture

By looking at the list of the first 25 prime numbers, we see several cases where consecutive prime numbers differ by 2. Examples are: 3 and 5; 11 and 13; 17 and 19; 29 and 31. Such pairs of prime numbers are said to be **twin primes**. How many twin primes exist? The answer is not known. The **Twin Prime Conjecture** states that there are infinitely many twin primes. As of January 1, 2002, this is still a conjecture as it has not been proven or disproven.

For some interesting information on prime numbers, visit the Web site *The Prime Pages* (http://www.utm.edu/research/primes/), where there is a link to The Largest Known Primes Web site. According to information at this site (http://www.utm.edu/research/primes/largest.html), as of December 1, 2001, the largest known twin primes are $\left(318032361 \times 2^{107001} \pm 1\right)$. Each of these prime numbers contains 32220 digits.

3. Goldbach's Conjecture

Given an even natural number, is it possible to write it as a sum of two prime numbers? For example:

$$4 = 2 + 2 \qquad 6 = 3 + 3 \qquad 8 = 5 + 3$$
$$78 = 37 + 41 \qquad 90 = 43 + 47 \qquad 138 = 67 + 71$$

One of the most famous unsolved problems in mathematics is a conjecture made by Christian Goldbach in a letter toLeonhard Euler in 1742. The conjecture, now known as **Goldbach's Conjecture**, is as follows:

> Every even integer greater than 2 can be expressed as the sum of two (not necessarily distinct) prime numbers.

As of January 1, 2002, it is not known if this conjecture is true or false, although most mathematicians believe it to be true.

Exercises 8.2

1. Prove the second and third parts of Theorem 8.11.

   (a) Let a be a nonzero integer, and let p be a prime number. If $p \mid a$, then $\gcd(a, p) = p$.

   (b) Let a be a nonzero integer, and let p be a prime number. If p does not divide a, then $\gcd(a, p) = 1$.

2. Prove the first part of Corollary 8.14.

   Let $a, b \in \mathbb{Z}$, and let p be a prime number. If $p \mid (ab)$, then $p \mid a$ or $p \mid b$. <u>Hint</u>: Consider two cases: (1) $p \mid a$; and (2) p does not divide a.

3. Use mathematical induction to prove the second part of Corollary 8.14.

   Let p be a prime number, let $n \in \mathbb{N}$, and let $a_1, a_2, \ldots, a_n \in \mathbb{Z}$. If $p \mid (a_1 a_2 \cdots a_n)$, then there exists a $k \in \mathbb{N}$ with $1 \leq k \leq n$ such that $p \mid a_k$.

4. **(a)** Let a and b be nonzero integers. If there exist integers x and y such that $ax + by = 1$, what conclusion can be made about $\gcd(a, b)$? Explain.

   (b) Let a and b be nonzero integers. If there exist integers x and y such that $ax + by = 2$, what conclusion can be made about $\gcd(a, b)$? Explain.

5. **(a)** Let $a \in \mathbb{Z}$. What is $\gcd(a, a + 1)$? That is, what is the greatest common divisor of two consecutive integers? Justify your conclusion.

   Hint: Exercise (4) might be helpful.

   (b) Let $a \in \mathbb{Z}$. What conclusion can be made about $\gcd(a, a + 2)$? That is, what conclusion can be made about the greatest common divisor of two integers that differ by 2? Justify your conclusion.

6. **(a)** Let $a \in \mathbb{Z}$. What conclusion can be made about $\gcd(a, a + 3)$? That is, what conclusion can be made about the greatest common divisor of two integers that differ by 3? Justify your conclusion.

   (b) Let $a \in \mathbb{Z}$. What conclusion can be made about $\gcd(a, a + 4)$? That is, what conclusion can be made about the greatest common divisor of two integers that differ by 4? Justify your conclusion.

7. **(a)** Let $a = 16$ and $b = 28$. Determine the value of $d = \gcd(a, b)$, and then determine the value of $\gcd\left(\dfrac{a}{d}, \dfrac{b}{d}\right)$.

   (b) Repeat Exercise (7a) with $a = 10$ and $b = 45$.

   (c) Let $a, b \in \mathbb{Z}$, not both equal to 0, and let $d = \gcd(a, b)$. Explain why $\dfrac{a}{d}$ and $\dfrac{b}{d}$ are integers. Then prove that $\gcd\left(\dfrac{a}{d}, \dfrac{b}{d}\right) = 1$.

   Hint: Start by writing d as a linear combination of a and b.

   This says that if you divide both a and b by their greatest common divisor, the result will be two relatively prime integers.

8. Are the following propositions true or false? Justify your conclusions.

   (a) Let $a, b, c \in \mathbb{Z}$. If $a \mid c$ and $b \mid c$, then $(ab) \mid c$.

   (b) Let $a, b, c \in \mathbb{Z}$. If $a \mid c$, $b \mid c$ and $\gcd(a, b) = 1$, then $(ab) \mid c$.

9. In Exercise (12) in Section 3.4, it was proven that if n is an odd integer, then $8 \mid (n^2 - 1)$. (This result was also proven in Exercise (15) in Section 7.4.) Now, prove the following proposition:

If n is an odd integer and 3 does not divide n, then $24 \mid (n^2 - 1)$.

10. **(a)** Prove the following proposition:

   Let $a, b, c \in \mathbb{Z}$. If $\gcd(a, b) = 1$ and $\gcd(a, c) = 1$, then $\gcd(a, bc) = 1$.

   (b) Use mathematical induction to prove the following proposition:

   Let $n \in \mathbb{N}$ and let $a, b_1, b_2, \ldots, b_n \in \mathbb{Z}$. If $\gcd(a, b_i) = 1$ for all $i \in \mathbb{N}$ with $1 \leq i \leq n$, then $\gcd(a, b_1 b_2 \cdots b_n) = 1$.

11. Is the following proposition true or false? Justify your conclusion.

   Let $a, b, c \in \mathbb{Z}$. If $\gcd(a, b) = 1$ and $c \mid (a + b)$, then $\gcd(a, c) = 1$ and $\gcd(b, c) = 1$.

12. Is the following proposition true or false? Justify your conclusion.

   If $n \in \mathbb{N}$, then $\gcd(5n + 2, 12n + 5) = 1$.

13. Let $y \in \mathbb{N}$. Use the Fundamental Theorem of Arithmetic to prove that there exists an odd natural number x and a nonnegative integer k such that $y = 2^k x$.

14. **(a)** Determine at least five different primes that are congruent to 3 modulo 4.

   (b) Prove that there are infinitely many primes that are congruent to 3 modulo 4.

15. **(a)** Let $n \in \mathbb{N}$. Prove that $2 \mid [(n + 1)! + 2]$.

   (b) Let $n \in \mathbb{N}$ with $n \geq 2$. Prove that $3 \mid [(n + 1)! + 3]$.

   (c) Let $n \in \mathbb{N}$. Prove that for each $k \in \mathbb{N}$ with $2 \leq k \leq (n + 1)$, $k \mid [(n + 1)! + k]$.

   (d) Use the result of Exercise (15c) to prove that for each $n \in \mathbb{N}$, there exist at least n consecutive composite natural numbers.

16. The Twin Prime Conjecture states that there are infinitely many twin primes, but it is not known if this conjecture is true or false. The answers to the following questions, however, can be determined.

   (a) How many pairs of primes p and q exist where $q - p = 3$? That is, how many pairs of primes are there that differ by 3? Prove that your answer is correct. <u>Note</u>: One such pair is 2 and 5.

(b) How many triplets of primes of the form p, $p + 2$, and $p + 4$ are there? That is, how many triplets of primes exist where each prime is 2 more than the preceding prime? Prove that your answer is correct. Note: One such triplet is 3, 5, and 7.

Hint: Try setting up cases using congruence modulo 3.

17. Prove the following proposition:

Let $n \in \mathbb{N}$. If $a \in \mathbb{Z}$ and $\gcd(a, n) = 1$, then for every $b \in \mathbb{Z}$, there exists an $x \in \mathbb{Z}$ such that $ax \equiv b \pmod{n}$.

Hint: One way is to start by writing 1 as a linear combination of a and n.

18. Prove the following proposition:

Let $m, n \in \mathbb{N}$. If m and n are twin primes other than the pair 3 and 5, then 36 divides $mn + 1$ and $mn + 1$ is a perfect square.

Hint: Look at several examples of twin primes. What do you notice about the number that is between the two twin primes? Set up cases based on this observation.

8.3   Linear Diophantine Equations

Preview Activity 1 (Integer Solutions for Linear Equations).

1. Does the linear equation $4x = 9$ have a solution that is an integer? Explain.

2. Does the linear equation $7x = -21$ have a solution that is an integer? Explain.

3. Prove the following proposition:

Let $a, b \in \mathbb{Z}$ with $a \neq 0$.

- If a does not divide b, then the equation $ax = b$ has no solution that is an integer.
- If a divides b, then the equation $ax = b$ has exactly one solution that is an integer.

4. Find integers x and y so that $2x + 6y = 253$ or explain why it is not possible to find such a pair of integers.

5. Find integers x and y so that $6x - 9y = 100$ or explain why it is not possible to find such a pair of integers.

Preview Activity 2 (Linear Equations in Two Variables).

1. Notice that $x = 2$ and $y = 1$ is a solution for the equation $3x + 5y = 11$, and that $x = 7$ and $y = -2$ is also a solution for the equation $3x + 5y = 11$.

   (a) Find two other pairs of integers x and y so that $x > 7$ and $3x + 5y = 11$. (Try to keep the integer values of x as small as possible.)

   (b) Find two other pairs of integers x and y so that $x < 2$ and $3x + 5y = 11$. (Try to keep the integer values of x as close to 2 as possible.)

   (c) Determine formulas (one for x and one for y) that will generate pairs of integers x and y so that $3x + 5y = 11$.

      <u>Hint</u>: The two formulas can be written in the form $x = 2 + km$ and $y = 1 + kn$, where k is an arbitrary integer and m and n are specific integers.

2. Notice that $x = 4$ and $y = 0$ is a solution for the equation $4x + 6y = 16$, and that $x = 7$ and $y = -2$ is a solution for the equation $4x + 6y = 16$.

   (a) Find two other pairs of integers x and y so that $x > 7$ and $4x + 6y = 16$. (Try to keep the integer values of x as small as possible.)

   (b) Find two other pairs of integers x and y so that $x < 4$ and $4x + 6y = 16$. (Try to keep the integer values of x as close to 4 as possible.)

   (c) Determine formulas (one for x and one for y) that will generate pairs of integers x and y so that $4x + 6y = 16$.

      <u>Hint</u>: The two formulas can be written in the form $x = 4 + km$ and $y = 0 + kn$, where k is an arbitrary integer and m and n are specific integers.

In the two preview activities for this section, we were interested only in integer solutions for certain equations. In such instances, we give the equation a special name.

> **Definition.** An equation whose solutions are required to be integers is called a **Diophantine equation**.

Diophantine equations are named in honor of the Greek mathematician Diophantus of Alexandria (circa 300 A.D.). Very little is known about Diophantus' life except that he probably lived in Alexandria in the early part of the fourth century A.D. and was probably the first to use letters for unknown quantities in arithmetic problems. His most famous work, *Arithmetica*, consists of approximately 130 problems and their solutions. Most of these problems involved solutions of equations in various numbers of variables. It is interesting to note that Diophantus did not restrict his solutions to the integers but recognized rational number solutions as well. Today, however, the solutions for a so-called Diophantine equation must be integers.

If a and b are integers with $a \neq 0$, then the equation

$$ax = b$$

is a **linear Diophantine equation in one variable**. The case for linear Diophantine equations was treated in Preview Activity 1, where the following theorem was proven.

Theorem 8.18. *Let $a, b \in \mathbb{Z}$ with $a \neq 0$.*

- *If a does not divide b, then the linear Diophantine equation $ax = b$ has no solution.*

- *If a divides b, then the linear Diophantine equation $ax = b$ has exactly one solution.*

A linear Diophantine equation in two variables can be defined in a manner similar to the definition for a linear Diophantine equation in one variable.

> **Definition.** Let a, b, and c be integers with $a \neq 0$ and $b \neq 0$. The Diophantine equation $ax + by = c$ is called a **linear Diophantine equation in two variables**.

The equations that were investigated in Preview Activity 2 were linear Diophantine equations.

The problem of determining all the solutions of a linear Diophantine equation has been completely solved. Before stating the general result, we will provide a few more examples.

Example 8.19 (Examples of Linear Diophantine Equations). The following
two examples are similar to the examples studied in Preview Activity 2.

1. It is easy to verify that $x = 2$ and $y = 1$ is a solution of the linear Diophantine equation

$$4x + 3y = 5.$$

The following table shows other solutions of this Diophantine equation.

x	y	x	y
2	−1	−1	3
5	−5	−4	7
8	−9	−7	11
11	−13	−10	15

It would be nice to determine the pattern that these solutions exhibit. If we consider the solution $x = 2$ and $y = 1$ to be the starting point, then we can see that the other solutions are obtained by adding 3 to x and subtracting 4 from y from the previous solution. This means that we can write these solutions to the equation as

$$x = 2 + 3k \qquad y = -1 - 4k,$$

where k can be any integer. We can verify that these equations give solutions of this equation as follows:

$$4x + 3y = 4\,(2 + 3k) + 3\,(-1 - 4k)$$
$$= (8 + 12k) + (-3 - 12k)$$
$$= 5.$$

We should note that we have not yet proven that these solutions are all of the solutions of the Diophantine equation $4x + 3y = 5$. This will be done later.

If the general form for a linear Diophantine equation is $ax + by = c$, then for this example, $a = 4$ and $b = 3$. Notice that for this equation, we started with one solution. We then obtained other solutions by adding $b = 3$ to x and subtracting $a = 4$ from y in the previous solution. Also, notice that $\gcd(3, 4) = 1$.

2. The following table shows some solutions of the linear Diophantine equation $6x + 9y = 12$.

x	y	x	y
2	0	−1	2
5	−2	−4	4
8	−4	−7	6
11	−6	−10	8

Following the pattern in this table, we can verify that the following equations produce solutions for the equation $6x + 9y = 12$:

$$x = 2 + 3k \qquad y = 0 - 2k,$$

where k can be any integer. Again, this does not prove that these are the only solutions.

If the general form for a linear Diophantine equation is $ax + by = c$, then for this example, $a = 6$ and $b = 9$. Notice that for this equation, we started with one solution. We then obtained other solutions by adding 3 to x and subtracting 2 from y in the previous solutions. Notice that these are not the values of b and a as in the previous example. However, notice that $\gcd(6, 9) = 3$ and that

$$\frac{b}{3} = \frac{9}{3} = 3, \qquad \frac{a}{3} = \frac{6}{2} = 2.$$

Activity 8.20 (Revisiting Preview Activity 2). Do the solutions for the linear Diophantine equations in Preview Activity 2 show the same pattern as the solutions for the linear Diophantine equations in Example 8.19? Explain.

The solutions for the linear Diophantine equations in Preview Activity 2 and Example 8.19 provide examples for the second part of the following theorem.

Theorem 8.21. *Let a, b, and c be integers with $a \neq 0$ and $b \neq 0$, and let $d = \gcd(a, b)$.*

1. *If d does not divide c, then the linear Diophantine equation $ax + by = c$ has no solution.*

2. *If d divides c, then the linear Diophantine equation $ax + by = c$ has infinitely many solutions. In addition, if (x_0, y_0) is a particular solution of this equation, then all the solutions of the equation are given by*

$$x = x_0 + \frac{b}{d}k, \qquad y = y_0 - \frac{a}{d}k$$

where $k \in \mathbb{Z}$.

Proof. The proof of Part (1) is included in the exercises.

For Part (2), we let a, b, and c be integers with $a \neq 0$ and $b \neq 0$, and let $d = \gcd(a, b)$. We also assume that $d \mid c$.

Since $d = \gcd(a, b)$, Theorem 8.8 tells us that d is a linear combination of a and b. So there exist integers s and t such that

$$d = as + bt. \tag{1}$$

Since $d \mid c$, there exists an integer m such that $c = dm$. We can now multiply both sides of Equation (1) by m and obtain

$$dm = (as + bt)m$$
$$c = a(sm) + b(tm).$$

This means that $x = sm$, $y = tm$ is a solution of $ax + by = c$, and we have proven that the Diophantine equation $ax + by = c$ has at least one solution.

Now, let $x = x_0$, $y = y_0$ be any particular solution of $ax + by = c$, let $k \in \mathbb{Z}$, and let

$$x = x_0 + \frac{b}{d}k \qquad y = y_0 - \frac{a}{d}k. \tag{2}$$

We now verify that for each $k \in \mathbb{Z}$, these equations produce a solution of $ax + by = c$.

$$ax + by = a\left(x_0 + \frac{b}{d}k\right) + b\left(y_0 - \frac{a}{d}k\right)$$
$$= ax_0 + \frac{ab}{d}k + by_0 - \frac{ab}{d}k$$
$$= ax_0 + by_0$$
$$= c.$$

This proves that the Diophantine equation $ax + by = c$ has infinitely many solutions.

We now show that every solution of this equation can be written in the form described in (2). So suppose that x and y are integers such that $ax + by = c$. Then

$$(ax + by) - (ax_0 + by_0) = c - c = 0,$$

and this equation can be rewritten in the following form:

$$a(x - x_0) = b(y_0 - y). \tag{3}$$

Dividing both sides of this equation by d, we obtain

$$\left(\frac{a}{d}\right)(x - x_0) = \left(\frac{b}{d}\right)(y_0 - y).$$

This implies that

$$\frac{a}{d} \mid \left(\frac{b}{d}\right)(y_0 - y).$$

However, by Exercise (7) in Section 8.2 , $\gcd\left(\dfrac{a}{d}, \dfrac{b}{d}\right) = 1$, and so by Theorem 8.12, we can conclude that $\dfrac{a}{d} \mid (y_0 - y)$. This means that there exists an integer k such that $y_0 - y = \dfrac{a}{d}k$, and solving for y gives

$$y = y_0 - \frac{a}{d}k.$$

Substituting this value for y in Equation (3) and solving for x yields

$$x = x_0 + \frac{b}{d}k.$$

This proves that every solution of the Diophantine equation $ax + by = c$ can be written in the form prescribed in (2). ■

The proof of the following corollary to Theorem 8.21 is included in the exercises.

Corollary 8.22. *Let a, b, and c be integers with $a \neq 0$ and $b \neq 0$. If a and b are relatively prime, then the linear Diophantine equation $ax + by = c$ has infinitely many solutions. In addition, if x_0, y_0 is a particular solution of this equation, then all the solutions of the equation are given by*

$$x = x_0 + bk \qquad y = y_0 - ak$$

where $k \in \mathbb{Z}$.

Example 8.23. $63x + 336y = 40$

For the linear Diophantine equation $63x + 336y = 40$, we can use the Euclidean algorithm to determine that $\gcd(63, 336) = 21$. Since 21 does not divide 40, Theorem 8.21 tells us that the Diophantine equation $63x + 336y = 40$ has no solution.

Remember that this means there is no ordered pair of integers (x, y) such that $63x + 336y = 40$. However, if we allow x and y to be real numbers, then there are real number solutions. In fact, we can graph the straight line whose equation is $63x + 336y = 40$ in the Cartesian plane. From the fact that there is no pair of integers x, y such that $63x + 336y = 40$, we can conclude that there is no point on the graph of this line in which both coordinates are integers.

Example 8.24. $144x + 225y = 27$

For this equation, we can use the Euclidean algorithm to determine that $\gcd(144, 225) = 9$. Since $9 \mid 27$, Theorem 8.21 tells us that the Diophantine equation $144x + 225y = 27$ has infinitely many solutions.

To write formulas that will generate all the solutions, we first need to find one solution for $144x + 225y = 27$. This can sometimes be done by trial and error, but there is a systematic way to find a solution. The first step is to use the Euclidean algorithm in reverse to write $\gcd(144, 225)$ as a linear combination of 144 and 225. See Section 8.1 to review how to do this. The result from using the Euclidean algorithm in reverse for this situation is

$$144 \cdot 11 + 225 \cdot (-7) = 9.$$

If we multiply both sides of this equation by 3, we obtain

$$144 \cdot 33 + 225 \cdot (-21) = 27.$$

This means that $x_0 = 33, y_0 = -21$ is a solution of the linear Diophantine equation $144x + 225y = 27$. We can now use Theorem 8.21 to conclude that all solutions of this Diophantine equation can be written in the form

$$x = 33 + \frac{225}{9}k \qquad y = -21 - \frac{144}{9}k,$$

where $k \in \mathbb{Z}$. Simplifying, we see that all solutions can be written in the form

$$x = 33 + 25k \qquad y = -21 - 16k,$$

where $k \in \mathbb{Z}$.

We can check this general solution as follows: Let $k \in \mathbb{Z}$. Then

$$144x + 225y = 144\,(33 + 25k) + 225\,(-21 - 16k)$$
$$= (4752 + 3600k) + (-4725 - 3600k)$$
$$= 27.$$

Activity 8.25 (Linear Congruences in One Variable). Let n be a natural number and let $a, b \in \mathbb{Z}$ with $a \neq 0$. A congruence of the form $ax \equiv b \pmod{n}$ is called a **linear congruence in one variable**.

This is called a linear congruence since the variable x occurs to the first power. A solution of a linear congruence in one variable is defined similarly to the solution of an equation. A solution is an integer that makes the resulting congruence true when the integer is substituted for the variable x. For example,

- The integer $x = 3$ is a solution for the congruence $2x \equiv 1 \pmod 5$ since $2 \cdot 3 \equiv 1 \pmod 5$ is a true congruence.

- The integer $x = 7$ is not a solution for the congruence $3x \equiv 1 \pmod 6$ since $3 \cdot 7 \equiv 1 \pmod 6$ is a not a true congruence.

1. Verify that $x = 2$ and $x = 6$ are the only solutions for the linear congruence $6x \equiv 4 \pmod 8$ with $0 \le x < 8$.

The following parts of this activity show that we can use the results of Theorem 8.21 to help find all solutions of the linear congruence $6x \equiv 4 \pmod 8$.

2. Use the definition of "congruence" to rewrite the congruence $6x \equiv 4 \pmod 8$ in terms of "divides."

3. Use the definition of "divides" to rewrite the result in Part (2) in the form of an equation. (To do this, an existential quantifier must be used.)

4. Use the results of Part (1) and Part (3) to write an equation that will generate all the solutions of the linear congruence $6x \equiv 4 \pmod 8$.

   Hint: Use Theorem 8.21. This can be used to generate solutions for x and the variable introduced in Part (3). In this case, we are interested only in the solutions for x.

Now let n be a natural number and let $a, c \in \mathbb{Z}$ with $a \neq 0$. A general linear congruence of the form $ax \equiv c \pmod{n}$ can be handled in the same way that we handled in $6x \equiv 4 \pmod 8$.

5. Use the definition of "congruence" to rewrite $ax \equiv c \pmod n$ in terms of "divides."

6. Use the definition of "divides" to rewrite the result in Part (5) in the form of an equation. (To do this, an existential quantifier must be used.)

7. Let $d = \gcd(a, n)$. State and prove a theorem about the solutions of the linear congruence $ax \equiv c \pmod n$ in the case where d does not divide c.

   Hint: Use Theorem 8.21.

8. Let $d = \gcd(a, n)$. State and prove a theorem about the solutions of the linear congruence $ax \equiv c \pmod n$ in the case where d divides c.

Exercises 8.3

1. Prove Part (1) of Theorem 8.21:

   Let a, b, and c be integers with $a \neq 0$ and $b \neq 0$, and let $d = gcd(a, b)$. If d does not divide c, then the linear Diophantine equation $ax + by = c$ has no solution.

2. Determine all solutions of the following linear Diophantine equations.

   (a) $9x + 14y = 1$
   (b) $18x + 22y = 4$
   (c) $48x - 18y = 15$
   (d) $12x + 9y = 6$

   (e) $200x + 49y = 10$
   (f) $200x + 54y = 21$
   (g) $10x - 7y = 31$
   (h) $12x + 18y = 6$

3. Prove Corollary 8.22.

   Let a, b, and c be integers with $a \neq 0$ and $b \neq 0$. If a and b are relatively prime, then the linear Diophantine equation $ax + by = c$ has infinitely many solutions. In addition, if (x_0, y_0) is a particular solution of this equation, then all the solutions of the equation are given by

$$x = x_0 + bk \qquad y = y_0 - ak,$$

where $k \in \mathbb{Z}$.

4. <u>A Balancing Problem</u>. A certain rare artifact is supposed to weigh exactly 25 grams. Suppose that you have an accurate balance scale and 500 each of 27 gram weights and 50 gram weights. Explain how to use Theorem 8.21 to devise a plan to check the weight of this artifact.

   <u>Hint</u>: Notice that $\gcd(50, 27) = 1$. Start by writing 1 as a linear combination of 50 and 27.

5. On the night of a certain banquet, a caterer offered the choice of two dinners, a steak dinner for \$25 and a vegetarian dinner for \$16. At the end of the evening, the caterer presented the host with a bill (before tax and tips) for \$1461. What is the minimum number of people who could have attended the banquet? What is the maximum number of people who could have attended the banquet?

6. The goal of this exercise is to determine all (integer) solutions of the linear Diophantine equation in three variables $12x_1 + 9x_2 + 16x_3 = 20$.

   (a) First, notice that $\gcd(12, 9) = 3$. Determine formulas that will generate all solutions for the linear Diophantine equation $3y + 16x_3 = 20$.

   (b) Explain why the solutions (for x_1 and x_2) of the Diophantine equation $12x_1 + 9x_2 = 3y$ can be used to generate solutions for $12x_1 + 9x_2 + 16x_3 = 20$.

   (c) Use the general value for y from Exercise (6a) to determine the solutions of $12x_1 + 9x_2 = 3y$.

   (d) Use the results from Exercises (6a) and (6c) to determine formulas that will generate all solutions for the Diophantine equation $12x_1 + 9x_2 + 16x_3 = 20$.

   <u>Note</u>: These formulas will involve two arbitrary integer parameters. Substitute specific values for these integers and then check the resulting solution in the original equation. Repeat this at least three times.

   (e) Check the general solution for $12x_1 + 9x_2 + 16x_3 = 20$ from Exercise (6d).

7. Use the method suggested in Exercise (6) to determine formulas that will generate all solutions of the Diophantine equation $8x_1 + 4x_2 - 6x_3 = 6$. Check the general solution.

8. Explain why the Diophantine equation $24x_1 - 18x_2 + 60x_3 = 21$ has no solution.

9. The purpose of this exercise will be to prove that the nonlinear Diophantine equation $3x^2 - y^2 = -2$ has no solution.

   (a) Explain why if there is a solution of the Diophantine equation $3x^2 - y^2 = -2$, then that solution must also be a solution of the congruence $3x^2 - y^2 \equiv -2 \pmod 3$.

   (b) If there is a solution to the congruence $3x^2 - y^2 \equiv -2 \pmod 3$, explain why there then must be an integer y such that $y^2 \equiv 2 \pmod 3$.

   (c) Use a proof by contradiction to prove that the Diophantine equation $3x^2 - y^2 = -2$ has no solution.

10. Use the method suggested in Exercise (9) to prove that the Diophantine equation $7x^2 + 2 = y^3$ has no solution.

Chapter 9

Topics in Set Theory

9.1 Functions Acting on Sets

Preview Activity 1 (Functions and Sets, Part 1).
For this preview activity, let $S = \{a, b, c, d\}$ and $T = \{s, t, u\}$. Define $f : S \to T$ by

$$f(a) = s \qquad f(b) = t \qquad f(c) = t \qquad f(d) = s.$$

1. Let $A = \{a, b, c\}$ and $B = \{a, d\}$. Notice that A and B are subsets of S. Use the roster method to describe each of the following two sets:

   (a) $\{f(x) \mid x \in A\}$ (b) $\{f(x) \mid x \in B\}$

2. Let $C = \{s, t\}$ and $D = \{s, u\}$. Notice that C and D are subsets of T. Use the roster method to describe each of the following two sets:

   (a) $\{x \in S \mid f(x) \in C\}$ (b) $\{x \in S \mid f(x) \in D\}$

Preview Activity 2 (Functions and Sets, Part 2).
Let $f : \mathbb{R} \to \mathbb{R}$ be defined by $f(x) = x^2$ for all $x \in \mathbb{R}$.

1. Determine $f(1)$, $f(2)$, $f(3)$, and $f(-1)$.

2. Let $A = \{1, 2, 3, -1\}$. Use the roster method to describe the set $\{f(x) \mid x \in A\}$.

3. (a) Find all $x \in \mathbb{R}$ such that $f(x) = 1$.

(b) Find all $x \in \mathbb{R}$ such that $f(x) = 4$.

(c) Find all $x \in \mathbb{R}$ such that $f(x) = 9$.

(d) Find all $x \in \mathbb{R}$ such that $f(x) = -1$.

4. Let $B = \{1, 4, 9\}$. Use the roster method to describe the set $\{x \in \mathbb{R} \mid f(x) \in B\}$.

Preview Activity 3 (Functions and Intervals).

Let $f : \mathbb{R} \to \mathbb{R}$ be defined by $f(x) = x^2$ for all $x \in \mathbb{R}$.

1. We will first determine where f maps the closed interval $[1, 2]$. That is, we will describe, in simpler terms, the set $\{f(x) \mid x \in [1, 2]\}$. This is the set of all images of the real numbers in the closed interval $[1, 2]$.

   (a) Draw a graph of the function f using $-3 \leq x \leq 3$.

   (b) On the graph, draw the vertical lines $x = 1$ and $x = 2$ from the x-axis to the graph. Label the points $P(1, f(1))$ and $Q(2, f(2))$ on the graph.

   (c) Now draw horizontal lines from the points P and Q to the y-axis. Use this information from the graph to describe the set $\{f(x) \mid x \in [1, 2]\}$ in simpler terms.

2. We will now determine all real numbers that map into the closed interval $[1, 4]$. That is, we will describe the set $\{x \in \mathbb{R} \mid f(x) \in [1, 4]\}$ in simpler terms. This is the set of all pre-images of the real numbers in the closed interval $[1, 4]$.

   (a) Draw a graph of the function f using $-3 \leq x \leq 3$.

   (b) On the graph, draw the horizontal lines $y = 1$ and $y = 4$ from the y-axis to the graph. Label all points where these two lines intersect the graph.

   (c) Now draw vertical lines from the points in Part (2) to the x-axis, and then use the resulting information to describe the set $\{x \in \mathbb{R} \mid f(x) \in [1, 4]\}$ in simpler terms. (You will need to describe this set as a union of two intervals.)

In our study of functions in Chapter 6, we focused on how a function "maps" individual elements of its domain to the codomain. We also studied the pre-image of an individual element in its codomain. For example, if $f : \mathbb{R} \to \mathbb{R}$ is defined by $f(x) = x^2$ for all $x \in \mathbb{R}$, then

- $f(2) = 4$. We say that f maps 2 to 4 or that 4 is the image of 2 under the function f.

- Since $f(x) = 4$ implies that $x = 2$ or $x = -2$, we say that the pre-images of 4 are 2 and -2.

For a function $f : S \to T$, the next step is to consider subsets of S or T and what corresponds to them in the other set. We did this in the Preview Activities. We will first give some definitions and then revisit the examples in the Preview Activities in light of these definitions.

Definition. Let $f : S \to T$. If $A \subseteq S$, then the **image of A under f** is the set $f(A)$ where

$$f(A) = \{f(x) \mid x \in A\}.$$

If there is no confusion as to which function is being used, we call $f(A)$ **the image of A**.

Definition. Let $f : S \to T$. If $C \subseteq T$, then the **pre-image of C under f** is the set $f^{-1}(C)$ where

$$f^{-1}(C) = \{x \in S \mid f(x) \in C\}.$$

If there is no confusion as to which function is being used, we call $f^{-1}(C)$ **the pre-image of C**. <u>Note</u>: The pre-image of the set C under f is frequently called the **inverse image of C under f**.

Notice that the set $f^{-1}(C)$ is defined whether or not the relation f^{-1} is a function.

Example 9.1. Let $S = \{a, b, c, d\}$ and $T = \{s, t, u\}$. Define $f : S \to T$ by

$$f(a) = s \qquad\qquad f(b) = t \qquad\qquad f(c) = t \qquad\qquad f(d) = s.$$

The following results are based on the examples in Preview Activity 1.

- Let $A = \{a, b, c\}$. Notice that $A \subseteq S$.

$$f(A) = \{f(x) \mid x \in A\} = \{s, t\}$$

- Let $B = \{a, d\}$. Notice that $B \subseteq S$.

$$f(B) = \{f(x) \mid x \in B\} = \{s\}$$

- Let $C = \{s, t\}$. Notice that $C \subseteq T$.

$$f^{-1}(C) = \{x \in S \mid f(x) \in C\} = \{a, b, c, d\}$$

- Let $D = \{s, u\}$. Notice that $C \subseteq T$.

$$f^{-1}(D) = \{x \in S \mid f(x) \in D\} = \{a, d\}$$

Example 9.2. Let $f : \mathbb{R} \to \mathbb{R}$ be defined by $f(x) = x^2$ for all $x \in \mathbb{R}$. The following results are based on the examples in Preview Activity 2 and Preview Activity 3.

- Let $A = \{1, 2, 3, -1\}$. Then $f(A) = \{1, 4, 9\}$.

- Let $B = \{1, 4, 9\}$. Then $f^{-1}(B) = \{-1, -2, -3, 1, 2, 3\}$.

The graphs in Figure 9.1 will be used for the following:

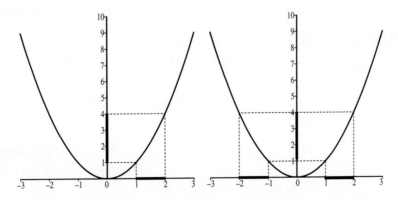

Figure 9.1: Graphs for Example 9.2

- Now let T be the closed interval $[1, 2]$. Then the graph on the left in Figure 9.1 illustrates the fact that the image of the set T is

$$f(T) = \{f(x) \mid x \in [1, 2]\}$$
$$= [1, 4].$$

- Let C be the closed interval $[1, 4]$. Then the graph on the right in Figure 9.1 illustrates the fact that the pre-image of the set C is

$$f^{-1}(C) = \{x \in \mathbb{R} \mid f(x) \in [1, 4]\} \quad = [-2, -1] \cup [1, 2].$$

We will now consider the following situation: Let S and T be sets and let $f : S \to T$. Also, let A and B be subsets of S and let C and D be subsets of T. In the remainder of this section and the exercises for this section, we will consider the following situations and answer the questions posed in each case.

- The set $A \cap B$ is a subset of S and so $f(A \cap B)$ is a subset of T. In addition, $f(A)$ and $f(B)$ are subsets of T. Hence, $f(A) \cap f(B)$ is a subset of T.

  Is there any relationship between $f(A \cap B)$ and $f(A) \cap f(B)$?

- The set $A \cup B$ is a subset of S and so $f(A \cup B)$ is a subset of T. In addition, $f(A)$ and $f(B)$ are subsets of T. Hence, $f(A) \cup f(B)$ is a subset of T.

  Is there any relationship between $f(A \cup B)$ and $f(A) \cup f(B)$?

- The set $C \cap D$ is a subset of T and so $f^{-1}(C \cap D)$ is a subset of S. In addition, $f^{-1}(C)$ and $f^{-1}(D)$ are subsets of S, and hence, $f^{-1}(C) \cap f^{-1}(D)$ is a subset of S.

  Is there any relationship between the sets $f^{-1}(C \cap D)$ and $f^{-1}(C) \cap f^{-1}(D)$?

- The set $C \cup D$ is a subset of T and so $f^{-1}(C \cup D)$ is a subset of S. In addition, $f^{-1}(C)$ and $f^{-1}(D)$ are subsets of S, and hence, $f^{-1}(C) \cup f^{-1}(D)$ is a subset of S.

  Is there any relationship between the sets $f^{-1}(C \cup D)$ and $f^{-1}(C) \cup f^{-1}(D)$?

These and other questions will be explored in the next activity.

Activity 9.3 (Set Operations and Functions Acting on Sets).

In Section 6.2, we introduced functions involving congruences. For example, if we let

$$\mathbb{Z}_8 = \{0, 1, 2, 3, 4, 5, 6, 7\},$$

then we can define $f : \mathbb{Z}_8 \to \mathbb{Z}_8$ by $f(x) = r$ where $(x^2 + 2) \equiv r \pmod 8$ and $r \in \mathbb{Z}_8$. Moreover, we shortened this notation to

$$f(x) = (x^2 + 2) \pmod 8.$$

Special Note for Those Who Have Studied Section 7.4

In Section 7.4, we used the notation $\mathbb{Z}_8$ to represent the set of all congruence classes of the integers for the equivalence relation of congruence modulo 8. That is,

$$\mathbb{Z}_8 = \{[0], [1], [2], [3], [4], [5], [6], [7]\}.$$

In this notation, we are in effect using one integer inside brackets to represent an entire congruence class. So, to make the notation somewhat easier to use, we are dropping the brackets and just writing the integer to represent the congruence class.

1. Verify that $f(0) = 2$, $f(1) = 3$, $f(2) = 6$, and $f(3) = 3$. Then determine $f(4)$, $f(5)$, $f(6)$, and $f(7)$.

We will now use the following subsets of $\mathbb{Z}_8$:

$$A = \{1, 2, 3\} \qquad B = \{3, 4, 6\} \qquad C = \{1, 2, 3\} \qquad D = \{3, 4, 5\}.$$

2. For each of the following, determine the two subsets of $\mathbb{Z}_8$ and then determine if there is a relationship between the two sets.

   (a) $f(A \cap B)$ and $f(A) \cap f(B)$
   (b) $f(A \cup B)$ and $f(A) \cup f(B)$
   (c) $f^{-1}(C \cap D)$ and $f^{-1}(C) \cap f^{-1}(D)$
   (d) $f^{-1}(C \cup D)$ and $f^{-1}(C) \cup f^{-1}(D)$

3. Now notice that $f(A)$ is a subset of the codomain, $\mathbb{Z}_8$. Consequently, $f^{-1}(f(A))$ is a subset of the domain, $\mathbb{Z}_8$. Is there any relation between A and $f^{-1}(f(A))$ in this case?

4. Now notice that $f^{-1}(C)$ is a subset of the domain, $\mathbb{Z}_8$. Consequently, $f(f^{-1}(C))$ is a subset of the codomain, $\mathbb{Z}_8$. Is there any relation between C and $f(f^{-1}(C))$ in this case?

Activity 9.4 (Set Operations and Functions Acting on Sets).
Define $f : \mathbb{R} \to \mathbb{R}$ by $f(x) = x^2 + 2$ for all $x \in \mathbb{R}$. For this activity, it will be useful to have a printed copy of the graph of the function f. Sketch a graph using $-3 \le x \le 3$ and $0 \le y \le 12$.

For this activity, we will use the following closed intervals:

$$A = [0,3] \qquad B = [-2,0] \qquad C = [2,6] \qquad D = [0,3]$$

1. For each of the following, determine the two subsets of $\mathbb{Z}_8$ and then determine if there is a relationship between the two sets.

   (a) $f(A \cap B)$ and $f(A) \cap f(B)$
   (b) $f(A \cup B)$ and $f(A) \cup f(B)$
   (c) $f^{-1}(C \cap D)$ and $f^{-1}(C) \cap f^{-1}(D)$
   (d) $f^{-1}(C \cup D)$ and $f^{-1}(C) \cup f^{-1}(D)$

2. Now notice that $f(A)$ is a subset of the codomain, $\mathbb{R}$. Consequently, $f^{-1}(f(A))$ is a subset of the domain, $\mathbb{R}$. Is there any relation between A and $f^{-1}(f(A))$ in this case?

3. Now notice that $f^{-1}(C)$ is a subset of the domain, $\mathbb{R}$. Consequently, $f(f^{-1}(C))$ is a subset of the codomain, $\mathbb{R}$. Is there any relation between C and $f(f^{-1}(C))$ in this case?

The examples in Activity 9.3 and Activity 9.4 were meant to illustrate general results about how functions act on sets. In particular, we investigated how the action of a function on sets interacts with the set operations of intersection and union. We will now state the theorems that these examples were meant to illustrate. Some of the proofs will be left as exercises.

Theorem 9.5. *Let $f : S \to T$ be a function and let A and B be subsets of S. Then*

  1. $f(A \cap B) \subseteq f(A) \cap f(B)$

  2. $f(A \cup B) = f(A) \cup f(B)$

Proof. We will prove Part (1). The proof of Part (2) is included in the exercises.

Assume that $f : S \to T$ is a function and let A and B be subsets of S. We will prove that $f(A \cap B) \subseteq f(A) \cap f(B)$ by proving that for all $y \in T$, if $y \in f(A \cap B)$, then $y \in f(A) \cap f(B)$.

We assume that $y \in f(A \cap B)$. This means that there exists an $x \in A \cap B$ such that $f(x) = y$. Since $x \in A \cap B$, we conclude that $x \in A$ and $x \in B$.

  • Since $x \in A$ and $f(x) = y$, we conclude that $y \in f(A)$.

- Since $x \in B$ and $f(x) = y$, we conclude that $y \in f(B)$.

Since $y \in f(A)$ and $y \in f(B)$, $y \in f(A) \cap f(B)$. This proves that if $y \in f(A \cap B)$, then $y \in f(A) \cap f(B)$. Hence $f(A \cap B) \subseteq f(A) \cap f(B)$. ∎

Theorem 9.6. *Let $f : S \to T$ be a function and let C and D be subsets of T. Then:*

1. $f^{-1}(C \cap D) = f^{-1}(C) \cap f^{-1}(D)$

2. $f^{-1}(C \cup D) = f^{-1}(C) \cup f^{-1}(D)$

Proof. We will prove Part (2). The proof of Part (1) is included in the exercises.

Assume that $f : S \to T$ is a function and that C and D are subsets of T. We will prove that $f^{-1}(C \cup D) = f^{-1}(C) \cup f^{-1}(D)$ by proving that each set is a subset of the other.

We start by letting x be an element of $f^{-1}(C \cup D)$. This means that $f(x)$ is an element of $C \cup D$. Hence,

$$f(x) \in C \text{ or } f(x) \in D.$$

In the case where $f(x) \in C$, we conclude that $x \in f^{-1}(C)$, and hence that $x \in f^{-1}(C) \cup f^{-1}(D)$. Likewise, in the case where $f(x) \in D$, we see that $x \in f^{-1}(D)$, and hence that $x \in f^{-1}(C) \cup f^{-1}(D)$. So, in both cases, $x \in f^{-1}(C) \cup f^{-1}(D)$, and we have proven that $f^{-1}(C \cup D) \subseteq f^{-1}(C) \cup f^{-1}(D)$.

We now let $t \in f^{-1}(C) \cup f^{-1}(D)$. This means that

$$t \in f^{-1}(C) \text{ or } t \in f^{-1}(D).$$

- In the case where $t \in f^{-1}(C)$, we conclude that $f(t) \in C$ and hence that $f(t) \in C \cup D$. This means that $t \in f^{-1}(C \cup D)$.

- Similarly, when $t \in f^{-1}(D)$, it follows that $f(t) \in D$ and hence that $f(t) \in C \cup D$. This means that $t \in f^{-1}(C \cup D)$.

These two cases prove that if $t \in f^{-1}(C) \cup f^{-1}(D)$, then $t \in f^{-1}(C \cup D)$. Therefore, $f^{-1}(C) \cup f^{-1}(D) \subseteq f^{-1}(C \cup D)$.

Since we have now proven that each of the two sets is a subset of the other set, we can conclude that $f^{-1}(C \cup D) = f^{-1}(C) \cup f^{-1}(D)$. ∎

Theorem 9.7. *Let $f : S \to T$ be a function and let A be a subset of S and let C be a subset of T. Then*

   1. $A \subseteq f^{-1}(f(A))$

   2. $f(f^{-1}(C)) \subseteq C$

Proof. We will prove Part (1). The proof of Part (2) is included in the exercises.

To prove Part (1), we need to prove that for all $a \in S$, if $a \in A$, then $a \in f^{-1}(f(A))$. So let $a \in A$. Then, by definition, $f(a) \in f(A)$.

Now, $f(A) \subseteq T$, and so $f^{-1}(f(A)) \subseteq S$. Notice that

$$f^{-1}(f(A)) = \{x \in S \mid f(x) \in f(A)\}.$$

Since $f(a) \in f(A)$, we use this to conclude that $a \in f^{-1}(f(A))$. This proves that if $a \in A$, then $a \in f^{-1}(f(A))$, and hence that $A \subseteq f^{-1}(f(A))$. ∎

Exercises 9.1

1. Let $f : S \to T$, let A and B be subsets of S, and let C and D be subsets of T. For $x \in S$ and $y \in T$, arefully explain what it means to say that

    (a) $y \in f(A \cap B)$ **(e)** $x \in f^{-1}(C \cap D)$

    (b) $y \in f(A \cup B)$ **(f)** $x \in f^{-1}(C \cup D)$

    (c) $y \in f(A) \cap f(B)$ **(g)** $x \in f^{-1}(C) \cap f^{-1}(D)$

    (d) $y \in f(A) \cup f(B)$ **(h)** $x \in f^{-1}(C) \cup f^{-1}(D)$

2. Let $f : \mathbb{R} \to \mathbb{R}$ by $f(x) = -2x + 1$. Let

    $A = [2, 5]$ $B = [-1, 3]$ $C = [-2, 3]$ $D = [1, 4]$.

    Find each of the following:

    (a) $f(A)$ **(d)** $f(f^{-1}(C))$

    (b) $f^{-1}(f(A))$ **(e)** $f(A \cap B)$

    (c) $f^{-1}(C)$ **(f)** $f(A) \cap f(B)$

(g) $f^{-1}(C \cap D)$ **(h)** $f^{-1}(C) \cap f^{-1}(D)$

3. Let $g : \mathbb{N} \times \mathbb{N} \to \mathbb{N}$ by $g(m, n) = 2^m 3^n$, let $A = \{1, 2, 3\}$, and let $C = \{1, 4, 6, 9, 12, 16, 18\}$. Find

(a) $g(A \times A)$ **(c)** $g^{-1}(g(A \times A))$

(b) $g^{-1}(C)$ **(d)** $g\left(g^{-1}(C)\right)$

4. **(a)** Let $S = \{1, 2, 3, 4\}$. Define $F : S \to \mathbb{N}$ by $F(x) = x^2$ for each $x \in S$. What is the range of the function F and what is $F(S)$? How do these two sets compare?

Now, let A and B be sets and let $f : A \to B$ be an arbitrary function from A to B.

(b) Explain why $f(A) = \text{range}(f)$.

(c) Define a function $g : A \to f(A)$ by $g(x) = f(x)$ for all x in A. Prove that the function g is a surjection.

5. Prove Part (2) of Theorem 9.5.

Let $f : S \to T$ be a function and let A and B be subsets of S. Then $f(A \cup B) = f(A) \cup f(B)$.

6. Prove Part (1) of Theorem 9.6.

Let $f : S \to T$ be a function and let C and D be subsets of T. Then $f^{-1}(C \cap D) = f^{-1}(C) \cap f^{-1}(D)$.

7. Prove Part (2) of Theorem 9.7.

Let $f : S \to T$ be a function and let $C \subseteq T$. Then $f\left(f^{-1}(C)\right) \subseteq C$.

8. Let $f : S \to T$ and let A and B be subsets of S. Prove or disprove each of the following:

(a) If $A \subseteq B$, then $f(A) \subseteq f(B)$.

(b) If $f(A) \subseteq f(B)$, then $A \subseteq B$.

9. Let $f : S \to T$ and let C and D be subsets of T. Prove or disprove each of the following:

(a) If $C \subseteq D$, then $f^{-1}(C) \subseteq f^{-1}(D)$.

(b) If $f^{-1}(C) \subseteq f^{-1}(D)$, then $C \subseteq D$.

10. Prove or disprove:

If $f : S \to T$ is a function and A and B are subsets of S, then $f(A) \cap f(B) \subseteq f(A \cap B)$.

<u>Note</u>: Part (1) of Theorem 9.5 states that $f(A \cap B) \subseteq f(A) \cap f(B)$.

11. Let $f : S \to T$ be a function, let $A \subseteq S$, and let $C \subseteq T$.

    (a) Part (1) of Theorem 9.7 states that $f^{-1}(f(A)) \subseteq A$. Give an example where $f^{-1}(f(A)) \subset A$.

    (b) Part (2) of Theorem 9.7 states that $C \subseteq f(f^{-1}(C))$. Give an example where $C \subset f(f^{-1}(C))$.

12. Is the following proposition true or false? Justify your conclusion with a proof or a counterexample.

If $f : S \to T$ is an injection and $A \subseteq S$, then $f^{-1}(f(A)) = A$.

13. Is the following proposition true or false? Justify your conclusion with a proof or a counterexample.

If $f : S \to T$ is a surjection and $C \subseteq T$, then $f(f^{-1}(C)) = C$.

14. Let $f : S \to T$. Prove that $f(A \cap B) = f(A) \cap f(B)$ for all subsets A and B of S if and only if f is an injection.

15. Let $f : S \to T$ and let A and B be subsets of S. Investigate the relationship between the sets $f(A - B)$ and $f(A) - f(B)$. Are the two sets equal? If not, is one a subset of the other? Are there any conditions on the function f that will ensure that the two sets are equal? Justify your conclusions.

9.2 Finite Sets

Preview Activity 1 (Equivalent Sets, Part 1).

1. Let $f : A \to B$. If necessary, review the material in Section 6.3 to complete the following:

    (a) The function f is an injection provided that ...

    (b) The function f is not an injection provided that ...

    (c) The function f is a surjection provided that ...

(**d**) The function f is not a surjection provided that ...

(**e**) The function f is a bijection provided that ...

Definition. Let A and B be subsets of a universal set U. The set A is **equivalent** to the set B provided that there exists a bijection from the set A onto the set B. In this case, we write $A \approx B$.

When $A \approx B$, we also say that the set A is in **one-to-one correspondence** with the set B and that the set A has the same **cardinality** as the set B.

<u>Note</u>: When A is not equivalent to B, we write $A \not\approx B$.

2. Use the definition of equivalent sets to justify your answers to the following questions:

(**a**) Is the set $A = \{1, 2, 3\}$ equivalent to the set $B = \{a, b, c\}$?

(**b**) Is the set $C = \{1, 2\}$ equivalent to the set $B = \{a, b, c\}$?

(**c**) Is the set $X = \{1, 2, 3, \ldots, 10\}$ equivalent to the set $Y = \{57, 58, 59, \ldots, 66\}$?

3. Let D^+ be the set of all odd natural numbers. Prove that the function $f : \mathbb{N} \to D^+$ defined by $f(x) = 2x - 1$, for all $x \in \mathbb{N}$, is a bijection and hence that $\mathbb{N} \approx D^+$.

4. Let r be an real number with $r > 0$. Let $(0, 1)$ and $(0, r)$ be the open intervals from 0 to 1 and 0 to r, respectively. Prove that the function $g : (0, 1) \to (0, r)$ by $g(x) = rx$, for all $x \in \mathbb{R}$, is a bijection and hence that $(0, 1) \approx (0, r)$.

Preview Activity 2 (Equivalent Sets, Part 2).

1. Review Theorem 6.18 in Section 6.4, Theorem 6.24 in Section 6.5, and Exercise (8) in Section 6.5.

2. Review the definitions of a reflexive relation on a set, a symmetric relation, a transitive relation, and an equivalence relation on a set in Section 7.2.

3. Now let U be a universal set and let A, B, and C be subsets of U. That is, A, B, and C are elements of the power set of U, which is denoted by $\mathcal{P}(U)$. Prove that $\approx$ is an equivalence relation on $\mathcal{P}(U)$ by proving each of the following three propositions.

    (a) Prove that for all $A \in \mathcal{P}(U)$, $A \approx A$. That is, prove that $\approx$ is a reflexive relation on $\mathcal{P}(U)$.

    (b) Prove that for all $A, B \in \mathcal{P}(U)$, if $A \approx B$, then $B \approx A$. That is, prove that $\approx$ is a symmetric relation.

    (c) Prove that for all $A, B, C \in \mathcal{P}(U)$, if $A \approx B$ and $B \approx C$, then $A \approx C$. That is, prove that $\approx$ is a transitive relation.

Equivalent Sets

In Preview Activity 1, we introduced the concept of equivalent sets. The motivation for this definition was to have a formal method for determining whether or not two sets "have the same number of elements." This idea was described in terms of a one-to-one correspondence (a bijection) from one set onto the other set. This idea may seem simple for finite sets, but as we will see, this idea has surprising consequences when we deal with infinite sets. (We will soon provide precise definitions for finite and infinite sets.)

Activity 9.8 (Examples of Equivalent Sets).

    Recall that the set A is **equivalent** to the set B provided that there exists a bijection from the set A onto the set B. In this case, we say that **A is equivalent to B** and write $A \approx B$.

    In Preview Activity 2, we proved that $\approx$ is an equivalence relation. Consequently, if $A \approx B$, we often say that A and B are **equivalent sets**.

1. Let $A = \{1, 2, 3\}$ and let $B = \{a, b, c\}$. To prove that $A \approx B$, we must prove that there exists a bijection from one set to the other set. Give an example of a bijection from A to B. (<u>Note</u>: At this point, we cannot simply say that the two sets have the same number of elements.)

2. Let $A = \{1, 2, 3, \ldots, 99, 100\}$ and let $B = \{351, 352, 353, \ldots, 449, 450\}$. Give an example of a bijection from the set A to the set B to prove that $A \approx B$.

3. Let E be the set of all even integers and let D be the set of all odd integers. Prove that $E \approx D$ by proving that the function $F : E \to D$, where $F(x) = x + 1$, for all $x \in E$, is a bijection.

4. Let $(0, 1)$ be the open interval of real numbers between 0 and 1. Similarly, if $b \in \mathbb{R}$ with $b > 0$, let $(0, b)$ be the open interval of real numbers between 0 and b.

Prove that $f : (0, 1) \rightarrow (0, b)$ by $f(x) = bx$, for all $x \in (0, 1)$, is a bijection and hence $(0, 1) \approx (0, b)$.

In Part (4) of Activity 9.8, notice that if $b > 1$, then $(0, 1)$ is a proper subset of $(0, b)$ and $(0, 1) \approx (0, b)$.

Also, in Part (3) of Preview Activity 1, we proved that the set D of all odd natural numbers is equivalent to $\mathbb{N}$, and we know that D is a proper subset of $\mathbb{N}$.

Although we have not defined the terms yet, we will see that one thing that will distinguish an infinite set from a finite set is that an infinite set can be equivalent to one of its proper subsets, whereas a finite set cannot be equivalent to one of its proper subsets.

Finite Sets

In Section 4.1, we defined the **cardinality** of a finite set A, denoted by $|A|$, to be the number of elements in the set A. Now that we know about functions and bijections, we can define this concept more formally and more rigorously. First, for each $k \in \mathbb{N}$, we define $\mathbb{N}_k$ to be the set

$$\mathbb{N}_k = \{1, 2, \ldots, k\}.$$

We will use the concept of **equivalent sets** introduced in Preview Activity 1 to define a finite set.

Definition. A set A is a **finite set** provided that $A = \emptyset$ or there exists a natural number k such that $A \approx \mathbb{N}_k$. A set is an **infinite set** provided that it is not a finite set.

If $A \approx \mathbb{N}_k$, we say that the set A has **cardinality k** (or **cardinal number k**), and we write card $(A) = k$.

In addition, we say that the empty set has **cardinality 0** (or **cardinal number 0**), and we write card $(\emptyset) = 0$.

Notice that by this definition, the empty set is a finite set. In addition, for each $k \in \mathbb{N}$, the identity function on $\mathbb{N}_k$ is a bijection and hence, by definition, the set $\mathbb{N}_k$ is a finite set with cardinality k.

Theorem 9.9. *Any set equivalent to a finite nonempty set A is a finite set and has the same cardinality as A.*

Proof. Suppose that A is a finite nonempty set, B is a set, and $A \approx B$. Then there exists a $k \in \mathbb{N}$ such that $A \approx \mathbb{N}_k$. We also have assumed that $A \approx B$ and since $\approx$ is an equivalence relation, we can use symmetry to conclude that $B \approx A$. Since $A \approx \mathbb{N}_k$, we can use the transitivity of $\approx$ to conclude that $B \approx \mathbb{N}_k$. Thus, B is finite and has the same cardinality as A. ∎

It may seem that we have done a lot of work to prove the "obvious" result in Theorem 9.9. The same may be true of the remaining results in this section, which give further results about finite sets. One of the goals is to make sure that the concept of cardinality for a finite set corresponds to our intuitive notion of the number of elements in the set. As important, another goal is to lay the groundwork for a more rigorous and mathematical treatment of infinite sets than we have encountered before. Along the way, we will see the mathematical distinction between finite and infinite sets.

The following two lemmas will be used to prove the theorem that states that every subset of a finite set is finite.

Lemma 9.10. *If A is a finite set and $x \notin A$, then $A \cup \{x\}$ is a finite set and $card\,(A \cup \{x\}) = card\,(A) + 1$.*

Proof. Let A be a finite set and assume $card\,(A) = k$, where $k = 0$ or $k \in \mathbb{N}$. Assume $x \notin A$.

If $A = \emptyset$, then $card\,(A) = 0$ and $A \cup \{x\} = \{x\}$, which is equivalent to $\mathbb{N}_1$. Thus, $A \cup \{x\}$ is finite with cardinality 1, which equals $card\,(A) + 1$.

If $A \neq \emptyset$, then $A \approx \mathbb{N}_k$, for some $k \in \mathbb{N}$. This means that $card\,(A) = k$, and there exists a bijection $f : A \to \mathbb{N}_k$. We will now use this bijection to define a function $g : A \cup \{x\} \to \mathbb{N}_{k+1}$ and then prove that the function g is a bijection.

So we define $g : A \cup \{x\} \to \mathbb{N}_{k+1}$ as follows: For each $t \in A \cup \{x\}$,

$$g\,(t) = \begin{cases} f\,(t) & \text{if } t \in A \\ k+1 & \text{if } t = x. \end{cases}$$

To prove that g is an injection, we let $x_1, x_2 \in A \cup \{x\}$ and assume $x_1 \neq x_2$.

- If $x_1, x_2 \in A$, then since f is a bijection, $f\,(x_1) \neq f\,(x_2)$, and this implies that $g\,(x_1) \neq g\,(x_2)$.

- If $x_1 = x$, then since $x_2 \neq x_1$, we conclude that $x_2 \neq x$ and hence $x_2 \in A$. So $g(x_1) = k + 1$, and since $f(x_2) \in \mathbb{N}_k$ and $g(x_2) = f(x_2)$, we can conclude that $g(x_1) \neq g(x_2)$.

This proves that the function g is an injection. The proof that g is a surjection is left as an exercise.

Since g is a bijection, we conclude that $A \cup \{x\} \approx \mathbb{N}_{k+1}$, and

$$\operatorname{card}(A \cup \{x\}) = k + 1.$$

Since $\operatorname{card}(A) = k$, we have proven that $\operatorname{card}(A \cup \{x\}) = \operatorname{card}(A) + 1$. ∎

Lemma 9.11. *For each natural number m, if $A \subseteq \mathbb{N}_m$, then A is finite and $\operatorname{card}(A) \leq m$.*

Proof. We will use a proof using induction on m. For each $m \in \mathbb{N}$, let $P(m)$ be, "if $A \subseteq \mathbb{N}_m$, then A is finite and $\operatorname{card}(A) \leq m$."

We first prove that $P(1)$ is true. If $A \subseteq \mathbb{N}_1$, then $A = \emptyset$ or $A = \{1\}$, both of which are finite and have cardinality less than or equal to the cardinality of $\mathbb{N}_1$. This proves that $P(1)$ is true.

For the inductive step, let $k \in \mathbb{N}$ and assume that $P(k)$ is true. That is, assume that if $A \subseteq \mathbb{N}_k$, then A is finite and $\operatorname{card}(A) \leq k$. We need to prove that $P(k+1)$ is true.

So assume that A is a subset of $\mathbb{N}_{k+1}$. Then $A - \{k+1\}$ is a subset of $\mathbb{N}_k$. Since $P(k)$ is true, $A - \{k+1\}$ is finite and

$$\operatorname{card}(A - \{k+1\}) \leq k.$$

There are two cases to consider: Either $k + 1 \in A$ or $k + 1 \notin A$.

If $k + 1 \notin A$, then $A = A - \{k+1\}$. Hence, A is finite and

$$\operatorname{card}(A) \leq k < k + 1.$$

If $k + 1 \in A$, then $A = (A - \{k+1\}) \cup \{k+1\}$. Hence, by Lemma 9.10, A is finite and

$$\operatorname{card}(A) = \operatorname{card}((A - \{k+1\})) + 1.$$

Since $\operatorname{card}(A - \{k+1\}) \leq k$, we can conclude that $\operatorname{card}(A) \leq k + 1$.

This means that we have proven the inductive step. Hence, by mathematical induction, for each $m \in \mathbb{N}$, if $A \subseteq \mathbb{N}_m$, then A is finite and $\operatorname{card}(A) \leq m$. ∎

The preceding two lemmas were proven to aid in the proof of the following theorem.

Theorem 9.12. *If S is a finite set and A is a subset of S, then A is finite and $card\,(A) \leq card\,(S)$.*

Proof. Let S be a finite set and assume that A is a subset of S. If $A = \emptyset$, then A is a finite set and $\text{card}\,(A) \leq \text{card}\,(S)$. So we assume that $A \neq \emptyset$.

Since S is finite, there exists a bijection $f : S \to \mathbb{N}_k$ for some $k \in \mathbb{N}$. In this case, $\text{card}\,(S) = k$. We need to show that A is equivalent to a finite set. To do this, we define $g : A \to f\,(A)$ by

$$g\,(x) = f\,(x) \text{ for each } x \in A.$$

Since f is an injection, we conclude that g is an injection. Now, let $y \in f\,(A)$. Then, there exists an $a \in A$ such that $f\,(a) = y$. But by the definition of g, this means that $g\,(a) = y$, and hence g is a surjection. This proves that g is a bijection.

Hence, we have proven that $A \approx f\,(A)$. Since $f\,(A)$ is a subset of $\mathbb{N}_k$, we use Lemma 9.11 to conclude that $f\,(A)$ is finite and $\text{card}\,(f\,(A)) \leq k$. In addition, by Theorem 9.9, A is a finite set and $\text{card}\,(A) = \text{card}\,(f\,(A))$. This proves that A is a finite set and $\text{card}\,(A) \leq \text{card}\,(S)$. ∎

Lemma 9.10 implies that adding one element to a finite set increases its cardinality by 1. It is also true that removing one element from a finite non-empty set reduces the cardinality by 1. The proof of Corollary 9.13 is left as Exercise (3).

Corollary 9.13. *If A is a finite set and $x \in A$, then $A - \{x\}$ is a finite set and $card\,(A - \{x\}) = card\,(A) - 1$.*

Corollary 9.14. *A finite set is not equivalent to any of its proper subsets.*

Proof. Let B be a finite set and assume that A is a proper subset of B. Since A is a proper subset of B, there exists an element x in $B - A$. This means that A is a subset of $B - \{x\}$. Hence, by Theorem 9.12,

$$\text{card}\,(A) \leq \text{card}\,(B - \{x\}).$$

Also, by Corollary 9.13

$$\text{card}\,(B - \{x\}) = \text{card}\,(B) - 1.$$

Hence, we may conclude that $\text{card}\,(A) \le \text{card}\,(B) - 1$ and that

$$\text{card}\,(A) < \text{card}\,(B)\,.$$

Theorem 9.9 implies that $B \not\approx A$. This proves that a finite set is not equivalent to any of its proper subsets. ∎

The Pigeonhole Principle

The last property of finite sets that we will consider in this section is often called the **Pigeonhole Principle**. The "pigeonhole" version of this property says, "If m pigeons go into r pigeonholes and $m > r$, then at least one pigeonhole has more than one pigeon."

In this situation, we can think of the set of pigeons as being equivalent to a set P with cardinality m and the set of pigeonholes as being equivalent to a set H with cardinality r. We can then define a function $f : P \to H$ that maps each pigeon to its pigeonhole. The Pigeonhole Principle states that this function is not an injection. (It is not one-to-one since there are at least two pigeons "mapped" to the same pigeonhole.)

Theorem 9.15 (The Pigeonhole Principle). *Let A and B be finite sets. If*
$\text{card}\,(A) > \text{card}\,(B)$, then any function $f : A \to B$ is not an injection.

Proof. Let A and B be finite sets. We will prove the contrapositive of the theorem, which is, if there exists a function $f : A \to B$ that is an injection, then $\text{card}\,(A) \le \text{card}\,(B)$.

So assume that $f : A \to B$ is an injection. As in Theorem 9.12, we define a function $g : A \to f\,(A)$ by

$$g\,(x) = f\,(x) \text{ for each } x \in A.$$

As we saw in Theorem 9.12, the function g is a bijection. But then $A \approx f\,(A)$ and $f\,(A) \subseteq B$. Hence,

$$\text{card}\,(A) = \text{card}\,(f\,(A)) \text{ and } \text{card}\,(f\,(A)) \le \text{card}\,(B).$$

Hence, $\text{card}\,(A) \le \text{card}\,(B)$, and this proves the contrapositive. Hence, if $\text{card}\,(A) > \text{card}\,(B)$, then any function $f : A \to B$ is not an injection. ∎

The Pigeonhole Principle has many applications in the branch of mathematics called "combinatorics." Some of these will be explored in the next activity and in the exercises.

Activity 9.16 (Using the Pigeonhole Principle). For this activity, we will consider subsets of $\mathbb{N}_{30}$ that contain eight elements.

1. One such set is $A = \{3, 5, 11, 17, 21, 24, 26, 29\}$. Notice that

$$\{3, 21, 24, 26\} \subseteq A \quad \text{and} \quad 3 + 21 + 24 + 26 = 74$$
$$\{3, 5, 11, 26, 29\} \subseteq A \quad \text{and} \quad 3 + 5 + 11 + 26 + 29 = 74.$$

   Use this information to find two disjoint subsets of A whose elements have the same sum.

2. Let $B = \{3, 6, 9, 12, 15, 18, 21, 24\}$. Find two disjoint subsets of B whose elements have the same sum. <u>Note</u>: By convention, if $T = \{a\}$, where $a \in \mathbb{N}$, then the sum of the elements in T is equal to a.

3. Now let C be any subset of $\mathbb{N}_{30}$ that contains eight elements.

   (a) How many subsets does C have?

   (b) The sum of the elements of the $\emptyset$ is 0. What is the maximum sum for any subset of C? Let M be this maximum sum.

   <u>Hint</u>: Make the elements of a subset of $\mathbb{N}_{30}$ with 8 elements as large as possible.

   (c) Now define a function $f : \mathcal{P}(C) \to \mathbb{N}_M$ so that for each $X \in \mathcal{P}(C)$, $f(X)$ is equal to the sum of the elements in X.

   Use the Pigeonhole Principle to prove that there exist two subsets of C whose elements have the same sum.

4. If the two subsets in Part (3c) are not disjoint, use the idea presented in Part (1) to prove that there exist two disjoint subsets of C whose elements have the same sum.

Exercises 9.2

1. Let A be a subset of some universal set U. Prove that if $x \in U$, then $A \times \{x\} \approx A$.

2. Let E^+ be the set of all even natural numbers. Prove that $\mathbb{N} \approx E^+$.

3. Prove Corollary 9.13.

   If A is a finite set and $x \in A$, then $A - \{x\}$ is a finite set and $\text{card}(A - \{x\}) = \text{card}(A) - 1$.

   <u>Hint</u>: One approach is to use the fact that $A = (A - \{x\}) \cup \{x\}$.

4. Let A and B be sets. Prove that

    (a) If A is a finite set, then $A \cap B$ is a finite set.

    (b) If $A \cup B$ is a finite set, then A and B are finite sets.

    (c) If $A \cap B$ is an infinite set, then A is an infinite set.

    (d) If A is an infinite set or B is an infinite set, then $A \cup B$ is an infinite set.

5. There are over 7 million people living in New York City. It is also known that the maximum number of hairs on a human head is less than 200,000. Use the Pigeonhole Principle to prove that there are at least two people in the city of New York with the same number of hairs on their heads.

6. Let S be a subset of $\mathbb{N}_{99}$ that contains 10 elements. Use the Pigeonhole Principle to prove that there exist two disjoint subsets of S whose elements have the same sum.

7. Prove the following propositions:

    (a) If A, B, C, and D are sets with $A \approx B$ and $C \approx D$, then $A \times C \approx B \times D$.

    (b) If A, B, C, and D are sets with $A \approx B$ and $C \approx D$ and if A and C are disjoint and B and D are disjoint, then $A \cup C \approx B \cup D$.

Hint: Since $A \approx B$ and $C \approx D$, there exist bijections $f : A \to B$ and $g : C \to D$. To prove that $A \times C \approx B \times D$, prove that $h : A \times C \to B \times D$ is a bijection, where $h(a, c) = (f(a), g(c))$, for all $(a, c) \in A \times C$.

If $A \cap C = \emptyset$ and $B \cap D = \emptyset$, then to prove that $A \cup C \approx B \cup D$, prove that the following function is a bijection: $k : A \cup C \to B \cup D$, where

$$k(x) = \begin{cases} f(x) & \text{if } x \in A \\ g(x) & \text{if } x \in C \end{cases}$$

8. Let $A = \{a, b, c\}$.

    (a) Construct a function $f : \mathbb{N}_5 \to A$ such that f is a surjection.

    (b) Use the function f to construct a function $g : A \to \mathbb{N}_5$ so that $f \circ g = I_A$, where I_A is the identity function on the set A. Is the function g an injection? Explain.

9. This exercise is a generalization of Exercise (8). Let m be a natural number, let A be a set and assume that $f : \mathbb{N}_m \to A$ is a surjection. Define $g : A \to \mathbb{N}_m$ as follows:

> For each $x \in A$, $g(x) = j$, where j is the least natural number in $f^{-1}(\{x\})$.

Prove that $f \circ g = I_A$, where I_A is the identity function on the set A and prove that g is an injection.

10. Let B be a finite, nonempty set and assume that $f : B \to A$ is a surjection. Prove that there exists a function $h : A \to B$ such that $f \circ h = I_A$ and h is an injection.

> Hint: Since B is finite, there exists a natural number m such that $\mathbb{N}_m \approx B$. This means there exists a bijection $k : \mathbb{N}_m \to B$. Now, let $h = k \circ g$, where g is the function constructed in Exercise (9).

9.3 Countable Sets

Preview Activity 1 (Introduction to Infinite Sets).
In Section 9.2, we defined a **finite set** to be the empty set or a set A such that $A \approx \mathbb{N}_k$ for some natural number k. We also defined an **infinite set** to be a set that is not finite.

One question is, "How do we know if a set is infinite?" One way to answer this is to use Corollary 9.14, which states that a finite set is not equivalent to any of its subsets. We can write this as a conditional statement as follows:

> If A is a finite set, then A is not equivalent to any of its proper subsets.

1. Write the contrapositive of the preceding conditional statement. Then explain how this statement can be used to determine if a set is infinite.

2. Let D^+ be the set of all odd natural numbers. In Preview Activity 1 from Section 9.2, we proved that $\mathbb{N} \approx D^+$.

   (a) Use this to explain carefully why $\mathbb{N}$ is an infinite set.
   (b) Is D^+ a finite set or an infinite set? Carefully explain how you know.

3. Let r be an real number with $r > 0$. Let $(0, 1)$ and $(0, r)$ be the open intervals from 0 to 1 and 0 to r, respectively. In Preview Activity 1 from Section 9.2, we proved that $(0, 1) \approx (0, r)$.

(a) Use a case where $0 < r < 1$ to explain why $(0, 1)$ is an infinite set.

(b) Use a case where $r > 1$ to explain why $(0, r)$ is an infinite set.

Preview Activity 2 (A function from $\mathbb{N}$ to $\mathbb{Z}$).

In this preview activity, we will define and explore a function $f : \mathbb{N} \to \mathbb{Z}$. We will start by defining $f(n)$ for the first few natural numbers n.

$$f(1) = 0$$
$$f(2) = 1 \qquad\qquad f(3) = -1$$
$$f(4) = 2 \qquad\qquad f(5) = -2$$
$$f(6) = 3 \qquad\qquad f(7) = -3$$

Notice that if we list the outputs of f in the order $f(1), f(2), f(3), \ldots,$ we create the following list of integers:

$$0, 1, -1, 2, -2, 3, -3, \ldots.$$

1. If the pattern suggested by the functional values we have defined continues, what is $f(8)$ and $f(9)$? What is $f(n)$ for n from 10 to 13?

2. If the pattern of outputs continues, does the function f appear to be an injection? Does f appear to be a surjection? (Formal proofs are not required.)

We will now attempt to determine a formula for $f(n)$ where $n \in \mathbb{N}$. We will actually determine two formulas: one for when n is even and one for when n is odd.

3. Look at the pattern of the values of $f(n)$ when n is even. What appears to be a formula for $f(n)$ when n is even?

4. Look at the pattern of the values of $f(n)$ when n is odd. What appears to be a formula for $f(n)$ when n is odd?

5. Use the work in Part (3) and Part (4) to complete the following: Define $f : \mathbb{N} \to \mathbb{Z}$, where

$$f(n) = \begin{cases} \text{?? if } n \text{ is even} \\ \\ \text{?? if } n \text{ is odd} \end{cases}$$

6. Use the formula in Part (5) to

    (a) Calculate $f(1)$ through $f(10)$. Are these results consistent with the pattern exhibited at the beginning of this preview activity?

    (b) Calculate $f(1000)$ and $f(1001)$.

    (c) Determine the value of n so that $f(n) = 1000$.

In this section, we will describe several infinite sets and define the cardinal number for so-called countable sets. Most of our examples will be subsets of some of our standard numbers systems such as $\mathbb{N}$, $\mathbb{Z}$, and $\mathbb{Q}$.

In Preview Activity 1, we saw how to use Corollary 9.14 to prove that a set is infinite. This corollary implies that if A is a finite set, then A is not equivalent to any of its proper subsets. By writing the contrapositive of this conditional statement, we can restate Corollary 9.14 in the following form:

Corollary 9.14 *If a set A is equivalent to one of its proper subsets, then A is infinite.*

Example 9.17. In Preview Activity 1, we used Corollary 9.14 to prove that

- The set of natural numbers, $\mathbb{N}$, is an infinite set.

- The open interval $(0, 1)$ is an infinite set.

Theorem 9.18. *Let A and B be sets.*

    1. If A is infinite and $A \approx B$, then B is infinite.

    2. If A is infinite and $A \subseteq B$, then B is infinite.

Activity 9.19 (Proof of Theorem 9.18). Write a proof for both parts of Theorem 9.18. For both parts, use a proof by contradiction. For each proof, a theorem in Section 9.2 will provide a contradiction.

Example 9.20 (Examples of Infinite Sets).

- Theorem 9.18 allows us to conclude that our standard number systems are infinite sets since they all have $\mathbb{N}$ as a subset. So $\mathbb{Z}$, $\mathbb{Q}$, and $\mathbb{R}$ are all infinite sets. In addition, the set of all positive rational numbers, $\mathbb{Q}^+$, and the set of all positive real numbers, $\mathbb{R}^+$, are infinite sets.

- Let D^+ be the set of all odd natural numbers. In Part (2) of Preview Activity 1, we proved that $D^+ \approx \mathbb{N}$. Therefore, by Theorem 9.18, D^+ is an infinite set. In a similar manner, the set E^+ of all even natural numbers is an infinite set. To do this, we can use the function $g : \mathbb{N} \to E^+$ defined by $g(x) = 2x$ for all $x \in \mathbb{N}$.

In Section 9.2, we used the set $\mathbb{N}_k$ as the standard set with cardinality k in the sense that a set is finite if and only if it is equivalent to $\mathbb{N}_k$ for some k. In a similar manner, we will use some infinite sets as standard sets for certain infinite cardinal numbers. The first set we will use is $\mathbb{N}$.

We will formally define what it means to say the elements of a set can be "counted" using the natural numbers. The elements of a finite set can be "counted" by defining a bijection (one-to-one correspondence) between the set and $\mathbb{N}_k$ for some natural number k. We will be able to "count" the elements of an infinite set if we can define a one-to-one correspondence between the set and $\mathbb{N}$.

Definition. The **cardinality of** $\mathbb{N}$ is denoted by $\aleph_0$. The symbol $\aleph$ is the first letter of the Hebrew alphabet, **aleph**. The subscript 0 is often read as "naught" (or sometimes as "zero" or "null"). So, we write

$$\operatorname{card}(\mathbb{N}) = \aleph_0$$

and say that the cardinality of $\mathbb{N}$ is "aleph naught."

Definition. A set A is **countably infinite** provided that $A \approx \mathbb{N}$. In this case, we write

$$\operatorname{card}(A) = \aleph_0.$$

A set that is countably infinite is sometimes called a **denumerable** set. A set is **countable** provided that it is finite or countably infinite. Otherwise, the set is **uncountable**.

Example 9.21 (Examples of Countably Infinite Sets).

1. In Preview Activity 1 from Section 9.2, we proved that $\mathbb{N} \approx D^+$ where, D^+ is the set of all odd natural numbers. To do this, we used the

function $f : \mathbb{N} \to D^+$ defined by $f(x) = 2x + 1$, for all $x \in \mathbb{N}$. This proves that

$$\text{card}\left(D^+\right) = \aleph_0.$$

2. In Example 9.20, we proved that $\mathbb{N} \approx E^+$, where E^+ is the set of all even natural numbers. Thus,

$$\text{card}\left(E^+\right) = \aleph_0.$$

3. At this point, if we wish to prove a set S is countably infinite, we must find a bijection from S to some set that is known to be countably infinite or a bijection from a set that is known to be countably infinite to S. For example, if S is the set of all natural numbers that are perfect squares, then the function

$$f : S \to \mathbb{N} \text{ defined by } f(n) = \sqrt{n} \text{ for all } n \in S$$

is a bijection. Therefore, $S \approx \mathbb{N}$ and $\text{card}(S) = \aleph_0$.

The fact that the set of integers is a countably infinite set is important enough to be called a theorem. The function we will use to establish that $\mathbb{N} \approx \mathbb{Z}$ was explored in Preview Activity 2.

Theorem 9.22. *The set $\mathbb{Z}$ of integers is countably infinite, and so $card(\mathbb{Z}) = \aleph_0$.*

Proof. To prove that $\mathbb{N} \approx \mathbb{Z}$, we will use the following function: $f : \mathbb{N} \to \mathbb{Z}$. where

$$f(n) = \begin{cases} \dfrac{n}{2} & \text{if } n \text{ is even} \\[2ex] \dfrac{1-n}{2} & \text{if } n \text{ is odd.} \end{cases}$$

From our work in Preview Activity 2, it appears that if n is an even natural number, then $f(n) > 0$, and if n is an odd natural number, then $f(n) \leq 0$. So it seems reasonable to use cases to prove that f is a surjection and that f is an injection.

To prove that f is a surjection, we let $y \in \mathbb{Z}$.

- If $y > 0$, then $2y \in \mathbb{N}$, and

$$f(2y) = \frac{2y}{2} = y.$$

- If $y \leq 0$, then $-2y \geq 0$ and $1 - 2y$ is an odd natural number. Hence,

$$f(1 - 2y) = \frac{1 - (1 - 2y)}{2} = \frac{2y}{2} = y.$$

These two cases prove that if $y \in \mathbb{Z}$, then there exists an $n \in \mathbb{N}$ such that $f(n) = y$. Hence, f is a surjection.

To prove that f is an injection, we let $m, n \in \mathbb{N}$ and assume that $f(m) = f(n)$. First note that if one of m and n is odd and the other is even, then one of $f(m)$ and $f(n)$ is positive and the other is less than or equal to 0. So if $f(m) = f(n)$, then both m and n must be even or both m and n must be odd.

- If both m and n are even, then

$$f(m) = f(n) \text{ implies that } \frac{m}{2} = \frac{n}{2}$$

and hence that $m = n$.

- If both m and n are odd, then

$$f(m) = f(n) \text{ implies that } \frac{1 - m}{2} = \frac{1 - n}{2}.$$

From this, we conclude that $1 - m = 1 - n$ and hence that $m = n$. This proves that if $f(m) = f(n)$, then $m = n$ and hence that f is an injection.

Since f is both a surjection and an injection, we see that f is a bijection, and therefore, $\mathbb{N} \approx \mathbb{Z}$. Hence, $\mathbb{Z}$ is countably infinite. ∎

The result in Theorem 9.22 can seem a bit surprising. It exhibits one of the distinctions between finite and infinite sets. If we add elements to a finite set, we will increase its size in the sense that the new set will have a greater cardinality than the old set. However, with infinite sets, we can add elements and not increase the size of the set. For example, there is a one-to-one correspondence between the elements of the sets $\mathbb{N}$ and $\mathbb{Z}$. We say that these sets have the same cardinality.

Following is a summary of some of the main examples dealing with the cardinality of sets that we have explored.

- The sets $\mathbb{N}_k$, where $k \in \mathbb{N}$, are examples of sets that are countable and finite.

- The sets $\mathbb{N}$, $\mathbb{Z}$, the set of all odd natural numbers, and the set of all even natural numbers are examples of sets that are countable and countably infinite.

- Although we have worked with uncountable sets, we have not proven that a given set is uncountable.

If we expect to find an uncountable set in our usual number systems, the rational numbers might be the place to start looking. The following activity illustrates some important properties of the rational numbers.

Activity 9.23 (Rational Numbers between Rational Numbers). Let $a, b \in \mathbb{Q}$ with $a < b$.

1. Prove that $c = \dfrac{a + b}{2}$ is a rational number and that $a < c < b$. Explain why this proves that there is a rational number between any two (unequal) rational numbers.

2. Now define $c_1 = \dfrac{a + b}{2}$, and for each $k \in \mathbb{N}$, define

$$c_{k+1} = \frac{c_k + b}{2}.$$

Prove that for each $k \in \mathbb{N}$, $a < c_k < c_{k+1} < b$. Use this to explain why the set $\{c_k \mid k \in \mathbb{N}\}$ is a countably infinite set where each element is a rational number between a and b.

Notice that are only finitely many integers between any two given (unequal) integers. The result in Activity 9.23 is that there are infinitely many rational numbers between any two given (unequal) rational numbers. This might suggest that the set $\mathbb{Q}$ of rational numbers is uncountable. Surprisingly, this is not the case. We start with a proof that the set of positive rational numbers is countable.

Theorem 9.24. *The set of positive rational numbers is countably infinite.*

Proof. We can write all the positive rational numbers in a two-dimensional array as shown in Figure 9.2.

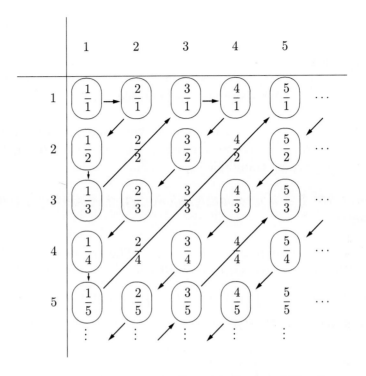

Figure 9.2: Counting the Positive Rational Numbers

The top row in Figure 9.2 represents the numerator of the rational number, and the left column represents the denominator. We follow the arrows in Figure 9.2 to define $f : \mathbb{N} \to \mathbb{Q}^+$. The idea is to start in the upper left corner of the table and move to successive diagonals as follows:

- We start with all fractions in which the sum of the numerator and denominator is 2 (only $\frac{1}{1}$). So $f(1) = \frac{1}{1}$.

- We next use those fractions in which the sum of the numerator and denominator is 3. So $f(2) = \frac{2}{1}$ and $f(3) = \frac{1}{2}$.

- We next use those fractions in which the sum of the numerator and denominator is 4. So $f(4) = \frac{1}{3}$, $f(5) = \frac{3}{1}$. We skipped $\frac{2}{2}$ since $\frac{2}{2} = \frac{1}{1}$. In this way, we will ensure that the function f is a one-to-one function.

We now continue with successive diagonals omitting fractions that are not in lowest terms. This process guarantees that the function f will be an injection and a surjection. Therefore, $\mathbb{N} \approx \mathbb{Q}^+$ and $\text{card}(\mathbb{Q}^+) = \aleph_0$. ■

Since $\mathbb{Q}^+$ is countable, it seems reasonable to expect that Q is countable. We will explore this soon. On the other hand, at this point, it may also seem reasonable to ask,

"Are there any uncountable sets?"

The answer to this question is yes, but we will wait until the next section to prove that certain sets are uncountable. We still have a few more issues to deal with concerning countable sets.

Theorem 9.25. *If A is a countably infinite set, then $A \cup \{x\}$ is a countably infinite set.*

Proof. Let A be a countably infinite set. The, there exists a bijection $f : \mathbb{N} \to A$. There are two cases to consider:

If $x \in A$, then $A \cup \{x\} = A$ and $A \cup \{x\}$ is countably infinite.

If $x \notin A$, define $g : \mathbb{N} \to A \cup \{x\}$ by

$$g(n) = \begin{cases} x & \text{if } n = 1 \\ f(n-1) & \text{if } n > 1. \end{cases}$$

The proof that the function g is a bijection is Exercise (3). Since g is a bijection, we have proven that $A \cup \{x\} \approx A$, and hence $A \cup \{x\}$ is countably infinite. ∎

Theorem 9.26. *If A is a countably infinite set and B is a finite set, then $A \cup B$ is a countably infinite set.*

Proof. Exercise (4). ∎

Theorem 9.26 says that if we add a finite number of elements to a countably infinite set, the resulting set is still countably infinite. In other words, the cardinality of the new set is the same of the cardinality of the original set. Finite sets behave very differently in the sense that if we add elements to a finite set, we will change the cardinality. What may even be more surprising than Theorem 9.26 is the result in Theorem 9.27.

Theorem 9.27. *If A and B are disjoint countably infinite sets, then $A \cup B$ is a countably infinite set.*

Proof. Let A and B be countably infinite sets and let $f : \mathbb{N} \to A$ and $g : \mathbb{N} \to B$ be bijections. Define $h : \mathbb{N} \to A \cup B$ by

$$
h(n) = \begin{cases} f\left(\dfrac{n+1}{2}\right) & \text{if } n \text{ is odd} \\[2ex] g\left(\dfrac{n}{2}\right) & \text{if } n \text{ is even} \end{cases}
$$

It is left as Exercise (5) to prove that the function h is a bijection. ∎

Theorem 9.28. *The set $\mathbb{Q}$ of all rational numbers is countable.*

Proof. Exercise (6). ∎

In Section 9.2, we proved that any subset of a finite set is finite (Theorem 9.12). A similar result should be expected for countable sets. We first prove that every subset of $\mathbb{N}$ is countable. For an infinite subset B of $\mathbb{N}$, the idea of the proof is to define a function $g : \mathbb{N} \to B$ as by removing the elements from B from smallest to the next smallest to the next smallest, and so on. We do this by defining the function g recursively as follows:

- Let $g(1)$ be the smallest natural number in B.

- Remove $g(1)$ from B and let $g(2)$ be the smallest natural number in $B - \{g(1)\}$.

- Remove $g(2)$ and let $g(3)$ be the smallest natural number in $B - \{g(1), g(2)\}$.

- We continue this process. The formal recursive definition of $g : \mathbb{N} \to B$ is included in the proof of Theorem 9.29.

Theorem 9.29. *Every subset of the natural numbers is countable.*

Proof. Let B be a subset of $\mathbb{N}$. If B is finite, then B is countable. So we next assume that B is infinite. We will next give a recursive definition of a function $g : \mathbb{N} \to B$ and then prove that g is a bijection.

- Let $g(1)$ be the smallest natural number in B.

- For each $n \in \mathbb{N}$, the set $B - \{g(1), g(2), \dots, g(n)\}$ is not empty since B is infinite. Define $g(n+1)$ to be the smallest natural number in $B - \{g(1), g(2), \dots, g(n)\}$.

The proof that the function g is a bijection is Exercise (10).   ∎

Corollary 9.30. *Every subset of a countable set is countable.*

Proof. Exercise (11).   ∎

Exercises 9.3

1. State whether each of the following is true or false.

    (a) If a set A is countably infinite, then A is infinite.

    (b) If a set A is countably infinite, then A is countable.

    (c) If a set A is uncountable, then A is not countably infinite.

    (d) If $A \approx \mathbb{N}_k$ for some $k \in \mathbb{N}$, then A is not countable.

2. Prove that each of the following sets is countably infinite.

    (a) The set F^+ of all natural numbers that are multiples of 5

    (b) The set F of all integers that are multiples of 5

    (c) $\left\{ \dfrac{1}{2^k} \mid k \in \mathbb{N} \right\}$ **(e)** $\mathbb{N} - \{4, 5, 6\}$

    (d) $\{n \in \mathbb{Z} \mid n \geq -10\}$ **(f)** $\{m \in \mathbb{Z} \mid m \equiv 2 \pmod{3}\}$

3. Complete the proof of Theorem 9.25 by proving the following:

Let A be a countably infinite set and $x \notin A$. If $f : \mathbb{N} \to A$ is a bijection, then g is a bijection, where $g : \mathbb{N} \to A \cup \{x\}$ by

$$g(n) = \begin{cases} x & \text{if } n = 1 \\ f(n-1) & \text{if } n > 1. \end{cases}$$

4. Prove Theorem 9.26.

If A is a countably infinite set and B is a finite set, then $A \cup B$ is a countably infinite set.

<u>Hint:</u> Let $\mathrm{card}\,(B) = n$ and use a proof by induction on n. Theorem 9.25 is the basis step.

5. Complete the proof of Theorem 9.27 by proving the following:

   Let A and B be disjoint countably infinite sets and let $f : \mathbb{N} \to A$ and $g : \mathbb{N} \to B$ be bijections. Define $h : \mathbb{N} \to A \cup B$ by

   $$h(n) = \begin{cases} f\left(\dfrac{n+1}{2}\right) & \text{if } n \text{ is odd} \\[2mm] g\left(\dfrac{n}{2}\right) & \text{if } n \text{ is even.} \end{cases}$$

   Then the function h is a bijection.

6. Prove Theorem 9.28.

   The set $\mathbb{Q}$ of all rational numbers is countable.

   Hint: Use Theorem 9.25 and Theorem 9.27.

7. Prove that if A is countably infinite and B is finite, then $A - B$ is countably infinite.

8. Define $f : \mathbb{N} \times \mathbb{N} \to \mathbb{N}$ as follows: For each $(m, n) \in \mathbb{N} \times \mathbb{N}$,

   $$f(m, n) = 2^{m-1}(2n - 1).$$

   (a) Prove that f is an injection. Hint: If $f(m, n) = f(s, t)$, there are three cases to consider: $m > s$, $m < s$, and $m = s$. Use laws of exponents to prove that the first two cases lead to a contradiction.

   (b) Prove that f is a surjection. Hint: You may use the fact that if $y \in \mathbb{N}$, then $y = 2^k x$, where x is an odd natural number and k is a non-negative integer. This is actually a consequence of the Fundamental Theorem of Arithmetic, Theorem 8.15. [See Exercise (13) in Section 8.2.]

   (c) Prove that $\mathbb{N} \times \mathbb{N} \approx \mathbb{N}$ and hence that $\operatorname{card}(\mathbb{N} \times \mathbb{N}) = \aleph_0$.

9. Use Exercise (8) to prove that if A and B are countably infinite sets, then $A \times B$ is a countably infinite set.

10. Complete the proof of Theorem 9.29 by proving that the function g defined in the proof is a bijection from $\mathbb{N}$ to B.

    Hints: To prove that g is an injection, it might be easier to prove that for all $r, s \in \mathbb{N}$, if $r \neq s$, then $g(r) \neq g(s)$. To do this, we may assume that $r < s$ since one of the two numbers must be less than the other. Then notice that $r \in \{g(1), g(2), \ldots, g(s-1)\}$.

To prove that g is a surjection, let $b \in B$ and notice that for some $k \in \mathbb{N}$, there will be k natural numbers in B that are less than b.

11. Prove Corollary 9.30, which states that every subset of a countable set is countable.

    Hint: Let S be a countable set and assume that $A \subseteq S$. There are two cases: A is finite or A is infinite. If A is infinite, let $f : S \to \mathbb{N}$ be a bijection and define $g : A \to f(S)$ by $g(x) = f(x)$, for each $x \in A$.

12. Use Corollary 9.30 to prove that the set of all rational numbers between 0 and 1 is countably infinite.

9.4   Uncountable Sets

Preview Activity 1 (The Game of Dodge Ball).
(From *The Heart of Mathematics: An Invitation to Effective Thinking* by Edward B. Burger and Michael Starbird, Key Publishing Company, ©2000 by Edward B. Burger and Michael Starbird.)

Dodge Ball is a game for two players. It is played on a game board such as the one shown in Figure 9.3.

Player One has a 6 by 6 array to complete and Player Two has a 1 by 6 row to complete. Each player has six turns as described below.

- Player One begins by filling in the first horizontal row of his or her table with a sequence of six X's and O's, one in each square in the first row.

- Then, Player Two places either an X or an O in the first box of his or her row. At this point, Player One has completed the first row and Player Two has filled in the first box of his or her row with one letter.

- The game continues with Player One completing a row with six letters (X's and O's), one in each box of the next row followed by Player Two writing one letter (an X or an O) in the next box of his or her row. The game is completed when Player One has completed all six rows and Player Two has completed all six boxes in his or her row.

Winning the Game

- Player One wins the game if any horizontal row in the 6 by 6 array is identical to the row that Player Two created. (Player One matches Player Two.)

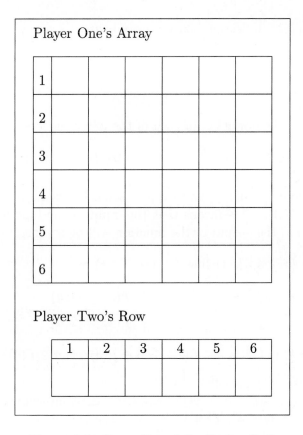

Figure 9.3: Game Board for Dodge Ball

- Player Two wins the game if Player Two's row of six letters is different than each of the six rows produced by Player One. (Player Two dodges Player One.)

There is a winning strategy for one of the two players. This means that there is plan by which one of the two players will always win. Which player has a winning strategy? Carefully describe this winning strategy.

Preview Activity 2 (Functions from a Set to Its Power Set).

Let A be a set. In Section 4.1, we defined the **power set** $\mathcal{P}(A)$ of A to be the set of all subsets of A. This means that

$$X \in \mathcal{P}(A) \text{ if and only if } X \subseteq A.$$

In Proposition 5.11 in Section 5.2, we proved that if a set A has n elements, then A has 2^n subsets or that $\mathcal{P}(A)$ has 2^n elements. Using our current notation for cardinality, this means that

$$\text{if card}(A) = n, \text{ then card}(\mathcal{P}(A)) = 2^n.$$

1. Determine the power set of each of the following sets:

   (a) $A = \{a, b\}$ **(b)** $B = \{a, b, c\}$

We are now going to define and explore some functions from a set A to its power set $\mathcal{P}(A)$. This means that the input of the function will be an element of A and the output of the function will be a subset of A.

2. Let $A = \{1, 2, 3, 4\}$. Define $f : A \to \mathcal{P}(A)$ by

$$f(1) = \{1, 2, 3\} \qquad\qquad f(3) = \{1, 4\}$$
$$f(2) = \{1, 3, 4\} \qquad\qquad f(4) = \{2, 4\}$$

   (a) Is $1 \in f(1)$? Is $2 \in f(2)$? Is $3 \in f(3)$? Is $4 \in f(4)$?
   (b) Determine $S = \{x \in A \mid x \notin f(x)\}$.
   (c) Notice that $S \in \mathcal{P}(A)$. Does there exist an element t in A such that $f(t) = S$? That is, is $S \in \text{range}(f)$?

3. Let $A = \{1, 2, 3, 4\}$. Define $f : A \to \mathcal{P}(A)$ by

$$f(x) = A - \{x\} \text{ for each } x \in A.$$

   (a) Determine $f(1)$. Is $1 \in f(1)$?
   (b) Determine $f(2)$. Is $2 \in f(2)$?
   (c) Determine $f(3)$. Is $3 \in f(3)$?
   (d) Determine $f(4)$. Is $4 \in f(4)$?

   (e) Determine $S = \{x \in A \mid x \notin f(x)\}$.
   (f) Notice that $S \in \mathcal{P}(A)$. Does there exist an element t in A such that $f(t) = S$? That is, is $S \in \text{range}(f)$?

4. Define $f : \mathbb{N} \to \mathcal{P}(\mathbb{N})$ by

$$f(n) = \mathbb{N} - \{n^2, n^2 - 2n\}, \text{ for each } n \in \mathbb{N}.$$

    (**a**) Determine $f(1)$, $f(2)$, $f(3)$, and $f(4)$. In each of these cases, determine if $k \in f(k)$.

    (**b**) Prove that if $n > 3$, then $n \in f(n)$. <u>Hint</u>: Prove that if $n > 3$, then $n^2 > n$ and $n^2 - 2n > n$.

    (**c**) Determine $S = \{x \in \mathbb{N} \mid x \notin f(x)\}$.

    (**d**) Notice that $S \in \mathcal{P}(\mathbb{N})$. Does there exist an element t in $\mathbb{N}$ such that $f(t) = S$? That is, is $S \in \text{range}(f)$?

We have seen examples of sets that are countably infinite, but we have not yet seen an example of an infinite set that is uncountable. We will do so in this section. The first example of an uncountable set will be the open interval of real numbers $(0, 1)$. The proof that this interval is uncountable uses a method similar to the winning strategy for Player Two in the game of Dodge Ball from Preview Activity 1. Before considering the proof, we need to review decimal expressions for real numbers.

Decimal Expressions for Real Numbers

In its decimal form, any real number a in the interval $(0, 1)$ can be written as $a = 0.a_1 a_2 a_3 a_4 \ldots$, where each a_i is an integer with $0 \le a_i \le 9$. For example,

$$\frac{5}{12} = 0.416666\ldots.$$

We often abbreviate this as $\dfrac{5}{12} = 0.41\bar{6}$ to indicate that the 6 is repeated. We can also repeat a block of digits. For example, $\dfrac{5}{26} = 0.19\overline{230769}$ to indicate that the block 230769 repeats. That is,

$$\frac{5}{26} = 0.19230769230769230769\ldots.$$

There is only one situation in which a real number can be represented as a decimal in more than one way. A decimal that ends with an infinite string of 9's is equal to one that ends with an infinite string of 0's. For example, $0.3199999\ldots$ represents the same real number as $0.3200000\ldots.$ Geometric series can be used to prove that a decimal that ends with an infinite string of 9's is equal to one that ends with an infinite string of 0's, but we will not do so here.

    We say that a decimal representation of real number a is in **normalized form** if and only if there is no k such that for all $n > k$, $a_n = 9$. That is, the decimal representation of a is in normalized form if and only if it does not end with an infinite string of 9's. One reason the normalized form is important is that

> *Two decimal numbers in normalized form are equal if and only if they have identical digits in each decimal position.*

Uncountable Subsets of $\mathbb{R}$

In the proof that follows, we will use only the normalized form for the decimal representations of real numbers in the interval $(0, 1)$.

Theorem 9.31. *The open interval $(0, 1)$ is an uncountable set.*

Proof. Since the interval $(0, 1)$ contains the infinite subset $\left\{\frac{1}{2}, \frac{1}{3}, \frac{1}{4}, \ldots\right\}$, we can use Theorem 9.18, to conclude that $(0, 1)$ is an infinite set. So $(0, 1)$ is either countably infinite or uncountable. We will show that $(0, 1)$ cannot be countably infinite.

Suppose that there is a function $f : \mathbb{N} \to (0, 1)$ that is an injection. We will show that f cannot be a surjection by showing that there exists an element in $(0, 1)$ that cannot be in the range of f. Writing the images of the elements of $\mathbb{N}$ in normalized form, we can write

$$f(1) = 0.a_{11}a_{12}a_{13}a_{14}a_{15}\ldots$$
$$f(2) = 0.a_{21}a_{22}a_{23}a_{24}a_{25}\ldots$$
$$f(3) = 0.a_{31}a_{32}a_{33}a_{34}a_{35}\ldots$$
$$f(4) = 0.a_{41}a_{42}a_{43}a_{44}a_{45}\ldots$$
$$f(5) = 0.a_{51}a_{52}a_{53}a_{54}a_{55}\ldots$$
$$\vdots$$
$$f(n) = 0.a_{n1}a_{n2}a_{n3}a_{n4}a_{n5}\ldots$$
$$\vdots$$

Notice the use of the double subscripts. The number a_{ij} is the jth digit to the right of the decimal point in the normalized decimal representation of $f(i)$.

We will now construct a real number $b = 0.b_1b_2b_3b_4b_5\ldots$ in $(0, 1)$ and in normalized form that is not in this list.

Informal Side Note: The idea is to start in the upper left corner and move down the diagonal in a manner similar to the winning strategy for Player

Two in the game in Preview Activity 1. At each step, we choose a digit that is not equal to the diagonal digit.

Start with a_{11} in $f(1)$. We want to choose b_1 so that $b_1 \neq a_{11}$ and $b_1 \neq 9$. (To ensure that we end up with a decimal that is in normalized form, we make sure that each digit is not equal to 9.) We then repeat this process with a_{22}, a_{33}, a_{44}, a_{55}, and so on.

We let b be the real number $b = 0.b_1b_2b_3b_4b_5\ldots$ where for each $k \in \mathbb{N}$

$$b_k = \begin{cases} 3 & \text{if } a_{kk} \neq 3 \\ 5 & \text{if } a_{kk} = 3 \end{cases}$$

Note: The choice of 3 and 5 is arbitrary. Other choices of distinct digits will also work.

Now, for each $n \in \mathbb{N}$, $b \neq f(n)$ since b and $f(n)$ are in normalized form and b and $f(n)$ differ in the nth decimal place.

This proves that any function from $\mathbb{N}$ to $(0,1)$ cannot be surjection and hence, there is no bijection from $\mathbb{N}$ to $(0,1)$. Therefore, $(0,1)$ is not countably infinite and hence must be an uncountable set. ■

The proof of Theorem 9.31 is often referred to as **Cantor's diagonal argument**. It is named after the mathematician Georg Cantor, who first published the proof in 1874.

Activity 9.32 (Dodge Ball and Cantor's Diagonal Argument).
Explain the connection between the winning strategy for Player Two in Dodge Ball (see Preview Activity 1) and the proof of Theorem 9.31 using Cantor's diagonal argument.

The open interval $(0,1)$ is our first example of an uncountable set. The cardinal number of $(0,1)$ is defined to be c, which stands for **the cardinal number of the continuum**. So the two infinite cardinal numbers we have seen are $\aleph_0$ for countably infinite sets and c.

Definition. A set A is said to have **cardinality** c provided that A is equivalent to $(0,1)$. In this case, we write $\text{card}(A) = c$ and say that the cardinal number of A is c.

The proof of the next theorem is included in Activity 9.34.

Theorem 9.33. *Let a and b be real numbers with $a < b$. The open interval (a, b) is uncountable and has cardinality $\mathbf{c}$.*

Activity 9.34 (Proof of Theorem 9.33).

1. In Part (4) of Preview Activity 1 in Section 9.2, we proved that if $r \in \mathbb{R}$ and $r > 0$, then the open interval $(0, 1)$ is equivalent to the open interval $(0, r)$. Now let a and b be real numbers with $a < b$. Find a function

$$f : (0, 1) \to (a, b)$$

   that is a bijection and conclude that $(0, 1) \approx (a, b)$.

   Hint: Find a linear function that passes through the points $(0, a)$ and $(1, b)$. Use this to define the function f. Make sure you prove that this function f is a bijection.

2. Let a, b, c, d be real numbers with $a < b$ and $c < d$. Prove that $(a, b) \approx (c, d)$.

Theorem 9.35. *The set of real numbers $\mathbb{R}$ is uncountable and has cardinality $\mathbf{c}$.*

Proof. Let $f : \left(-\dfrac{\pi}{2}, \dfrac{\pi}{2}\right) \to \mathbb{R}$ be defined by $f(x) = \tan x$, for each $x \in \mathbb{R}$. The function f is a bijection and hence, $\left(-\dfrac{\pi}{2}, \dfrac{\pi}{2}\right) \approx \mathbb{R}$. So by Theorem 9.33, $\mathbb{R}$ is uncountable and has cardinality $\mathbf{c}$. ∎

Cantor's Theorem

We have now seen two different infinite cardinal numbers, $\aleph_0$ and $\mathbf{c}$. It can seem surprising that there is more than one infinity. A reasonable question at this point is, "Are there any other infinite cardinal numbers?" The astonishing answer is that there are, and in fact, there are infinitely many different infinite cardinal numbers. The basis for this fact is the following theorem which states that a set is not equivalent to its power set. The proof is due to Georg Cantor (1845-1918).

Note: The idea for this proof was explored in Preview Activity 2.

Theorem 9.36 (Cantor's Theorem). *For every set A, A and $\mathcal{P}(A)$ do not have the same cardinality.*

Proof. Let A be a set. If $A = \emptyset$, then $\mathcal{P}(A) = \{\emptyset\}$, which has cardinality 1. Therefore, $\emptyset$ and $\mathcal{P}(\emptyset)$ do not have the same cardinality.

Suppose that $A \neq \emptyset$, and let $f : A \to \mathcal{P}(A)$. We will show that f cannot be a surjection, and hence there is no bijection from A to $\mathcal{P}(A)$. This will prove that A is not equivalent to $\mathcal{P}(A)$. Define

$$S = \{x \in A \mid x \notin f(x)\}.$$

Assume that there exists a t in A such that $f(t) = S$. Now, either $t \in S$ or $t \notin S$.

- If $t \in S$, then $t \in \{x \in A \mid x \notin f(x)\}$. By the definition of S, this means that $t \notin f(t)$. However, $f(t) = S$ and so we conclude that $t \notin S$. But now we have $t \in S$ and $t \notin S$. This is a contradiction.

- If $t \notin S$, then $t \notin \{x \in A \mid x \notin f(x)\}$. By the definition of S, this means that $t \in f(t)$. However, $f(t) = S$ and so we conclude that $t \in S$. But now we have $t \notin S$ and $t \in S$. This is a contradiction.

So in both cases we have arrived at a contradiction. This means that there does not exist a t in A such that $f(t) = S$. Therefore, any function from A to $\mathcal{P}(A)$ is not a surjection and hence not a bijection. Hence, A and $\mathcal{P}(A)$ do not have the same cardinality. ∎

Corollary 9.37. $\mathcal{P}(\mathbb{N})$ *is an infinite set that is not countably infinite.*

Proof. Since $\mathcal{P}(\mathbb{N})$ contains the infinite subset $\{\{1\}, \{2\}, \{3\} \ldots\}$, we can use Theorem 9.18, to conclude that $\mathcal{P}(\mathbb{N})$ is an infinite set. By Cantor's Theorem (Theorem 9.36), $\mathbb{N}$ and $\mathcal{P}(\mathbb{N})$ do not have the same cardinality. Therefore, $\mathcal{P}(\mathbb{N})$ is not countable and hence is an uncountable set. ∎

Some Final Comments about Uncountable Sets

1. We have now seen that any open interval of real numbers is uncountable and has cardinality $\mathbf{c}$. In addition, $\mathbb{R}$ is uncountable and has cardinality $\mathbf{c}$. Now, Corollary 9.37 tells us that $\mathcal{P}(\mathbb{N})$ is uncountable. A question that can be asked is,

"Does $\mathcal{P}(\mathbb{N})$ have the same cardinality as $\mathbb{R}$?"

The answer is yes, although we are not in a position to prove it yet. A proof of this fact uses the following theorem, which is known as the Cantor-Schröder-Bernstein Theorem.

Theorem 9.38 (Cantor-Schröder-Bernstein). *Let A and B be sets. If there exist injections $f_1 : A \to B$ and $f_2 : B \to A$, then $A \approx B$.*

In the statement of this theorem, notice that it is not required that the function f_2 is the inverse of the function f_1. We will not prove the Cantor-Schröder-Bernstein Theorem, but the following item will indicate another use of this theorem.

2. The Cantor-Schröder-Bernstein Theorem can also be used to prove that the closed interval $[0, 1]$ is equivalent to the open interval $(0, 1)$. See Exercise (4).

3. Another question that was posed earlier is,

"Are there other infinite cardinal numbers other than $\aleph_0$ and $\boldsymbol{c}$?"

Again, the answer is yes, and the basis for this is Cantor's Theorem (Theorem 9.36).

We can start with card $(\mathbb{N}) = \aleph_0$. We then define the following infinite cardinal numbers:

card $(\mathcal{P}(\mathbb{N})) = \aleph_1$.   card $(\mathcal{P}(\mathcal{P}(\mathcal{P}(\mathbb{N})))) = \aleph_3$.

card $(\mathcal{P}(\mathcal{P}(\mathbb{N}))) = \aleph_2$. $\vdots$

Cantor's Theorem tells us that these are all different cardinal numbers. In fact, there is a reasonably easy way to "order" these cardinal numbers in such a way that

$$\aleph_0 < \aleph_1 < \aleph_2 < \aleph_3 < \cdots.$$

Keep in mind, however, that even though these are different cardinal numbers, Cantor's Theorem does not tell us that these are the only cardinal numbers.

4. In Comment (1), we indicated that $\mathcal{P}(\mathbb{N})$ and $\mathbb{R}$ have the same cardinality. Combining this with the notation in Comment (3), this means that

$$\aleph_1 = \boldsymbol{c}.$$

However, this does not necessarily mean that $\boldsymbol{c}$ is the "next largest" cardinal number after $\aleph_0$. A reasonable question is, "Is there an infinite set with cardinality between $\aleph_0$ and c?" Rewording this in terms of the real number line, the question is, "On the real number line, is there an infinite set of points that is not equivalent to the entire line and also not equivalent to the set of natural numbers?" This question was asked by Cantor, but he was unable to find any such set. He conjectured that no such set exists. That is, he conjecture that c is really the next cardinal number after $\aleph_0$. This conjecture has come to be known as the **Continuum Hypothesis**. Stated somewhat more formally, the Continuum Hypothesis is

$$\text{There is no set } X \text{ such that } \aleph_0 < \text{card}\,(X) < c.$$

The question of whether the Continuum Hypothesis is true or false is one of the most famous problems in modern mathematics.

Through the combined work of Kurt Gödel in the 1930s and Paul Cohen in 1963, it has been proven that the Continuum Hypothesis cannot be proven or disproven from the standard axioms of set theory. This means that either the Continuum Hypothesis or its negation can be added to the standard axioms of set theory without creating a contradiction.

Activity 9.39 (The Closed Interval $[0,1]$).

We have seen that card $((0,1)) = c$ and card $(\mathbb{R}) = c$. It would seem reasonable to expect that if we add the endpoints to the open interval $(0,1)$, we would not change the cardinality. That is, it might be reasonable to expect that card $([0,1]) = c$. We have, in fact, indicated that the Cantor-Schröder-Bernstein Theorem (Theorem 9.38) can be used to prove that this is true. This may seem a bit unsatisfactory since we have not proven the Cantor-Schröder-Bernstein Theorem. In this activity, we will prove that card $([0,1]) = c$ by using appropriate bijections.

1. Let $f : [0,1] \to [0,1)$ by

$$f(x) = \begin{cases} \dfrac{1}{n+1} & \text{if } x = \dfrac{1}{n} \text{ for some } n \in \mathbb{N} \\ x & \text{otherwise.} \end{cases}$$

(a) Determine $f(0)$, $f(1)$, $f\left(\dfrac{1}{2}\right)$, $f\left(\dfrac{1}{3}\right)$, $f\left(\dfrac{1}{4}\right)$, and $f\left(\dfrac{1}{5}\right)$.

(b) Sketch a graph of the function f. *Hint*: Start with the graph of $y = x$ for $0 \le x \le 1$. Remove the point $(1, 1)$ and replace it with the point $\left(1, \dfrac{1}{2}\right)$. Next, remove the point $\left(\dfrac{1}{2}, \dfrac{1}{2}\right)$ and replace it with the point $\left(\dfrac{1}{2}, \dfrac{1}{3}\right)$. Continue this process of removing points on the graph of $y = x$ and replacing them with the points determined from the information in Part (1a). Stop after repeating this four or five times so that pattern of this process becomes apparent.

(c) Explain why the function f is a bijection.

(d) Prove that $[0, 1] \approx [0, 1)$.

2. Let $g : [0, 1) \to (0, 1)$ by

$$
g(x) = \begin{cases}
\dfrac{1}{2} & \text{if } x = 0 \\[2mm]
\dfrac{1}{n + 1} & \text{if } x = \dfrac{1}{n} \text{ for some } n \in \mathbb{N} \\[2mm]
x & \text{otherwise}
\end{cases}
$$

(a) Follow the procedure suggested in Part (1) to sketch a graph of g.

(b) Explain why the function g is a bijection.

(c) Prove that $[0, 1) \approx (0, 1)$.

3. Prove that $[0, 1]$ and $[0, 1)$ are both uncountable and have cardinality c.

Exercises 9.4

1. Use an appropriate bijection to prove that each of the following sets has cardinality c.

   (a) $(0, \infty)$
   (b) (a, ∞), for any $a \in \mathbb{R}$
   (c) $\mathbb{R} - \{0\}$
   (d) $\mathbb{R} - \{a\}$, for any $a \in \mathbb{R}$

2. Is the set of irrational numbers countable or uncountable? Prove that your answer is correct.

3. Prove that if A is uncountable and $A \subseteq B$, then B is uncountable.

4. The goal of this exercise is to use the Cantor-Schröder-Bernstein Theorem to prove that the cardinality of the closed interval $[0, 1]$ is c.

    (a) Find an injection $f : (0, 1) \to [0, 1]$.

    (b) Find an injection $h : [0, 1] \to (-1, 2)$.

    (c) Use the fact that $(-1, 2) \approx (0, 1)$ to prove that there exists an injection $g : [0, 1] \to (0, 1)$. <u>Note</u>: It is only necessary to prove that the injection g exists. It is not necessary to determine a specific formula for $g(x)$.

    (d) Use the Cantor-Schröder-Bernstein Theorem to conclude that $[0, 1] \approx (0, 1)$, and hence that the cardinality of $[0, 1]$ is c.

5. In Activity 9.39 and Exercise (4), we proved that the closed interval $[0, 1]$ is uncountable and has cardinality c. Now let $a, b \in \mathbb{R}$ with $a < b$. Prove that $[a, b] \approx [0, 1]$ and hence that $[a, b]$ is uncountable and has cardinality c.

6. Do two uncountable sets always have the same cardinality? Justify your conclusion.

Appendix A

Guidelines for Writing Mathematical Proofs

One of the most important forms of mathematical writing is writing mathematical proofs. The writing of mathematical proofs is an acquired skill and takes a lot of practice. Throughout the textbook, we have introduced various guidelines for writing proofs. These guidelines are in Sections 1.1, 1.2, 3.1, 3.2, 3.3, and 5.1.

Following is a summary of all the writing guidelines introduced in the text. This summary contains some standard conventions that are usually followed when writing a mathematical proof.

1. **Begin with a carefully worded statement of the theorem or result to be proven.** The statement should be a simple declarative statement of the problem. Do not simply rewrite the problem as stated in the textbook or given on a handout. Problems often begin with phrases such as "Show that" or "Prove that." This should be reworded as a simple declarative statement of the theorem. Then skip a line and write "Proof" in boldface font (when using a word processor). Begin the proof on the same line. Make sure that all paragraphs can be easily identified. Skipping a line between paragraphs or indenting each paragraph can accomplish this.

   As an example, an exercise in a text might read, "Prove that if x is an odd integer, then x^2 is an odd integer." This could be started as follows:

   Theorem. If x is an odd integer, then x^2 is an odd integer.

   Proof: We assume that x is an odd integer ...

2. **Begin the proof with a statement of the assumptions.** This is illustrated in the example in Part (1). Follow the statement of the assumptions with a statement of what will be proven.

   Theorem If x is an odd integer, then x^2 is an odd integer.

   Proof: We assume that x is an odd integer, and we will prove that x^2 is an odd integer.

3. **Use the pronoun "we."** If a pronoun is used in a proof, the usual convention is to use "we" instead of "I." The idea is that the author and the reader are proving the theorem together.

4. **Use italics for variables.** When using a word processor to write mathematics, the word processor needs to be capable of producing the appropriate mathematical symbols and equations. The mathematics that is written with a word processor should look like typeset mathematics. This means that variables need to be italicized, boldface is used for vectors, and regular font is used for mathematical terms such as the names of the trigonometric functions and logarithmic functions. The use of italics is illustrated in the example in Part (2).

5. **Use complete sentences and proper paragraph structure.** Good grammar is an important part of any writing. Therefore, conform to the accepted rules of grammar. Pay careful attention to the structure of sentences. Write proofs using **complete sentences** but avoid run-on sentences. Part of good grammar is correct spelling. Always use a spell checker when using a word processor.

6. **Keep the reader informed.** Sometimes a theorem is proven by proving the contrapositive or by using a proof by contradiction. If either proof method is used, this should be indicated within the first few lines of the proof. This also applies if the result is going to be proven using mathematical induction.

   Examples:

   - We will prove this result by proving the contrapositive of the statement.
   - We will prove this statement using a proof by contradiction.
   - We will assume to the contrary that . . .
   - We will use mathematical induction to prove this result.

In addition, make sure the reader knows the status of every assertion that is made. That is, make sure it is clearly stated whether an assertion is an assumption of the theorem, a previously proven result, a well-known result, or something from the reader's mathematical background.

7. **Display important equations and mathematical expressions.** Equations and manipulations are often an integral part of the exposition. Do not write equations, algebraic manipulations, or formulas in one column with reasons given in another column (as is often done in geometry texts). Important equations and manipulations should be displayed. This means that they should be centered with blank lines before and after the equation or manipulations. If several steps are shown together, the equals signs should be aligned, and if one side of an equation does not change, it should not be repeated. For example,

Using algebra, we obtain

$$
\begin{aligned}
x \cdot y &= (2m + 1)(2n + 1) \\
&= 4mn + 2m + 2n + 1 \\
&= 2(2mn + m + n) + 1.
\end{aligned}
$$

Since m and n are integers, we conclude that ...

8. **Equation numbering guidelines** If it is necessary to refer to an equation later in a proof, that equation should be centered and displayed, and it should be given a number. The number for the equation should be written in parentheses on the same line as the equation at the right-hand margin.

Example:

Since x is an odd integer, there exists an integer n such that

$$
x = 2n + 1. \tag{1}
$$

Later in the proof, there may be a line such as

Then, using the result in Equation (1), we obtain . . .

Please note that we should only number those equations we will be referring to later in the proof. Also, note that the word "Equation" begins with a capital "E" when we are referring to an equation by number.

9. **Do not begin a sentence with a mathematical symbol.** In addition to not beginning a sentence with a mathematical symbol, in formal writing in mathematics, we do not use the special symbols $\forall$ (for all), $\exists$ (there exists), $\ni$ (such that), or $\therefore$ (therefore).

10. **Tell the reader when the proof has been completed.** Perhaps the best way to do this is to say outright that, "This completes the proof." Although it may seem repetitive, a good alternative is to finish a proof with a sentence that states precisely what has been proven. In any case, it is usually good practice to use some "end of proof symbol" such as ■.

11. **Write a first draft of your proof and then revise it.** Remember that a proof is written so that readers are able to read and understand the reasoning in the proof. Be clear and concise. Include details but do not ramble. Do not be satisfied with the first draft of a proof. Read it over and refine it. Just like any worthwhile activity, learning to write mathematics well takes practice and hard work. This can be frustrating. Everyone can be sure that there will be some proofs that are difficult to construct, but remember that proofs are a very important part of mathematics. So work hard and have fun.

Appendix B

Answers and Hints for Selected Exercises

Section 1.1

1. Sentences (a), (d), and (h) are propositions. Sentence (f) is a proposition if we are assuming that n is a prime number means that n is an integer.

2.

	Hypothesis	Conclusion
a.	n is a prime number	n^2 has three positive divisors.
b.	a is an irrational number and b is an irrational number	$a \cdot b$ is an irrational number.
c.	p is a prime number	$p = 2$ or p is an odd number.

3. Statements (a), (c), and (d) are true.

4. (a) True when $a \neq 3$. (b) True when $a = 3$.

Section 1.2

1. (a)

Step	Know	Reason
P	m is an even integer.	Hypothesis
$P1$	There exists an integers k such that $m = 2k$.	Definition of an even integer
$P2$	$m + 1 = 2k + 1$	Algebra
$Q1$	There exists an integer q such that $m + 1 = 2q + 1$	Substitution of $k = q$
Q	$m + 1$ is an odd integer.	Definition of an odd integer

3. (c)

Step	Know	Reason
P	x and y are odd integers.	Hypothesis
$P1$	There exist integers m and n such that $x = 2m + 1$ and $y = 2n + 1$.	Definition of an odd integer
$P2$	$x + y = (2m + 1) + (2n + 1)$	Substitution
$P3$	$x + y = 2m + 2n + 2$ $x + y = 2(m + n + 1)$	Algebra
$P4$	$(m + n + 1)$ is an integer	Closure properties of the integers
$Q1$	There exists an integer q such that $x + y = 2q$	$q = m + n + 1$
Q	$x + y$ is an even integer.	Definition of an even integer

7. (a) Prove that they are not zero and their quotient is equal to 1.

(d) Prove that two of the sides have the same length. Prove that the triangle has two congruent angles. Prove that an altitude of the triangle is a perpendicular bisector of a side of the triangle.

9. (a) Some examples are $-5, -2, 1, 4, 7, 10$.

10. (a) Let a and b be integers and assume that a and b are both congruent to 1 modulo 3. Then, there exist integers m and n such that $a = 3m + 1$ and $b = 3n + 1$. Now show that

$$a + b = 3(m + n) + 2$$

The closure properties of the integers imply that $m + n$ is an integer. Therefore, the last equation tells us that $a + b \equiv 2 \pmod 3$.

Hence, we have proven that if a and b are both congruent to 1 modulo 3, then $a + b$ is congruent to 2 modulo 3.

Section 2.1

1. (a) $\{\frac{1}{2}, -2\}$ (d) $\{1, 2, 3, 4\}$ (e) $\{-0.5, 4.5\}$

2. (b) and (c)

3. (a) $\{x \in \mathbb{Z} \mid x \geq 5\}$ (e) $\{x \in \mathbb{R} \mid x^2 > 10\}$

4. (a), (d), (i), and (j) are true statements. (f), (g), and (k) are false statements.

Section 2.2

1. The statement is not false. When the hypothesis is false, the conditional statement is true.

2. (a) P and Q are false. (b) $P \wedge Q$ is false.   (c) $P \vee Q$ is false.

4. Statements (b) and (d) have the same truth table.

6. The two statements have the same truth table.

8. (c) The integer x is even only if x^2 is even.

9. (e) The integer x^2 is even is sufficient for the integer x to be even.

Section 2.3

1. (a) $a \neq 5$ or $a^2 = 25$ (b) It is raining or Laura is playing golf.

2. (a) $a = 5$ and $a^2 \neq 25$

   (b) It is not raining and Laura is not playing golf.

3. (a) We will not win the first game or we will not win the second game.

   (c) You mow the lawn and I will not pay you $20.

   (f) You graduate from college, and you will not get a job and you will not go to graduate school.

Section 2.4

1. (a) $x = 0$ is a counterexample.
    (c) $x = \frac{\pi}{2}$ is a counterexample.
    (d) Both solutions for the equation are irrational.

2. (a) There exists a rational number x such that $x > \sqrt{2}$.
    $(\forall x \in \mathbb{Q}) \left(x \leq \sqrt{2}\right)$.
    For each rational number x, $x \leq \sqrt{2}$.
    (c) For each integer x, x is even or x is odd.
    $(\exists x \in \mathbb{Z}) \left(x \text{ is odd and } x \text{ is even}\right)$.
    There exists an integer x such that x is odd and x is even.
    (e) For each integer x, if x^2 is odd, then x is odd.
    $(\exists x \in \mathbb{Z}) \left(x^2 \text{ is odd and } x \text{ is even}\right)$.
    There exists an integer x such that x^2 is odd and x is even.
    (h) There exists a real number x such that $\cos(2x) = 2(\cos x)$.
    $(\forall x \in \mathbb{R}) \left(\cos(2x) \neq 2(\cos x)\right)$.
    For each real number x, $\cos(2x) \neq 2(\cos x)$.

3. (a) There exist integers m and n such that $m > n$.
    (e) There exists an integer n such that for each integer m, $m^2 > n$.

4. (a) $(\forall m)(\forall n)(m \leq n)$.
    For all integers m and n, $m \leq n$.
    (e) $(\forall n)(\exists m)(m \leq n)$.
    For each integer n, there exists an integer m such that $m^2 \leq n$.

Section 3.1

1. Both are true statements. Remember that to prove that $a \mid (b+c)$, you need to prove that there exists an integer q such that $b + c = a \cdot q$.

2. First, try some examples. If you want to prove the statement is true, what do you need to do in order to prove that n^3 is odd? Notice that if n is odd, then by Theorem 1.3, n^2 is odd. Now use the fact that $n^3 = n \cdot n^2$.

5. In both cases, make sure you first try some examples. How do you prove that an integer is an odd integer? For part (b), the following algebra may be useful.

$$4(2m+1)^2 + 7(2m+1) + 6 = 6m^2 + 30m + 17$$
$$= 2\left(3m^2 + 15m + 8\right) + 1.$$

8. Let $n \in \mathbb{N}$. For $a, b \in \mathbb{Z}$, you need to prove that if $a \equiv b \pmod{n}$, then $b \equiv a \pmod{n}$. Remember that for $x, y \in \mathbb{Z}$, $x \equiv y \pmod{n}$ if and only if $n \mid (x - y)$.

10. <u>Another hint</u>: $(4n + 3) - 2(2n + 1) = 1$.

11. The assumptions mean that $n \mid (a - b)$ and that $n \mid (c - d)$. Use these divisibility relations to obtain an expression that is equal to a and to obtain an expression that is equal to c. Then use algebra to rewrite the resulting expressions for $a + c$ and $a \cdot c$.

Section 3.2

1. (a) Let n be an even integer. Since n is even, there exists an integer k such that $n = 2k$. Now use this to prove that n^3 must be even.

   (b) Prove the contrapositive.

   (c) Explain why Parts (a) and (b) prove this.

   (d) Explain why Parts (a) and (b) prove this.

2. (a) The contrapositive is, Let a and b be integers. If $ab \equiv 0 \pmod 6$, then $a \equiv 0 \pmod 6$ or $b \equiv 0 \pmod 6$.

4. (a) If $a \equiv 2 \pmod 5$, then there exists an integer k such that $a - 2 = 5k$. Then, $a^2 = (2 + 5k)^2 = 4 + 20k + 25k^2$. This means that $a^2 - 4 = 5\left(4k + 5k^2\right)$.

7. Remember that there are two conditional statements associated with this biconditional statement. Be willing to consider the contrapositive of one of these conditional statements.

8. (b) Since 4 divides a, there exist an integer n such that $a = 4n$. Using this, we see that $b^3 = 16n^2$. This means that b^3 is even and hence by Exercise (1), b is even. So there exists an integer m such that $b = 2m$. Use this to prove that m^3 must be even and hence by Exercise (1), m is even.

Section 3.3

1. (a) $P \vee C$

2. (a) Let r be a real number such that $r^2 = 18$. We will prove that r is irrational using a proof by contradiction. So we assume that r is a rational number.

3. What is the contrapositive of the proposition? Why does it seem like the contrapositive will not be a good approach? Try a proof by contradiction.

4. One of the propositions is true and the other is false.

8. Recall that $\log_2 y$ is the real number a such that $2^a = y$. That is, $a = \log_2 y$ means that $2^a = y$. If we assume that a is rational, then there exist integers m and n, with $n \neq 0$, such that $a = \dfrac{m}{n}$.

10. What happens if you expand $(\sin \theta + \cos \theta)^2$? Don't forget your trigonometric identities.

Section 3.4

1. Use the fact that $n^2 + n = n(n+1)$.
2. (b) Factor $n^3 - n$.

3. Do not use the quadratic formula. Try a proof by contradiction. <u>Hint</u>: If there exists a solution of the equation that is an integer, then we can conclude that there exists an integer n such that $n^2 + n - u = 0$.

4. Use the Division Algorithm. What are the possible remainders when n is divided by 4 if n is odd?

6. (a) Use the definition of congruence.
   (b) Let $a \in \mathbb{Z}$. Corollary 3.22 tell us that if $a \not\equiv 0 \pmod 3$, then $a \equiv 1 \pmod 3$ or $a \equiv 2 \pmod 3$.
   (c) Part (b) tells us we can use a proof by cases using the following two cases: (1) $a \equiv 1 \pmod 3$; (2) $a \equiv 2 \pmod 3$.

8. Remember that $3 \mid k$ if and only if $k \equiv 0 \pmod 3$. This proposition is a special case of the proposition in Exercise (5c) with $d = 3$. The general result in Exercise (5c) is false, but this special case is true.

10. (a) 21   (b) 43

Section 3.5

1. Use $x = \dfrac{p+q}{2}$. You must still prove the inequalities are true.

3. Write p and q as quotients of integers and then use the rule for division of fractions. Pay close attention to which integers are nonzero.

5. Show that $\delta = 0.005$ will work or show that any real number δ with $\delta < 0.005$ will work.

8. The propositions in (a) and (c) are true.

Section 4.1

1. (a) $A = B$ (c) $C \neq D$ (e) $A \not\subseteq D$
    (b) $A \subseteq B$ (d) $C \subseteq D$

3. | A | $\subset, \subseteq, \neq$ | B | $\emptyset$ | $\subset, \subseteq, \neq$ | A |
    |---|---|---|---|---|---|
    | 5 | $\in$ | C | $\{5\}$ | $\subset, \subseteq, \neq$ | C |
    | $|A|$ | $=$ | $|D|$ | A | $\in$ | $\mathcal{P}(B)$ |

5. (a) The set $\{a, b\}$ is a not a subset of $\{a, c, d, e\}$ since $b \in \{a, b\}$ and $b \notin \{a, c, d, e\}$.

6. (c) $(A \cup B)^c = \{2, 8, 10\}$ (h) $(A \cap C) \cup (B \cap C) = \{3, 6, 9\}$
    (d) $A^c \cap B^c = \{2, 8, 10\}$ (n) $(A \cup B) - D = \{1, 3, 5, 7, 9\}$
    (e) $(A \cup B) \cap C = \{3, 6, 9\}$

7. (a) The interval (a, b) is a proper subset of $(a, b]$. $[(b \in (a, b]$ and $b \notin (a, b).]$
    (e) $\{x \in \mathbb{R} \mid |x| > 2\} = (-\infty, 2) \cup (2, \infty)$.

Section 4.2

1. (a) If $x \in A$, then there exists an integer k such that $x = 9k$. This means that $x = 3(3k)$ and since $3k \in \mathbb{Z}$, we see that $x \in B$.
    (b) The set B is not a subset of A since there exist elements (such as 3 and 6) that are in B but not in A.
2. (a) The set A is a subset of B. A proof is required. The idea is that if $x \in A$, then $-2 < x < 2$. Since $x < 2$, we conclude that $x \in B$.
    (b) The set B is not a subset of A. Give an example of a real number that is in B but not in A.
4. (c) $A \subseteq B$ (d) $B \not\subseteq A$

6. For (a) and (b), start by letting x be an element of the set on the left side of the subset symbol.
9. (a) Let $x \in A \cap C$. Then $x \in A$ and $x \in C$. Since we are assuming $A \subseteq B$, we see that $x \in B$ and $x \in C$. This proves that $A \cap C \subseteq B \cap C$.

Section 4.3.

1. (a) Let $x \in (A^c)^c$. Then $x \notin A^c$, which means $x \in A$. Hence, $(A^c)^c \subseteq A$. Now prove that $A \subseteq (A^c)^c$.

   (c) Let $x \in U$. Then $x \notin \emptyset$ and so $x \in \emptyset^c$. Therefore, $U \subseteq \emptyset^c$. Now prove that $\emptyset^c \subseteq U$.

2. We still need to prove that if $B^c \subseteq A^c$, then $A \subseteq B$.

5. (a) $A - (B \cup C) = (A - B) \cap (A - C)$.

Section 4.4.

1. (a) $A \times B = \{(1,a),(1,b),(1,c),(1,d),(2,a),(2,b),(2,c),(2,d)\}$

   (b) $B \times A = \{(a,1),(b,1),(c,1),(d,1),(a,2),(b,2),(c,2),(d,2)\}$

   (e) $A \times (B \cap C) = \{(1,a),(1,b),(2,a),(2,b)\}$

3. Start of proof that $A \times (B \cap C) \subseteq (A \times B) \cap (A \times C)$:

   Let $u \in A \times (B \cap C)$. Then there exists $x \in A$ and there exists $y \in B \cap C$ such that $u = (x,y)$. Since $y \in B \cap C$, we know that $y \in B$ and $y \in C$. So we have
   $u = (x,y)$, where $x \in A$ and $y \in B$. This means that $u \in A \times B$.
   $u = (x,y)$, where $x \in A$ and $y \in C$. This means that $u \in A \times C$.

4. Start of proof that $(A \cup B) \times C \subseteq (A \times C) \cup (B \times C)$:

   Let $u \in (A \cup B) \times C$. Then there exists $x \in A \cup B$ and there exists $y \in C$ such that $u = (x,y)$. Since $x \in A \cup B$, we know that $x \in A$ or $x \in B$.

Section 5.1

1. The sets in Parts (a) and (b) are inductive.

2. A finite nonempty set is not inductive (why?) but the empty set is inductive (why?).

3. (a) For each $n \in \mathbb{N}$, let $P(n)$ be, $1 + 2 + 3 + \cdots + n = \frac{n(n+1)}{2}$. Verify that $P(1)$ is true. The key to the inductive step is that if $P(k)$ is true, then

$$1 + 2 + 3 + \cdots + k + (k+1) = (1 + 2 + 3 + \cdots + k) + (k+1)$$
$$= \frac{k(k+1)}{2} + (k+1).$$

Now use algebra to complete the proof that if $P(k)$ is true, then $P(k+1)$ is true.

5. The conjecture is that for each $n \in \mathbb{N}$, $\sum\limits_{j=1}^{n} (2j-1) = n^2$. The key to the inductive step is that

$$\sum_{j=1}^{k+1} (2j-1) = \sum_{j=1}^{k} (2j-1) + [2(k+1)-1]$$

$$= \sum_{j=1}^{k} (2j-1) + [2k+1].$$

6. (a) The key to the inductive step is that if $4^k = 1 + 3m$, then $4^k \cdot 4 = 4(1+3m)$, which implies that

$$4^{k+1} - 1 = 3(1+4m).$$

11. For the inductive step, the following trigonometric identities are useful:
$$\cos(\alpha + \beta) = \cos \alpha \cos \beta - \sin \alpha \sin \beta.$$
$$\sin(\alpha + \beta) = \sin \alpha \cos \beta + \cos \alpha \sin \beta.$$

Section 5.2

1. (a) If $P(k)$ is true, then $3^k > 1 + 2^k$. Multiplying both sides of this inequality by 3 gives

$$3^{k+1} > 3 + 3 \cdot 2^k.$$

Now, since $3 > 1$ and $3 \cdot 2^k > 2^{k+1}$, we see that $3 + 3 \cdot 2^k > 1 + 2^{k+1}$ and hence, $3^{k+1} > 1 + 2^{k+1}$. Thus, if $P(k)$ is true, then $P(k+1)$ is true.

3. Let $P(n)$ be the predicate, "$8^n \mid (4n)!$." Verify that $P(0), P(1), P(2)$, and $P(3)$ are true. Now, find a counterexample or use induction to prove the proposition is true.

4. Let $P(n)$ be, "The natural number n can be written as a sum of natural numbers, each of which is a 2 or a 3." Verify that $P(4), P(5), P(6)$, and $P(7)$ are true.

Now assume that $k \in \mathbb{N}, k \geqslant 5$ and that $P(4), P(5), \ldots, P(k)$ are true. Then notice that

$$k + 1 = (k - 1) + 2.$$

Since $k - 1 \geqslant 4$, we have assumed that $P(k - 1)$ is true. Use this to complete the inductive step.

6. If $n \geq 5$, then $n^2 < 2^n$. For the inductive step, we assume that $k^2 < 2^k$ and that $k \geq 5$. With these assumptions, prove that

$$(k + 1)^2 = k^2 + 2k + 1 < 2^k + 2k + 1.$$

Now use the assumption that $k > 4$ to prove that $2k + 1 < k^2$ and combine this with the assumption that $k^2 < 2^k$.

Section 5.3

4. (a) Let $P(n)$ be, "f_{4n} is a multiple of 3." Since $f_4 = 3$, $P(1)$ is true. If $P(k)$ is true, then there exists an integer m such that $f_{4k} = 3m$. Use the following:

$$\begin{aligned}
f_{4(k+1)} &= f_{4k+4} \\
&= f_{4k+3} + f_{4k+2} \\
&= (f_{4k+2} + f_{4k+1}) + (f_{4k+1} + f_{4k}) \\
&= f_{4k+2} + 2f_{4k+1} + f_{4k}
\end{aligned}$$

5. (a) $a_2 = \sqrt{6}$, $a_3 = \sqrt{\sqrt{6} + 5} \approx 2.729$, $a_4 \approx 2.780$, $a_5 \approx 2.789$, $a_6 \approx 2.791$

6. (a) $a_3 = 7$, $a_4 = 15$, $a_5 = 31$, $a_6 = 63$
   (b) Think in terms of powers of 2.

7. (a) $a_3 = \dfrac{3}{2}$, $a_4 = \dfrac{7}{4}$, $a_5 = \dfrac{37}{24}$, $a_6 = \dfrac{451}{336}$

8. (a) $a_4 = 3$, $a_5 = 5$, $a_6 = 9$, $a_7 = 17$

Section 6.1

1. (b) The pre-images of 0 are 0 and 2. The pre-images of 4 are $\dfrac{2 - \sqrt{20}}{2}$ and $\dfrac{2 + \sqrt{20}}{2}$. (Use the quadratic formula.)
   (c) There are no pre-images of -2.

(e) range $(f) = \{y \in \mathbb{R} \mid y \geqslant -1\}$

3. Only (a) can be used to represent a function from A to B. Why?

5. (b) The only pre-image of 5 is 2. There are no pre-images of 4.

(c) The range of the function f is the set of all odd integers.

(d) The graph of the function f consists of an infinite set of discrete points.

6. (a) $d(1) = 1$, $d(2) = 2$, $d(3) = 2$, $d(4) = 3$, $d(8) = 4$, $d(9) = 3$

(c) The only natural numbers n such that $d(n) = 2$ are the prime numbers. The set of pre-images of the natural number 2 is the set of prime numbers.

(e) $d(2^0) = 1$, $d(2^1) = 2$, $d(2^2) = 3$, $d(2^3) = 4$

(f) The idea for the inductive step is that the divisors of 2^{k+1} are 2^{k+1} and the divisors of 2^k.

7. (a) $f(-3, 4) = 9$, $f(-2, -7) = -23$

(b) $\{(m, n) \in \mathbb{Z} \times \mathbb{Z} \mid m = 4 - 3n\}$

9. (b) $\mathrm{dom}\,(F) = \{x \in \mathbb{R} \mid x > \frac{1}{2}\}$, range $(F) = \mathbb{R}$

(d) $\mathrm{dom}\,(g) = \{x \in \mathbb{R} \mid x \neq 2 \text{ and } x \neq -2\}$,
range $(g) = \{y \in \mathbb{R} \mid y > 0\} \cup \{y \in \mathbb{R} \mid y \leqslant -1\}$

Section 6.2

1. (a) $f(0) = 0$, $f(1) = 1$, $f(2) = 1$, $f(3) = 1$, $f(4) = 1$

(c) $f = \{(0, 0), (1, 1), (2, 1)\,(3, 1), (4, 1)\}$

3. (a) $\langle a_n \rangle$, where $a_n = \dfrac{1}{n^2}$ for each $n \in \mathbb{N}$. The domain is $\mathbb{N}$.

(d) $\langle a_n \rangle$, where $a_n = \cos(n\pi)$ for each $n \in \mathbb{N}$. The domain is $\mathbb{N}$. This is equal to the sequence in Part (c).

5. (b) For example, $S(3) = \{1, 3\}$, $S(8) = \{1, 2, 4, 8\}$,
$S(15) = \{1, 3, 5, 15\}$

(c) For example, $S(2) = \{1, 2\}$, $S(3) = \{1, 3\}$, $S(5) = \{1, 5\}$,
$S(31) = \{1, 31\}$

6. Start of the inductive step: Let $P(n)$ be "A convex polygon with n sides has $\dfrac{n(n-3)}{2}$ sides." Let $k \in D$ and assume that $P(k)$ is true. We assume that a convex polygon with k sides has $\dfrac{k(k-3)}{2}$ diagonals. Now let Q be convex polygon with $(k+1)$ sides. Let v be one of the

$(k + 1)$ vertices of P and let u and w be the two vertices adjacent to v. By drawing the line segment from u to w and omitting the vertex v, we form a convex polygon with k sides. Now complete the inductive step.

7. (a) $\det \begin{bmatrix} 3 & 5 \\ 4 & 1 \end{bmatrix} = -17$, $\det \begin{bmatrix} 1 & 0 \\ 0 & 7 \end{bmatrix} = 7$, and $\det \begin{bmatrix} 3 & -2 \\ 5 & 0 \end{bmatrix} = 10$

Section 6.3

4. Let $F : \mathbb{R} \to \mathbb{R}$ be defined by $F(x) = 5x + 3$ for all $x \in \mathbb{R}$. Let $x_1, x_2 \in \mathbb{R}$ and assume that $F(x_1) = F(x_2)$. Then, $5x_1 + 3 = 5x_2 + 3$. Show that this implies that $x_1 = x_2$, and hence F is an injection.

Now let $y \in \mathbb{R}$. Then, $\dfrac{y - 3}{5} \in \mathbb{R}$. Prove that $F\left(\dfrac{y - 3}{5}\right) = y$. Thus, F is a surjection and hence F is a bijection.

5. (a) The function f is an injection and is not a surjection.

(b) The function F is an injection and is a surjection.

9. The birthday function is not an injection since there are two different people with the same birthday. The birthday function is a surjection since for each day of the year, there is a person that was born on that day.

11. (a) The function f is an injection and a surjection.

(b) The function g is an injection and is not a surjection.

Section 6.4

2. (a) $F(x) = (g \circ f)(x)$, $f(x) = e^x$, $g(x) = \cos x$

(b) $G(x) = (g \circ f)(x)$, $f(x) = \cos x$, $g(x) = e^x$

3. (a) $[(h \circ g) \circ f](x) = \sqrt[3]{\sin(x^2)}$; $[h \circ (g \circ f)](x) = \sqrt[3]{\sin(x^2)}$

4. (a) For each $x \in A$, $(f \circ I_A)(x) = f(I_A(x)) = f(x)$. Therefore, $f \circ I_A = f$.

5. Start of a proof: Let A, B, and C be nonempty sets and let $f : A \to B$ and $g : B \to C$. Assume that f and g are both injections. Let $x, y \in A$ and assume that $(g \circ f)(x) = (g \circ f)(y)$.

Section 6.5

2. (b) $f^{-1} = \{(c, a), (b, b), (d, c), (a, d)\}$

    (d) $(f^{-1} \circ f)(x) = x = (f \circ f^{-1})(x)$. This illustrates Corollary 6.26.

4. (a) Let $x, y \in A$ and assume that $f(x) = f(y)$. Apply g to both sides of this equation to prove that $(g \circ f)(x) = (g \circ f)(y)$. Since $g \circ f = I_A$, this implies that $x = y$ and hence that f is an injection.

    (b) Start by assuming that $f \circ g = I_B$, and then let $y \in B$. You need to prove there exists an $x \in A$ such that $f(x) = y$.

5. (a) $x = \dfrac{1}{2}(\ln y + 1)$ (b) $g : \mathbb{R}^+ \to \mathbb{R}$ by $g(y) = \dfrac{1}{2}(\ln y + 1)$

6. (a) The inverse of f is not a function.

    (b) The inverse of g is a function.

Section 7.1

1. (a) The set $A \times B$ contains nine ordered pairs. The set $A \times B$ is a relation from A to B since $A \times B$ is a subset of $A \times B$.

    (b) The set R is a relation from A to B since $R \subseteq A \times B$.

    (c) $\operatorname{dom}(R) = A$, $\operatorname{range}(R) = \{p, q\}$

    (d) $R^{-1} = \{(p, a), (q, b), (p, c), (q, a)\}$

2. Only the statement in Part (b) is true.

3. (a) The domain of D consists of the female citizens of the United States whose mother is a female citizen of the United States.

    (b) The range of D consists of those female citizens of the United States who have a daughter that is a female citizen of the United States.

4. (a) $(S, T) \in R$ means that $S \subseteq T$.

    (b) The domain of the subset relation is $\mathcal{P}(U)$.

    (c) The range of the subset relation is $\mathcal{P}(U)$.

    (d) $R^{-1} = \{(T, S) \in \mathcal{P}(U) \times \mathcal{P}(U) \mid S \subseteq T\}$.

    (e) The relation R is not a function from $\mathcal{P}(U)$ to $\mathcal{P}(U)$ since any proper subset of U is a subset of more than one subset of U.

6. (a) $\{x \in \mathbb{R} \mid (x, 6) \in S\} = \{-8, 8\}$
       $\{x \in \mathbb{R} \mid (x, 9) \in S\} = \{-\sqrt{19}, \sqrt{19}\}$

    (e) The relation S is not a function from $\mathbb{R}$ to $\mathbb{R}$.

Section 7.2

1. The relation R is not reflexive on A and is not symmetric. However, it is transitive since the conditional statement "If x R y and y R z, then x R z" is a true conditional statement.

2. The relation R is not reflexive on A, is symmetric, and is not transitive.

5. (b) $C = \{-5, 5\}$

Section 7.3

1. (b) $[a] = [b] = \{a, b\}$; $[c] = \{c\}$; $[d] = [e] = \{d, e\}$

2. The equivalence classes are $\{0, 1, 2, \ldots, 9\}$, $\{10, 11, 12, \ldots, 99\}$, $\{100, 101, 102, \ldots, 999\}$, $\{1000\}$.

3. The congruence classes for the relation of congruence modulo 5 on the set of integers are:

$[0] = \{5n \mid n \in \mathbb{Z}\}$ $[3] = \{5n + 3 \mid n \in \mathbb{Z}\}$

$[1] = \{5n + 1 \mid n \in \mathbb{Z}\}$

$[2] = \{5n + 2 \mid n \in \mathbb{Z}\}$ $[4] = \{5n + 4 \mid n \in \mathbb{Z}\}$

Section 7.4

1. (a)

$\oplus$	$[0]$	$[1]$
$[0]$	$[0]$	$[1]$
$[1]$	$[1]$	$[0]$

$\odot$	$[0]$	$[1]$
$[0]$	$[0]$	$[0]$
$[1]$	$[0]$	$[1]$

(b)

$\oplus$	$[0]$	$[1]$	$[2]$
$[0]$	$[0]$	$[1]$	$[2]$
$[1]$	$[1]$	$[2]$	$[0]$
$[2]$	$[2]$	$[0]$	$[1]$

$\odot$	$[0]$	$[1]$	$[2]$
$[0]$	$[0]$	$[0]$	$[0]$
$[1]$	$[0]$	$[1]$	$[2]$
$[2]$	$[0]$	$[2]$	$[1]$

3. (a) $[x] = [1]$ or $[x] = [3]$ (e) $[x] = [2]$ or $[x] = [3]$

(g) The equation has no solution.

4. The statement in (a) is false. The statement in (b) is true.

Section 8.1

1. (a) $\gcd(21, 28) = 7$ (c) $\gcd(58, 63) = 1$
   (b) $\gcd(-21, 28) = 7$ (d) $\gcd(0, 12) = 12$

2. (a) <u>Hint</u>: Prove that $d \mid [(a + 1) - a]$.

4. (a) $\gcd(36, 60) = 12$ $12 = 36 \cdot 2 + 60 \cdot (-1)$
   (b) $\gcd(901, 935) = 17$ $17 = 901 \cdot 27 + 935 \cdot (-26)$

Section 8.2

1. The only natural number divisors of a prime number p are 1 and p.

2. Use cases: (1) p divides a; (2) p does not divide a. In this case, use the fact that $\gcd(a, p) = 1$ to write the number 1 as a linear combination of a and p.

3. A hint for the inductive step: Write $p \mid (a_1 a_2 \cdots a_m) a_{m+1}$. Then look at two cases: (1) $p \mid a_{m+1}$; (2) p does not divide a_{m+1}.

4. (a) $\gcd(a, b) = 1$. Why?
   (b) $\gcd(a, b) = 1$ or $\gcd(a, b) = 2$. Why?

7. (a) $\gcd(16, 28) = 4$. Also, $\dfrac{16}{4} = 4$, $\dfrac{28}{4} = 7$, and $\gcd(4, 7) = 1$.

Section 8.3

2. (a) $x = -3 + 14k$, $y = 2 - 9k$
   (b) $x = -1 + 11k$, $y = 1 + 9k$
   (c) No solution
   (d) $x = 2 + 3k$, $y = -2 - 4k$

4. There are several possible solutions to this problem, each of which can be generated from the solutions of the Diophantine equation $27x + 50y = 25$.

5. This problem can be solved by finding all solutions of a linear Diophantine equation in x and y, where both x and y are positive. The mininum number of people attending the banquet is 66.

6. (a) $y = 12 + 16k$, $x_3 = -1 - 3k$
   (c) $x_1 = y + 3n$, $x_2 = -y + 4n$

Section 9.1

1. (a) There exists an $x \in A \cap B$ such that $f(x) = y$.

    (d) There exists an $a \in A$ such that $f(a) = y$ or there exists a $b \in B$ such that $f(b) = y$.

    (f) $f(x) \in C \cup D$

    (h) $f(x) \in C$ or $f(x) \in D$

2. (b) $f^{-1}(f(A)) = [2, 5]$.   (e) $f(A \cap B) = [-5, -3]$

    (d) $f(f^{-1}(C)) = [-2, 3]$ (f) $f(A) \cap f(B) = [-5, -3]$

3. (a) $g(A \times A) = \{6, 12, 18, 24, 36, 54, 72, 108, 216\}$

    (b) $g^{-1}(C) = \{(1, 1), (2, 1), (1, 2)\}$

4. (a) $\text{range}(F) = F(S) = \{1, 4, 9, 16\}$

5. To prove $f(A \cup B) \subseteq f(A) \cup f(B)$, start by letting $y \in f(A \cup B)$. How do you prove that $y \in f(A) \cup f(B)$?

6. Prove that each set is a subset of the other set.

Section 9.2

1. Use $f : A \times \{x\} \to A$ by $f(a) = (a, x)$, for all $(a, x) \in A \times \{x\}$.

3. Notice that $A = (A - \{x\}) \cup \{x\}$. Use Theorem 9.12 to conclude that $A - \{x\}$ is finite. Then use Lemma 9.10.

7. (a) Remember that two ordered pairs are equal if and only if their corresponding coordinates are equal. So if $h(a_1, c_1) = h(a_2, c_2)$, then $(f(a_1), g(c_1)) = (f(a_2), g(c_2))$. We can then conclude that $f(a_1) = f(a_2)$ and $g(c_1) = g(c_2)$.

Section 9.3

1. All except Part (d) are true.

2. (e) Either define an appropriate bijection or use Corollary 9.30 to conclude that $\mathbb{N} - \{4, 5, 6\}$ is countable. Prove that $\mathbb{N} - \{4, 5, 6\}$ cannot be finite.

    (f) $\{m \in \mathbb{Z} \mid m \equiv 2 \pmod{3}\} = \{3k + 2 \mid k \in \mathbb{Z}\}$

4. For each $n \in \mathbb{N}$, let $P(n)$ be "If card $(B) = n$, then $A \cup B$ is a countably infinite set.

   Note that if card $(B) = k + 1$ and $x \in B$, then card $(B - \{x\}) = k$. Apply the inductive assumption to $B - \{x\}$.

5. Notice that if $h(n) = h(m)$, then since A and B are disjoint, either $h(n)$ and $h(m)$ are both in A or are both in B.

   Also, if $y \in A \cup B$, then there are only two cases to consider: $y \in A$ or $y \in B$.

Section 9.4

1. (a) $f : (0, \infty) \to \mathbb{R}$ by $f(x) = \ln x$ for all $x \in (0, \infty)$
   (b) Find a bijection from $(0, \infty)$ to (a, ∞).

Appendix C

List of Symbols

Symbol	Meaning	Page
$\rightarrow$	conditional statement	7, 39
$\mathbb{R}$	set of real numbers	11
$\mathbb{Q}$	set of rational numbers	11
$\mathbb{Z}$	set of integers	11
$\mathbb{N}$	set of natural numbers	27
$y \in A$	y is an element of A	28
$z \notin A$	z is not an element of A	28
$\{ \ \mid \ \}$	set builder notation	31
$\forall$	universal quantifier	32
$\exists$	existential quanitifer	32
$\emptyset$	the empty set	33
$\wedge$	conjunction	39
$\vee$	disjunction	39
$\neg$	negation	39
$\leftrightarrow$	biconditional statement	43
$\equiv$	logically equivalent	46
$m \mid n$	m divides n	73
$a \equiv b \ (\text{mod } n)$	a is congruent to b modulo n	77
$A = B$	A equals B (set equality)	126, 129
$A \subseteq B$	A is a subset of B	126, 129
$A \nsubseteq B$	A is not a subset of B	126, 129
$A \subset B$	A is a proper subset of B	129

Symbol	Meaning	Page
$\mathcal{P}(A)$	power set of A	131
$\lvert A\rvert$	cardinality of a finite set A	132
$A \cap B$	intersection of A and B	133
$A \cup B$	union of A and B	133
A^c	complement of A	133
$A - B$	set difference of A and B	133
$A \times B$	Cartesian product of A and B	163, 164
(a, b)	ordered pair	164
$\mathbb{R} \times \mathbb{R}$	Cartesian plane	165
$\mathbb{R}^2$	Cartesian plane	165
$n!$	n factorial	187
$f_1, f_2, f_3, \ldots$	Fibonacci numbers	199
$s(n)$	sum of the divisors of n	212
$f : A \to B$	function from A to B	214
$\text{dom}(f)$	domain of the function f	215
$\text{codom}(f)$	codmain of the function f	215
$f(x)$	image of x under f	215
$\text{range}(f)$	range of the function f	216
$d(n)$	number of divisors of n	223
I_A	identity function on the set A	233
χ_A	characteristic function of the set A	233
p_1, p_2	projection functions	233
$\det(A)$	determinant of A	238
A^T	transpose of A	239
$g \circ f : A \to C$	composition of functions f and g	257
f^{-1}	the inverse of the function f	268
Sin	the restricted sine function	276
Sin^{-1}	the inverse sine function	276
$\text{dom}(R)$	domain of the relation R	283
$\text{range}(R)$	range of the relation R	283
$x\,R\,y$	x is related to y	285
$x\,\not\!R\,y$	x is not related to y	285
$x \sim y$	x is related to y	286
$x \not\sim y$	x is not related to y	286
R^{-1}	the inverse of the relation R	287
$[a]$	equivalence class of a	305
$[a]$	congruence class of a	306
$\mathbb{Z}_n$	the integers modulo n	315
$[a] \oplus [c]$	addition in $\mathbb{Z}_n$	317
$[a] \odot [c]$	multiplication in $\mathbb{Z}_n$	317

Symbol	Meaning	Page
$\gcd(a, b)$	greatest common divisor of a and b	327, 329
$f(A)$	image of A under the function f	364
$f^{-1}(C)$	pre-image of C under the funciton f	364
$A \approx B$	A is equivalent to B	373
	A and B have the same cardinality	
$\mathbb{N}_k$	$\mathbb{N}_k = \{1, 2, \ldots, k\}$	375
$\operatorname{card}(A) = k$	cardinality of A is k	375
$\aleph_0$	cardinality of $\mathbb{N}$	385
c	cardinal number of the continuum	399

Index